Detection and Prevention of Leaks from Dams

Detection and Prevention of Leaks from Dams

ANTONIO PLATA BEDMAR
Head, Isotope Hydrology Section, CEDEX, CETA, Madrid, Spain
Former Technical Officer at the IAEA, Vienna, Austria

LUIS ARAGUÁS ARAGUÁS
Consultant, Isotope Hydrology Section, CEDEX, CETA, Madrid, Spain
Former Technical Officer at the IAEA, Vienna, Austria

CRC Press
Taylor & Francis Group
Boca Raton London New York

CRC Press is an imprint of the
Taylor & Francis Group, an **informa** business

Library of Congress Cataloging-in-Publication Data

CRC Press
Taylor & Francis Group
6000 Broken Sound Parkway NW, Suite 300
Boca Raton, FL 33487-2742

ISBN-13: 978-9-058-09355-4 (hbk)

Publisher's Note

The publisher has gone to great lengths to ensure the quality of this reprint but points out that some imperfections in the original copies may be apparent.

Cover Design: Studio Jan de Boer, Amsterdam

Typesetting: Charon Tec Pvt. Ltd, Chennai, India

Visit the Taylor & Francis Web site at
http://www.taylorandfrancis.com

and the CRC Press Web site at
http://www.crcpress.com

Contents

Preface

Most of the reservoirs constructed in the world present seepage or leakage, which are located at the geological formations of the dam site, dam foundations or the dam body itself. Usually, a well-constructed dam is not affected by this problem, although leakage can be generated during the reservoir exploitation due to differential sitting down processes, seismic events, etc. Experience shows that reservoir leakage is frequently linked to important economical repercussions. Repairing works are usually very costly and, in most cases, they are planned and executed by engineers without having a sufficient knowledge on the real nature of the stated problem. It leads frequently to failures of such works and/or to unnecessary increments of their cost. A previous investigation oriented to the establishment of the local groundwater flow patterns of the affected area can produce significant benefits to this respect. Most of the techniques that are necessary to this purpose are not familiar to engineers responsible for the dam management and maintenance. Main goal in writing this book has been to cover this lack of information.

Due to the diversity of the involved techniques and the multidisciplinar character of the investigations, the collaboration of specialised working groups is necessary to achieve the pursued objectives. The present book can provide to dam engineers information on the possibilities offered by these techniques to solve their particular problem and, on the other hand, can provide guidance to specialised working groups with limited experience in tackling these kind of problems. Working groups, which are familiar in the application of tracer techniques (natural and artificial tracers) for hydrological purposes, are the most suitable ones to conduct this kind of studies. However, in specific cases, the collaboration of working groups specialised in geophysical methods may also be necessary. Apart from that, it has to be pointed out that some of the techniques described in Chapter 10 for the localisation of leakage and seepage zones at the bottom of the reservoir need further development.

Special mention deserves the application of the techniques described in this book during the feasibility, pre-construction and construction stages of the dam. These techniques can provide a valuable contribution to the evaluation of the suitability of a dam site. On the other hand, the prevention and prediction of future problems can provide important benefits due to the fact that repairing works use to be much cheaper before or at a previous stage of the dam construction. A detailed information of the flow pattern of dam site groundwater will surely help in the planning of the grouting and impermeabilisation works during the dam construction. Chapter 11 provides guidance for the planning and execution of the investigation.

In the same way, the application of natural and artificial tracer techniques prior to the first filling of the reservoir will provide information of the main characteristics of the local aquifers. This information will be basic for the investigation of the replacement of the aquifer water by reservoir water during the first filling and during successive years. Only with this previous investigation, the relationship between aquifers and reservoir can be understood in a detailed way. Guidance for these studies is given in Chapter 11.

Chapter 12 is devoted to the presentation of different case studies of reservoir and lake leakage which will help decision-makers, project planners, field personnel and other environmental professionals to a better understanding of the possibilities offered by the techniques described in the book for the study of particular problems.

Madrid, November 2001
Antonio Plata Bedmar
Luis Araguás Araguás

CHAPTER 1

General considerations

1.1 INTRODUCTION: UNDERSTANDING LEAKAGE AND SEEPAGE IN RELATION TO DAM SAFETY

Practically all reservoirs constructed all over the world show, to some extent, water losses of variable importance. In spite of the studies commonly carried out before and during dam construction, it is not always possible to predict precisely the hydraulic behaviour of the dam body and/or the geological formations in the vicinity of the dam site. Therefore, when a dam is constructed, a certain probability of leakage and/or seepage is assumed. In some instances, the rate of leakage or seepage is accepted as long as there is no concern from the safety point of view. However, it is a difficult task to assess the risks and associated consequences derived from excessive leakage and seepage.

In many instances, unforeseen remedial costs have to be assumed to control the leaks. Clearly, the potential effects derived from uncontrolled leaks can have a major impact on the stability of the dam or the complementary structures, as well as on the surrounding geological formations (dam abutments). In some cases, under arid and semiarid conditions, the value of the escaping water may be of higher relevance, providing additional value to the studies on the origin and characteristics of the leakage.

From the safety point of view, dealing with the risks derived from leakage and seepage represents a major concern, due to the difficulties inherent in characterising their properties. In many instances, these aspects represent the major component of Dam Safety Programs (DSP). These programs require the contribution and coordination of many experts and professionals in a number of disciplines during the life cycle of the structure, starting with the studies dealing with the site selection and continuing during the design of the structure and during its full operational cycle. The DSPs include monitoring and surveillance tasks in order to ensure the safety of the structure with regard to the properties and lives of people threatened, should the dam fail. Additionally, comprehensive reviews, including in-depth evaluation of the performance of the dam, need to be conducted every 5–10 years, as well as the evaluation of the need for remedial measures. These reviews usually include specific tasks to identify failure modes and to assess the overall risk. In many instances, these operations provide the only opportunity to consider all aspects that may affect the safety of the dam, including those derived from leakage.

The proper assessment of the risks derived from leakage and seepage usually constitutes the major part of DSPs for many large dams due to the lack of a proper hydrogeological characterisation in the vicinity of the dam. To illustrate the relevance of these studies, we should consider the potential events that can lead to the failure of earth dams.

It is expected that dams properly designed and constructed should present limited water losses through the dam body. Failure of a dam can result according to the following categories:

1) Piping at the dam toe, at contacts of dam structures with the terrain or at the surrounding or underlying formations.
2) Blowout derived from the development of high uplift pressures on the base of the dam.
3) Excessive leakage through the dam body or along the surrounding geological formations linked to erosion processes.
4) Overtopping of the dam due to insufficient spillway capacity. Usually this failure is associated with a very rapid increase of the water level derived from a cyclonal situation.
5) Slope failures on the downstream or upstream faces of the dam associated with inadequate breakwater.
6) Movement of materials within the foundations or the abutments along planes of geological weakness (sitting-down processes).

Except in the case of overtopping, the role of underground circulation of water and/or local groundwater derived from nearby aquifers might be highly relevant to the safety of the dam and all related programs. Unfortunately, in many instances, not enough relevance is given to the recommendations requesting additional hydrogeological studies and monitoring programs.

Frequently, the need for reducing the risk of failure or for controlling water losses derived from leakage leads to the initiation of very costly remedial works, in most cases, with insufficient hydrological information. These repair works are undertaken focussing almost exclusively on engineering aspects, avoiding the preliminary works required for a detailed study on the nature of the problem. In many instances, the results obtained in stopping or reducing the leaks are not those anticipated. Several cases are known from all continents where despite the large amount of money invested in remedial works (grouting, drainage curtains, etc.), no positive results were achieved.

Very often, the lack of an appropriate previous hydrological investigation, e.g., at the construction stage, as well as the lack of appropriate monitoring programs, leads to decisions on the remedial works on short notice and with limited hydrological information. Additionally, the engineers responsible for the dam construction and management are not familiar with the possibilities offered by many hydrogeological techniques recently developed. The present book intends to fill this gap, providing a description of the available techniques and documenting examples where positive results have been obtained.

The literature available on dam leak studies is extremely scarce. As will be shown throughout this book, practically all reported studies in this field deal with partial aspects of the leakage problem, that is, they describe the results or possibilities offered by one or two particular techniques. It is not easy to find comprehensive studies that integrate the use of several techniques and conclude with practical recommendations for the solution of the leakage problem.

The present book is based on the experience gained by the authors through the studies carried out at the following reservoirs:

BRAZIL: Botafogo, Poços de Caldas and Tapacurá.
CHILE: Colbún, Lake Laja and Yeso.
COLOMBIA: Chingaza.
COSTA RICA: Arenal.
CUBA: Cautillo, Cuyaguateje and Herradura.
CYPRUS: Takna.
DOMINICAN REPUBLIC: Aguacate, Chacuey, Maguaca and Sabaneta.
ECUADOR: Agoyán and Lake San Marcos.
ETHIOPIA: Koka.
INDONESIA: Sermo.
MOROCCO: Aoulouz.
NAMIBIA: Oanob.
NICARAGUA: Lakes Asososca and Managua.
PANAMA: La Sierpe.
SPAIN: Algar, Arquillo de San Blas, Atazar, Bellús, Beninar, Buenas Hierbas, Canelles, Cetransa landfill, Contreras, El Tejo, Escalona, Francisco Abellán, Giribaile, Guadalhorce, Lagunas de Ruidera, Montanejos, Peñarroya, Rules, Santa Liestra, Tous, Uldecona (La Cenia) and mining tail reservoirs of Aznalcóllar, Riotinto, Almagrera, Fertiberia, Cala and Filón Sur.
SUDAN: Rosieres and Sennar.
TUNISIA: Joumine.
VENEZUELA: La Honda.

1.2 LEAKAGE AND SEEPAGE

Water losses frequently appear below the dam foundation as well as at the abutments. Water losses through the dam body are less frequent. Water can escape from the reservoir either as leakage or as seepage. Leakage can be described as concentrated water losses derived from structural or construction deficiencies such as those found at contact surfaces between the dam body and the terrain, lack of or deficient isolation of alluvial materials, activation of perched old karstification levels when flooded by the reservoir water, etc. In the latter case, groundwater usually flows through concentrated pathways and emerges downstream, forming springs.

On the other hand, the presence of limited seepage through and under the dam embankment is a natural, expected process that basically is evaluated by investigating the permeability of the geological formations in the vicinity of the dam. In the case of earth dams, the limited flow of water across the dam body is considered a positive feature, permitting the development of pore pressures within their mass. However, it is a difficult task to evaluate precisely the hydraulic behaviour of the structure and the surrounding geological formations. On the other hand, the amount of seepage is not the main indicator of the risk associated with a potential dam failure.

Therefore, seepage can be described as spread infiltration produced at the reservoir bottom through permeable or partially permeable granular materials. Seepage rates frequently can be predicted if the hydraulic parameters and the geometrical dimensions of the involved geological formation are known. Under these conditions, a flow net can be modelled that permits the definition of the groundwater flow pattern and the evaluation of the water losses for different water levels in the reservoir. The escaping water either can be incorporated into the downstream aquifer without evidence in the surface or can emerge as a diffuse discharge in a wide area downstream.

1.3 MECHANISMS OF FAILURE

Piping has accounted for the majority of the catastrophic earth dam failures involving loss of life and large property damage. Development of flow nets within the dam body may result in the development of cavities and transport of material, leading to a catastrophic failure. Seepage takes place in the form of well-defined water veins with concentrated emerging flow. The leak gradually enlarges the permeable channel until failure occurs. The erosion starts at the water emergence point and progresses back towards the reservoir. To develop such a failure mode, the following conditions must occur:

1) There must be a well-defined flow path.
2) There must be an unprotected exit, allowing the transport of material.
3) The material in or under the structure must support a "roof" of nonerodible material, which therefore does not collapse, allowing the development of the pipe.
4) The material along the flow path must be erodible, that is, it presents adequate grain size to be transported to the exit point.

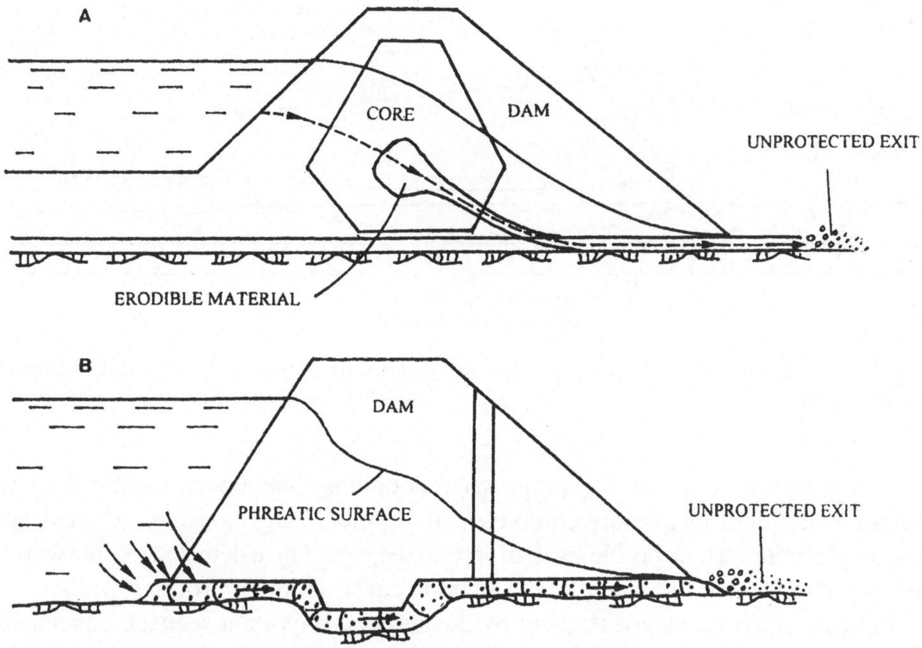

Fig. 1.1. Illustration of two different mechanisms leading to piping.

Figure 1.1 illustrates two examples of piping generation. In the first case, piping is occurring through the materials of the dam body due to inadequate compaction (density gradation due to inadequate thickness of material layers) or heterogeneity of the materials. The same mechanism is valid for piping occurring through an eventual pervious layer of alluvial materials located at the abutments. The alternation of pervious and impervious layers can create optimal conditions for the formation of a nonerodible roof below which water can flow with relatively high velocity. The same situation may exist in the case of sandy and clayey layers embedded between layers of consolidated rocks. If the flow velocity through the permeable layer is high enough, seepage erosion can take place.

In the case of Fig. 1.1B, piping is taking place through the contact of the concrete structure of the dam foundation and the terrain. The same sort of piping may also occur through the contact of concrete channels or iron pipes installed for water discharge because of the difficulty of performing appropriate compaction of the terrain in zones located close to these structures.

Most cases of catastrophic failure of earth dams are related to piping because the process may develop unnoticed for a long period of time and suddenly cause the collapse of the structure in a matter of hours, as has commonly occurred in many countries. In many cases, the process remained unnoticed for long periods despite the implementation of regular monitoring programs and safety evaluations. In some

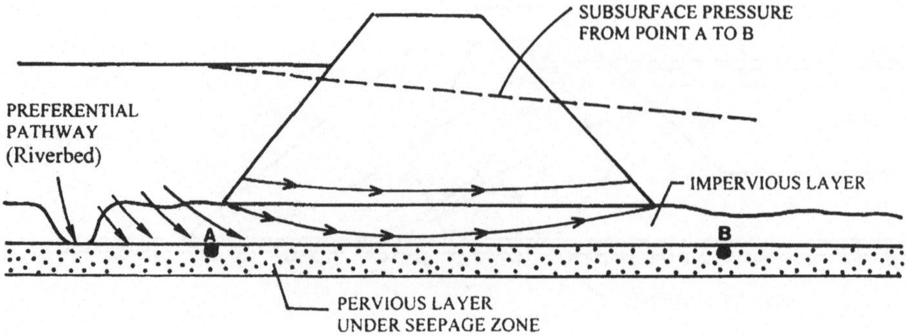

Fig. 1.2. Illustration of the mechanism for blowout failure derived from excessive uplift at toe of a dam.

cases, piping is accompanied by progressive sloughing. The concentrated leaks may remain small until just before complete failure. Obviously, concentrated leakage can create more serious problems than spread seepage. Once detected, repair works are easier to launch, for example, through the construction of grouting walls.

In some of the cases investigated by the authors, important seepage developed both at earth-filled dams and concrete dams. This was the case of Herradura dam in Cuba, where the problem was derived from an inadequate compaction of the dam body materials (Plata et al., 1999). In the case of concrete dams, the origin of the leakage can be the poor quality of the concrete or the joints. This can be derived from (1) an inappropriate ratio between cement and gravel or sand, (2) a lack of or deficient concrete vibration (Tapacurá dam, Brazil), (3) washout of the cement by local water flows (Sabaneta dam, Dominican Republic), or (4) crack formation due to sitting-down processes (El Atazar dam, Spain).

Figure 1.2 illustrates the failure derived from blowout processes. The problem is then created by excessive uplift pressure derived from the presence of a high permeability layer below the dam. An excessive water pressure is transmitted to the downstream slope of the dam so that the forces derived from the weight of the dam materials (gravity forces) and from friction of successive layers (upstream-oriented forces) can be lower than the uplift pressure until some kind of collapse takes place. This mechanism was responsible for the failure produced at Colorado dam in Chile, which is described in Chapter 12. Most of the water losses from the reservoir were located at a riverbed that was dug over the permeable sandy layer.

1.4 NATURE AND OBJECTIVES OF THE STUDIES ON DAM LEAKAGE

As previously indicated, planning for a reasonable probability of success in the reparation works of leaks from a reservoir requires a detailed knowledge of the hydraulic behaviour of the affected geological formations. Such knowledge cannot be obtained exclusively from hydrological and geological data available at a regional scale. The hydraulic parameters of the geological formations and, therefore, the groundwater flow pattern in the vicinity of the dam, may present drastic changes within short distances. This is especially true when dealing with consolidated rocks. For this reason, the investigation of leaks demands the use of direct techniques that provide information on the precise groundwater flow pattern prevailing in an area, usually in small dimensions. This information cannot be inferred exclusively from the analysis of general geological and hydrogeological data. These direct techniques include, among others, the use of natural and artificial tracers.

The detailed investigation of groundwater flows related to a particular leakage is always a difficult task. As previously stated, the combined use of complementary techniques is necessary. For obvious reasons, the starting point for this type of study is the geological and hydrogeological background information of the area.

An important characteristic of the studies performed for the investigation of leaks from lakes and reservoir is their low cost in comparison with those derived from the repairing works. The ratio of costs/benefits is in almost all cases absolutely negligible.

In principle, studies related to leaks from reservoirs can be undertaken at different stages during the life cycle of a dam. Table 1.1 shows the expected contribution of the techniques described in this book at different stages of the life cycle and the benefits derived from such application. In general terms, the studies can be conducted at the following stages:

1) Before the dam construction, as a part of the feasibility and reconnaissance studies that are usually performed.
2) During the dam construction, when relevant geological structures remain uncovered or become accessible through galleries or boreholes drilled for different purposes.
3) During the operational phase, after the dam construction, once the leakage or seepage problem has been identified and a solution has to be proposed or considered.

The ideal situation is attained when all potential problems related to the leaks have been solved during stages 1 and 2. Usually, the associated costs to reduce the permeability of selected geological formations are much lower when they are performed during the dam construction. However, this procedure is not always possible because, at this stage, the geological structures located at higher altitudes are usually unsaturated and the evaluation and prediction of their hydraulic behaviour after impoundment are difficult to achieve. This is true particularly in the case of fractured consolidated rocks.

The studies carried out before and during the dam construction should have as the main goal to provide information on the following topics:

1) Identification at the dam site of the permeable geological structures, formations or strata through which water may escape from the reservoir once it has

Table 1.1. Relationship between the contribution of the techniques described in this book and the expected benefits at different stages during the life cycle of the structure.

Life cycle stage	Purpose of application	Expected benefit
Site selection or assessment of existing dam site	Identification of structural features (faults and fractures)	Allow better potential for identification of possible flow paths
	Permeability tests during site investigation	Provide more realistic measure than current practice
	Baseline hydrological data collection	Adds baseline parameters
	Interconnection of water sources—e.g. springs to river	More confidence in site selection
	Groundwater dynamics	Can provide better baseline data for site
	Reservoir slope stability evaluation	Recognise recent movements
	Unsaturated zone studies	Early recognition of potential problems
	Get baseline of groundwater downstream—to determine changes due to reservoir	Better understanding of system dynamics; adds baseline parameters
Design	Help define instrumentation and monitoring requirements	More definitive identification of potential problems
	Surveillance program requirements	As above
Construction	Additional site assessment	More rapid and precise assessments — save time and money
	Investigate specific features	Most tracer techniques not affected by infrastructure
	Compaction—nuclear probes	Standard practice nowadays for moisture and density control
	Quality control/evaluation of grouting	Provides additional information on effectiveness of grouting
	Install pre-injection tubes at suspect area	Enable leakage detection during operation

First filling and initial operation	Evaluate performance	Allows comparison with baseline data on a definitive basis
	Rate of change in some indicators (dynamics of the system)	Better "detectability" of potential problems
	Origin of water at leakage point/spring	Higher confidence level in performance
	Evaluate adequacy of surveillance program and monitoring	
	Flow path determination, including source	
	Understanding of pathways from analysis of mixing processes	
	Identification of inflow point(s) in dam or reservoir	
Operation	Early warning of piping potential	Improve defensibility against law suits (liability)
	Measure velocity in flow path to estimate possibility of erosion	Not possible with other techniques. Improved risk assessment
	Evaluate conduit flow/seepage, track the direction of flow	Determine erosion potential
	Determine permeability in all layers, by water labelling	Improved precision compared to current practice
	Check effectiveness of cutoff/grouting, etc.	More appropriate remedial action
	Assess dissolution of material from abutment	
	"Optimising" design of repair works	More appropriate remedial action

been filled. Particularly important might be the localisation of eventual paleo-riverbeds at the dam site, which can act as a bypass between the reservoir and the discharge zone downstream. The techniques recommended for this purpose are described in Chapters 3 and 8.

2) Possible existence of springs located either at the dam site or downstream from the dam. The investigations on the origin of the emerging water in these springs are very important, even if their flow rate is very low. It is possible that the emerging water derives from the infiltration of river water within the zone that will be flooded by the future reservoir. If this is the case, it is very likely that the flow rate of the springs will drastically increase after impoundment. This is frequently observed in karstic regions. Springs located at the riverside can be fed by surface water derived from the same river that infiltrates at a certain distance upstream.
3) Exhaustive inventory of all water points that could be related to the future reservoir. Depending on the local geology, the inventory should cover an area of 10–20 km downstream of the dam site. The chemical and isotopic characterisation of these waters should be performed, according to the indications given in Chapters 5 and 6. This characterisation will allow the establishment of the baseline for water emerging downstream and will permit follow up on eventual changes in flow rates and chemical compositions produced after impoundment. In this case, it is possible to study the replacement of the locally derived groundwater by water derived from the reservoir as a function of time. Even if this characterisation cannot be performed due to the lack of technical means, it is recommended that water samples from the most relevant points be collected and kept in waterproof bottles or containers. These samples can be analysed later on, for example, once a leakage problem has been identified. The information on the chemical and isotopic contents prevailing before the dam construction can be of great help for a better understanding of the stated leakage problem.

Obviously, the main goal of the studies performed with the reservoir under hydraulic load is the localisation and characterisation of the groundwater flows that are responsible for eventual springs emerging downstream. Equally important is the study of the changes of the potentiometric levels observed in aquifers located downstream. The infiltration of the reservoir water can occur at the following zones:

1) Through the local geological formations located at the dam site. In this case, the eventual presence of faults, as well as other structural accidents, such as open diaclases, fractures, dolines, etc., can play a relevant role. Sometimes, water infiltration occurs through perched permeable sandy strata connected to ancient alluvial terraces. This was the case found at the Chacuey reservoir in the Dominican Republic.
2) Through the geological formations hosting the dam foundation. More frequently, these units are formed by remains of alluvial materials that either were not

fully removed during the excavation works or were not properly made watertight. This occurred at the Maguaca reservoir in the Dominican Republic.
3) Through the contact surfaces of concrete structures with the terrain. This was the case of the Sabaneta reservoir (Dominican Republic), where a large leak was identified at the contact surface between the morning glory and the terrain.
4) In the case of earth-filled dams, through the dam body due to either poor quality of the clayey materials used for its construction or to deficient compaction of these materials. This case was reported at the La Herradura dam, Cuba.
5) In the case of concrete dams, as a consequence of the poor quality of the concrete.
6) Also in the case of concrete dams, through fissures created as a result of a differential settling of various blocks of the dam. This was the case of the Atazar reservoir located near Madrid, Spain.

The main conclusion derived from the previous statements is that all cases of dam leakage show their particular characteristics. Therefore, a common methodology that can be generally applied for the investigation of all cases of dam leakage does not exist. On the contrary, each case has to be analysed separately in such a way that the most adequate techniques will be applied, according to the particular geological and hydraulic settings. An initial planning of the activities to be undertaken is necessary. But this planning must not be rigid because the results obtained during the course of the study will surely provide important guidelines for the implementation of new actions and experiments to be conducted in the investigation.

1.5 RISK ANALYSIS

As mentioned above, a well-constructed dam usually should not present a relevant leakage or seepage through the dam body. Obviously, this statement applies preferentially to concrete dams, where leakage through the dam body usually only occurs in the case of fractures associated with sitting-down or sliding problems of the foundation. Leakage and seepage may occur both through fractures of consolidated rock formations and through permeable granular materials. Due to the high hydraulic gradients usually prevailing at the dam site, even low permeability formations can transmit significant amounts of water.

Dam collapse is usually linked to the transport of solid particles, from the dam body, from the foundation material or from the surrounding formations. This transport only takes place when specific conditions exist, which involve the formation of water flows with a velocity higher than the critical velocity. This parameter can be defined as the minimum velocity needed to initiate the transport of the finest solid particles existing in the medium, as long as water is flowing.

According to Lawrence von Thun (1996), three different terms have to be defined for the risk evaluation. These are hazard, consequences and risk. Hazard is the term used to indicate the probability of harm occurring. For example, the term hazard accounts for loads such as the probability of the occurrence of an earthquake, flood or tornado, as well as potential harms linked to a specific facility or structure. Hazard is not specifically quantified and is not used for risk analysis. Consequences are the adverse outcomes estimated for a failure of the structure, for example, loss of life, harm to properties, etc.

Two kinds of risk can be differentiated: risk cost and risk of loss of life. These are given by the product of the load probability, failure probability for a given load, and consequence of a failure scenario, according to the following expressions:

$$\begin{aligned} \textit{Risk cost} &= \textit{Load likelihood} \times \textit{Failure likelihood} \ (\textit{given a load}) \\ &\quad \times \textit{Economic loss} \ (\textit{given the failure}). \\ \textit{Risk of loss of life} &= \textit{Load likelihood} \times \textit{Failure likelihood} \ (\textit{given a load}) \\ &\quad \times \textit{Estimated life loss} \ (\textit{given the failure}). \end{aligned}$$

All parameters involved in these equations are described in probability terms. Obviously, the load likelihood is directly related to the local meteorological conditions, such as the probability of a tornado, hurricane or big storm occurring, which can lead to an uncontrolled rise of the water inflow in the reservoir. The loss of life heavily depends on the available warning time, that is to say, the time elapsed between the detection of the problem and the moment the local authorities are informed of the fact, and the arrival of the flood wave to the population at risk. Separate calculations can be made for static (for example, piping), hydrologic and seismic conditions.

Risk analysis with quantitative information can be efficiently and effectively added to dam safety evaluations. If the available data are not very precise, a qualitative risk assessment can still be performed (McCann and Castro, 1998).

1.6 TYPICAL FEATURES OF GEOLOGICAL FORMATIONS

The analysis of groundwater flow needs to take into account the geologic settings and the hydraulic properties of the different types of rocks forming aquifers. As previously stated, groundwater flows through interconnected void spaces, pores, cracks, fissures, large-scale fractures or cavities developed as a result of dissolution. Therefore, groundwater movement is controlled by differences in fluid pressure and elevation (i.e., the hydraulic gradient) as well as by the properties of the geological medium.

The main hydraulic properties of a porous medium are *porosity* and *hydraulic conductivity*. The first term refers to the fraction of the void space per unit volume of rock. *Effective porosity* considers the fraction of interconnected voids through which groundwater flows. These parameters are dimensionless, although they are usually expressed as a percentage or as a fraction. *Primary porosity* refers to the void space between grains as the rock was originally formed. *Secondary porosity* considers additional increase in porosity due to dissolution or fracturing of the mineral network. Nonfractured igneous rocks show porosity values as low as 0.0001, while some clayey rocks may show values as high as 0.6. Typical values of porosity for common aquifers range between 0.2 and 0.3.

Hydraulic conductivity measures the ability of a fluid to move through the interconnected voids and directly derives from Darcy's law. This property is therefore controlled by the geological medium and by the fluid characteristics. This parameter has a dimension of "length/time," e.g., m/day. The range of hydraulic conductivity values found in nature is extremely high, being one of the few properties showing eleven or twelve orders of magnitude, as shown in Fig. 1.3.

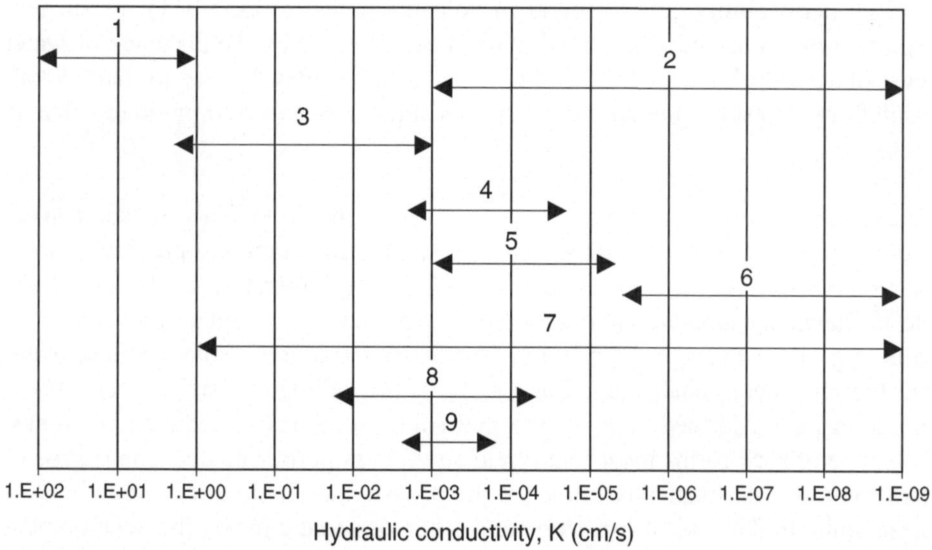

1. Clean gravel, very coarse sand.
2. Silty clay, silt, clayey fine sand, clay, rock-flour, etc.
3. Clean gravel and sand mixes, clean sand, coarse to medium coarse sand.
4. Medium coarse sand and silt mixtures, fine sand.
5. Sandstone.
6. Shale.
7. Basalt.
8. Coal.
9. Peat.

Fig. 1.3. Ranges of hydraulic conductivity for various types of common rocks and sediments.

From the geologic and hydraulic point of view, the main types of rocks are the following:

Igneous rocks Igneous rocks are formed by solidification of magma at some depth (*intrusive rocks*, such as granites) or by rapid cooling of magma that erupts and flows across the earth surface (*extrusive* or *volcanic rocks*, such as andesites or basalts). Because of the different cooling rates, the size of the minerals may present large variability among rocks of a similar chemical composition. The main mineral groups in igneous rocks include quartz, feldspars and ferromagnesian minerals (pyroxene, hornblende and olivine). These rocks are classified according to the proportion of these basic minerals. Most of the intrusive rocks are characterized by low values of primary porosity and permeability. However, in the case of intense fracturation, aquifers of some relevance may be developed. Extrusive (volcanic) rocks show a wide range of values of both porosity and permeability. Hydraulic properties may be modified by weathering and intense fracturation.

High permeability can be found in volcanic rocks, especially in recent geological formations and if the cooling has been affected by the presence of water (case of the Lake Laja, Chile). In other cases, galleries may develop in shallow volcanic flows. However, medium to low permeability values are common in volcanic rocks.

Sedimentary rocks Basically, two processes form sedimentary rocks. *Clastic sedimentary rocks* are formed by compaction and cementation of previously weathered, eroded and transported rock particles (conglomerates, sandstone, shale, etc.). *Chemical sedimentary rocks* are those formed by precipitation of dissolved salts (e.g., limestones, evaporites) or by the accumulation of biochemical material (corals, shells, coal, etc.). The nature of the geological environment where the deposits are formed controls the hydraulic properties of sedimentary rocks. This type of rock forms most high-yield aquifers in porous media. The degree of cementation may drastically influence the porosity and hydraulic conductivity of these units. In the case of limestone and other carbonatic rocks, the development of cavities leads to the formation of karstic terrains, where groundwater plays an important role in shaping the landscape.

The presence of perched karstic aquifers or karstic features, such as dolines or other dissolution structures, may play an important role during impoundment because preferential pathways are activated, permitting the generation of important flow rates in the abutments. These dissolution phenomena may also be valid for areas with evaporitic rocks, such as massive gypsum, which allow the rapid development of karstic features. However, the hydraulic properties of these rocks may be controlled by the presence of clayey horizons.

As shown in Fig. 1.3, the range of hydraulic conductivity in sedimentary rocks is larger than the typical values found in igneous rocks. Most of the major aquifers worldwide are formed on sedimentary rocks, and therefore most dams constructed over this type of rock require some type of work to reduce the permeability of the geological units on the abutments and/or foundations.

Unconsolidated sediments These geological units usually correspond to recent sediments (Quaternary) covering harder rocks. They originate from in situ weathering of a parent rock—igneous, sedimentary or metamorphic rocks (developing a regolith)—or by transportation by fluvial or glacial processes (gravels, tills, loess, fluvial deposits, etc.). Usually, these sediments show high porosity values (30–40%) due to the lack of cementation, forming shallow aquifers of high yields. These materials are characterized by much larger horizontal permeability values, allowing an easy transport of water along a given lithological unit. A common feature found in many dams is difficulty at the time of removing unconsolidated sediments because they usually represent a candidate formation for serious leaks. If possible, these deposits are removed from the dam site.

Metamorphic rocks Metamorphic rocks are formed by mineral crystallization due to the high temperatures and pressures found in tectonically active areas. The magnitude of the increase in these two parameters controls the mineral phases formed and the hydraulic properties of the new rock. A major division of these rocks is based upon the development of foliation (such as in the case of slates, schists or gneisses). Nonfoliated rocks include marble and quartzite. In most cases, primary porosity of these rocks is found to be low, but due to fracturing, a significant increase in porosity creates aquifers of some importance.

1.7 AVAILABLE TECHNIQUES

As previously indicated, obtaining geological and hydrogeological information of the area involved in the leakage study should always be the initial step of the investigation. Usually, this information will lead to the establishment of the initial hypothesis about the investigation, on which the initial experimental program has to be based. Practical experience demonstrates that this sort of study requires the combined use of diverse techniques. No technique should be disregarded a priori because sometimes the key information to solve the problem is found by applying the most unexpected technique.

The techniques commonly used for the investigation of leakage at reservoirs and lakes are the following:

1) Water balance of the reservoir or the lake.
2) Relationship between the water level in the reservoir and the flow rate of the springs located downstream, which is called by some authors the filtration law.
3) Techniques based on the measurement of piezometric levels at the boreholes existing at the dam site and at the borders of the reservoir.

4) Permeability tests carried out in boreholes and testing holes drilled in the ground.
5) Geophysical well-logging techniques, such as:
   a) Measurement of natural gamma rays profiles.
   b) Measurement of vertical flows in boreholes using commercially available measuring systems.
   c) Profiles of self-potential and resistivity in the case of boreholes without internal casing.
   d) In some cases, ultrasonic techniques, magnetic induction, etc.
6) Geophysical techniques applied on the ground, the most important of which are, apart from natural gamma radiation, self-potential and resistivity.
7) Use of the temperature as a natural tracer of water for the determination of the elevation in the reservoir from which water infiltrates into the terrain.
8) Techniques based on the measurement of physical and chemical parameters in boreholes and springs, including:
   a) Measurement of temperature and conductivity profiles in boreholes, allowing the evaluation of the possible relationship with those made at the reservoir.
   b) Distribution of temperature, conductivity and other chemical parameters in other springs and discharge points in the area.
   c) Chemical analysis of major ionic components, that is, Cl^-, $SO_4^=$, HCO_3^-, NO_3^-, Na^+, K^+, Ca^{++} and Mg^{++}. In some cases, the measurement of SiO_2, Fe(total) and other minor components may be of interest due to the presence of these compounds in the reservoir waters from, for example, the discharge of sewage.
9) Techniques based on the measurement of environmental isotopes naturally occurring in water, such as the stable isotopes deuterium and oxygen-18, carbon isotopes, and the radioactive isotope of hydrogen, tritium.
10) Techniques based on the use of artificial tracers, including:
    a) Measurement of flow properties along a vertical profile in boreholes.
    b) Interconnection experiments between the reservoir (or boreholes) and the discharge points (or boreholes) located downstream.
    c) Infiltration experiments inside the reservoir, with the aim of identifying the main infiltration area.
11) Detection and measurement of flow in the reservoir using floating devices and float drogues.
12) Identification of areas of preferential infiltration in the reservoir using specially designed devices, known as infiltrometers.

In the following chapters, the theoretical considerations of the techniques are presented, emphasising their advantages and limitations. Based on the experience of the authors, there is not a single institute or research centre in the world equipped and routinely applying all mentioned techniques. Most of the studies carried out about dam leakage are conducted using a limited number of techniques (mostly one or two), and very frequently the techniques used are not the most appropriate to deal with the leakage problem observed. When this happens, the probability of success is rather low. The material presented in this book has been compiled with the aim of filling this gap, providing information on the possibilities offered by the different techniques available to the user. It is expected that a more comprehensive working plan can be adopted, according to the specific circumstances of the seepage or leakage problem.

REFERENCES

Lawrence von Thun, J., 1996. Understanding seepage and piping failures—the no. 1 Dam safety problem in the West. *ASDSO Western Regional Conference, Lake Tahoe, Nevada, April, 1996.*

McCann, M.W. Jr. and Castro, G., 1998. A framework for applying and conducting risk-based analysis for dams. *Proceedings of the 1998 USCOLD Annual Meeting, Buffalo, New York.*

Plata, A., Rodríguez, B., González, B., Romero, Z. and Llanes, A., 1999. Estudio de fugas en la presa La Herradura, Cuba. In *Estudios de fugas en embalses*, International Atomic Energy Agency, Vienna (in Spanish).

CHAPTER 2

Water balance, leakage law and piezometry

2.1 WATER BALANCE OF A RESERVOIR

Frequently, the existence of water losses derived from a reservoir is suspected, but the losses cannot be quantified because either no water is emerging downstream of the reservoir or the existing springs do not represent the total amount of the leak. The latter situation occurs when part of the water losses is incorporated into the groundwater flow of a local aquifer or emerges at the bottom of a second reservoir located downstream. The last case was observed in the Canelles reservoir, Spain, where most of the escaping water was emerging at the bottom of the Santana reservoir.

In these cases, the quantification of the water losses by means of the lake water balance referred to a certain period of time can be very convenient. Obviously, this water balance requires the control of all water inputs and outputs during a given period of time. Water losses are computed as the difference between the total inflow and outflow, which will be reflected in the water level of the reservoir. The balance equation can be expressed in the following way (Hutchinson, 1975):

$$[(V_{in})_{SW} + (V_{in})_{GW} + (V_{in})_{PREC}] - [V_{LEAKS} + (V_{out})_{EVAP}] = \textit{Change of volume} \tag{2.1}$$

where V_{in} and V_{out} represent, respectively, the total water volumes of the inflows and the outflows during the whole period of time during which the water balance is performed, and the subscripts *SW*, *GW*, *PREC* and *EVAP* represent, respectively, surface water, groundwater, precipitation and evaporation. The term $(V_{in})_{PREC}$ corresponds to the rainfall directly falling on the surface of the reservoir.

Obviously, the determination of all these parameters entails important uncertainties whose magnitude depends on the local prevailing conditions of the water flows. Particularly difficult to determine might be the contribution to the inflow of small intermittent streams such as ravines or creeks that flow only during the rainfall periods. The water balance of the reservoir can only be estimated if the total uncertainty is small compared with the magnitude of the water losses. This condition is not fulfilled in most cases, especially when water losses are small compared with the total inflow and outflow. The right selection of the period of time during which the water balance is going to be carried out is also a critical factor. In most cases, the dry season is chosen in order to keep small the contributions derived from local precipitation.

In some cases, mass-balance equations similar to equation (2.1) can be introduced for certain chemical or isotopic components of water that reach the reservoir in the form of a pulse. This may occur both as a result of normal seasonal variations in the concentration of a pollutant in the incoming water or associated with a concrete discharge of wastewater into the river. For example, the concentration of some conservative components, such as the ions Cl^-, $SO_4^=$, and NO_3^- or the isotopic deviation of oxygen-18 ($\delta^{18}O$, see Chapter 6), can be used, depending on the prevailing conditions. The balance equation is expressed as the product of the different fluxes or masses (water volumes) included in equation (2.1) by their respective concentrations. These concentrations need to be determined experimentally. This equation (or equations, if more than one component is determined), combined with equation (2.1), allows the suppression of one (or more than one) parameter whose measurement might be difficult, costly or not possible at all.

In Annex 3, the theoretical fundamentals for the use of the stable isotopes deuterium and oxygen-18 to compute balances in lakes and reservoirs are described. For further information on this topic, the following references can be consulted: Hutchinson (1975); Merlivat (1970); Nir (1973); and Zuber (1983).

Water balance calculations may be particularly interesting in the case of shallow lakes or reservoirs where the chemical and/or isotopic composition of water can undergo important changes throughout the year due to evaporation. This is the case of reservoir Koka in Ethiopia, which is used for hydropower production. The average water column of this reservoir at full capacity is about 8 m, and its surface extends for about 250 km^2. The reservoir is located at an elevation of about 1500 m.a.s.l. Due to this fact and to the high yearly average temperature, about 28°C, the effective evaporation rate reaches up to about 2.5 m/y. On the other hand, more than 90% of the local precipitation takes place during the months of July and August. The rest of the year is extremely dry. Under these circumstances, if the mass balance is made during the dry season, the terms $(V_{in})_{SW}$ and $(V_{in})_{PREC}$ can be neglected. The water balance shown in equation (2.1), as well as the electrical conductivity EC and $\delta^{18}O$ balances, were recommended to be conducted in this case by the senior author on the occasion of a technical mission undertaken under the framework of a technical cooperation project of the International Atomic Energy Agency (IAEA). The following equations can be written for these balances:

$$(V_{in})_{GW} - V_{LEAKS} + (V_{out})_{EVAP} = dV_L/dt \tag{2.2}$$

$$(C_{in})_{GW}(V_{in})_{GW} - C_L V_{LEAKS} = (dC_L/dt)(dV_L/dt) \tag{2.3}$$

$$(\delta^{18}O_{in})_{GW}(V_{in})_{GW} - (\delta^{18}O)_L V_{LEAKS} + (\delta^{18}O)_E(V_{out})_{EVAP}$$
$$= (d(\delta^{18}O)_L/dt)(dV_L/dt) \qquad (2.4)$$

where the subscripts L and E represent, respectively, the lake and the water escaping from the reservoir by evaporation. Obviously, the term C_E (electrical conductivity of evaporating water) is equal to zero because no salt content is associated with the evaporating water.

The usefulness of these balances requires (1) the continuous control of the flow rate of the discharge through the electrical power plant, (2) the periodic measurement of the water level at the reservoir, and (3) the periodic determination of the electrical conductivity and $\delta^{18}O$ values of the lake water. The value of $(\delta^{18}O)_E$ can be calculated from the value of $(\delta^{18}O)_L$, the surface water temperature of the lake and the relative humidity of the local atmosphere (see Chapter 6 and Annex 3). The approximate values of $(C_{in})_{GW}$ and $(\delta^{18}O_{in})_{GW}$ can be obtained by analysing water samples collected from springs and boreholes derived from the local recharge. Therefore, the previous equations can provide quantitative information on (1) the flow rate of groundwater contribution $(V_{in})_{GW}$, (2) the combined flow rate of the leaks V_{LEAKS} and (3) the evaporation rate $(V_{out})_{EVAP}$. According to the available information, water losses from the reservoir are taking place exclusively under full reservoir conditions, which usually occur at the end of the rainy season. The electrical conductivity of reservoir water is around 180 μS/cm at the end of the rainy season and about 350 μS/cm at the end of the dry season. The respective typical $\delta^{18}O$ values are −3.5‰ and −0.5‰ (see Chapter 6). The high variability of these parameters allows the performance of the balances with relatively high precision.

2.2 EFFECTS RELATED TO THE WATER LEVEL AT THE RESERVOIR

2.2.1 Leakage law

The importance of establishing the relationship between the water level at the reservoir and the flow rate of the springs located downstream is pretty obvious. This relationship can provide valuable information on the elevation interval inside the reservoir where the infiltration of the emerging water at the springs is taking place. Some authors call this relationship the *leakage law*. As shown later, the study of the relationship between the water level at the reservoir and the piezometric levels existing in the boreholes located in the surrounding geological formations is also very important. The piezometric variations of the water table will provide valuable information on the response of the local aquifers to the change of water level in the reservoir.

The method requires the periodic measurement of the flow rate of the springs for different values of the water level at the reservoir. Obviously, the frequency of the measurements should be chosen according to the water level variation at the

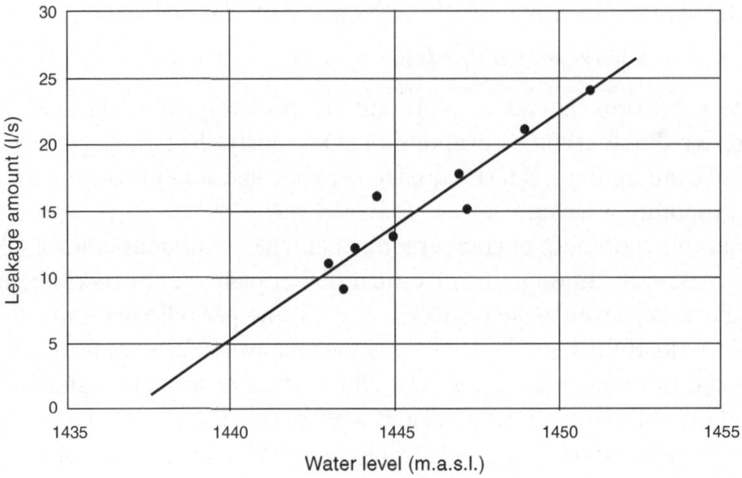

Fig. 2.1. Leakage law obtained for the reservoir Oanob in Namibia.

reservoir. If various springs are fed by independent water flows, it is very important to separately perform the measurement of their respective flow rates, because the response of each spring to the water level variation at the reservoir can be very different.

The chemical and isotopic contents, including the stable isotopes deuterium and oxygen-18, as well as tritium (see Chapter 6) and sometimes also the temperature of the emerging waters, can provide valuable information for the identification of springs fed by different groundwater flows. Even springs that are fed by the same groundwater flow can show considerable changes in the chemical composition. This situation may happen, for example, when the transit time of water between the reservoir and the springs is different and, as a result of it, the amount of soluble material dissolved by groundwater along its pathway can show large variations from one spring to another.

The results obtained using this approach can be plotted in diagrams, such as those shown in Figs. 2.1 and 2.2 for two different case studies. When all the water flow emerging at a given spring is coming from the same elevation of reservoir, the points corresponding to the flow rate variation in function of the water level tend to be aligned along a straight line. The flow rate of the spring is then proportional to the water level, that is, to the hydraulic head. The extrapolation of this straight line to a value of the flow rate equal to zero provides the approximate elevation at which the seepage or infiltration of the reservoir water is taking place.

The leakage amount Q is approximately governed by Darcy's law:

$$Q = \Delta h \cdot k \cdot S \cdot p \tag{2.5}$$

where Δh is the hydraulic gradient between the reservoir and the spring, expressed as a fraction of the unity, k is the hydraulic conductivity or permeability, S is the

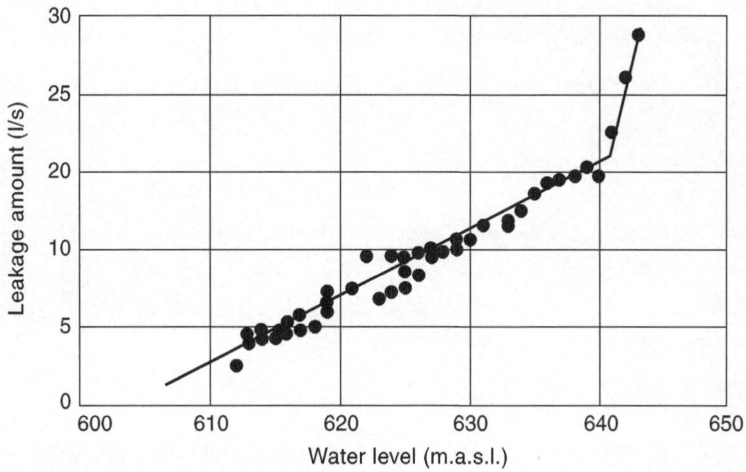

Fig. 2.2. Leakage law obtained for the reservoir Sabaneta in the Dominican Republic.

cross section of the leakage flow and p is the average porosity of the involved rock formation.

Usually, the flow rate of any leak is mainly governed by the minimum average values of parameters k, S and p along the underground flow path of the escaping water. In the case of concentrated leakage, these parameters are kept approximately constant, and, therefore, the flow rate Q is proportional to Δh. However, Darcy's equation will not account very precisely for the leakage process when the groundwater flow is of turbulent character, as happens very frequently.

The diagram presented in Fig. 2.1 corresponds to the Oanob reservoir in Namibia (see Figs. 2.3 and 2.4). The intercept of the straight line with the ordinate, corresponding to a flow rate equal to zero, occurs at an elevation of 1437 m.a.s.l. This elevation is only 12 m lower than the maximum elevation of the water level at the reservoir, therefore indicating that the infiltration responsible for the leakage is taking place at the uppermost layer of the reservoir.

Sometimes the straight line relating the spring flow rate and the water level exhibits a change of slope, such as the one shown in Fig. 2.2, which corresponds to the Sabaneta reservoir in the Dominican Republic. Such a slope change is interpreted as being produced by the activation of a second infiltration zone at the reservoir, located more or less at the elevation of the water level prevailing at the moment when the slope change is observed. In the case of this figure, a second infiltration zone was identified at the contact between the terrain and the concrete structure of the morning glory (spillway). The flow rate increase was observed at an elevation of 640 m.a.s.l., which coincided with the elevation of a terrace that allowed the reservoir water to reach the above-mentioned contact.

Obviously, in the case of a diffuse or dispersed seepage at the reservoir (seepage distributed over a wide range of elevations), the correlation between the flow rate

Fig. 2.3. General view of the Oanob dam (Namibia).

Fig. 2.4. Measuring the flow rate of a leak at the Oanob reservoir (Namibia) using rhodamine WT as a tracer. The seepage zone is located at the uppermost layer of the reservoir.

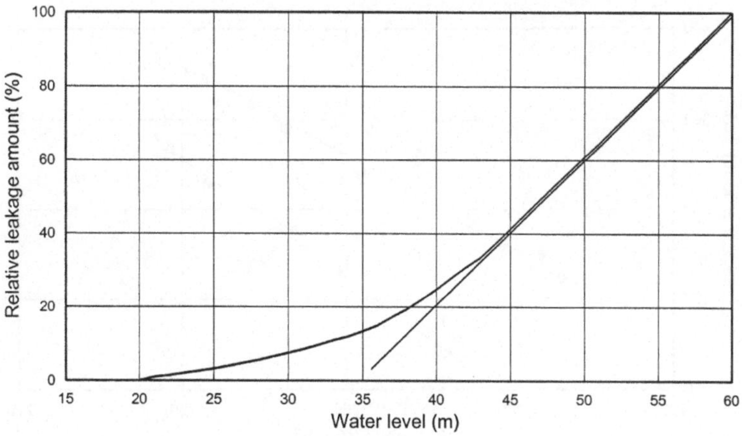

Fig. 2.5. Leakage law in case of diffuse infiltration at a reservoir between 20 and 43 m of water column.

and the water level is not defined by a straight line. A curve is then obtained, with a slope that increases at higher water levels due to the parallel increase of the surface of the seepage area. This case is illustrated in Fig. 2.5, where the leakage zone is located between 20 and 43 m of water level at the reservoir.

2.2.2 Response velocity of springs and boreholes to level changes at the reservoir

The time response in the flow rate of a spring as a function of the water level at the reservoir (velocity of response) provides valuable information on the characteristics of the hydraulic connection existing between the reservoir and the spring. The response velocity depends on the average permeability of the rock formation involved in the leakage and, on the other hand, on the water volume stored in the geological formation between the reservoir and the spring. This water volume produces the smoothness and the delay of the flow rate peaks that are transmitted to the spring as a consequence of the level change in the reservoir.

The hydrographs observed at the springs are different when the water level at the reservoir is increasing, with respect to those obtained when it is decreasing. In the first case, a fraction of the escaping water is filling up empty cavities of the terrain. As a consequence, the hydrograph undergoes a certain delay, the magnitude of which is giving by the ratio between the total volume of the cavities and the escaping flow rate. In the second case, the water stored in the cavities has to be drained through the spring when the water level at the reservoir decreases. As a result of this drainage, the flow rate of the springs decreases also with a certain delay. In both cases, the delay observed in the hydrograph increases when the total volume of the cavities is more important.

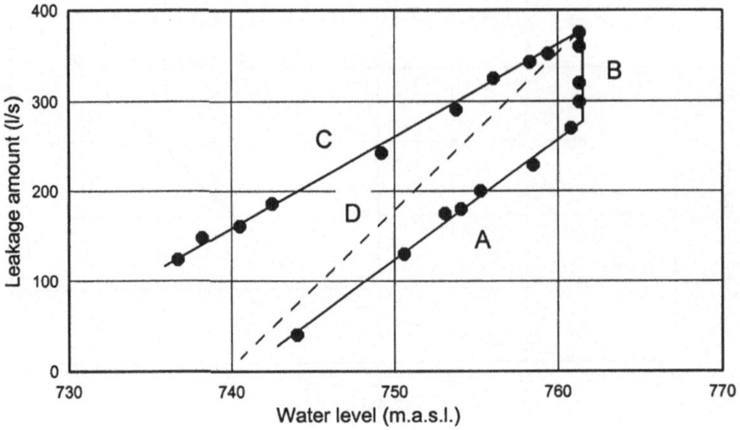

Fig. 2.6. Variation of the flow rate of the leaks at the PUNA reservoir, Africa, as a function of the water level. Line *A* was determined during the filling of the reservoir, while line *C* was measured during emptying. Line *B* shows the increase in the flow rate produced with the maximum level due to the progressive filling of the rock cavities. Line *D* represents the approximate flow rate variation that would be expected in the case of very slow filling and emptying of the reservoir.

The influence of the filling velocity of the reservoir is clearly shown in Fig. 2.6, which corresponds to the results obtained by the senior author at the PUNA reservoir in Africa.[1] During the filling of the reservoir, the straight line *A* was obtained for the variation of part of the emerging flow rate as a function of the water level. Due to the favourable meteorological conditions, the filling of the reservoir was very rapid. For this reason, the flow rate of the emerging water continued increasing even after the water level at the reservoir reached the maximum elevation of 761.3 m.a.s.l., increasing approximately from 280 l/s to 380 l/s (line *B*). This flow rate became stable once all cavities were filled up. During the emptying of the reservoir, which was also completed in a short time, the line *C* was obtained. The difference between this line and line *A* reflects the delay observed in the hydrographs obtained under decreasing conditions. Obviously, if the filling and emptying processes had taken place very slowly, a line similar to the one marked with the character *D* would have been obtained. The difference between this line and line *C* is a consequence of the emptying process of the cavities. The volume of these cavities was, in this case, very large because it was mainly due to several galleries that were dug in the terrain during the exploration of an old abandoned mine.

[1] The name PUNA is fictitious and represents the initials of "Publishing non authorised". For reasons beyond our control, authorities and officers responsible for the construction and management of this dam did not authorise the publication of the results obtained in the investigation, which was carried out under the framework of an IAEA's technical cooperation project (Vienna, Austria).

The above described results clearly show that the flow rate of the water emerging at a given spring for the same water level at the reservoir can be very different, depending on whether the water level is ascending or descending. The difference between the flow rates obtained in both conditions strongly depends upon the velocity of the level change at the reservoir, as well as on the volume of the cavities of the rock formation in the stretch involved in the level change. A rough estimation of this volume seems, in principle, to be possible, as shown in the next section.

2.2.3 Time evolution of the leakage

According to the statements made in the previous paragraphs, the convenience of introducing time as an additional parameter in studies dealing with the leakage law of the reservoir is pretty obvious. Usually, diagrams such as the one shown in Fig. 2.7 for the PUNA reservoir are produced, in which the variation, both of the water level in the reservoir and the flow rate in the springs, is analysed as a function of time.

In cases similar to the one described above, where the water level at the reservoir increases and decreases very rapidly, the results of the investigation may be used to estimate the minimum volume of the cavities forming the groundwater storage through which the springs are connected to the reservoir. This is achieved by comparing the flow rates emerging at the springs under rapid filling conditions

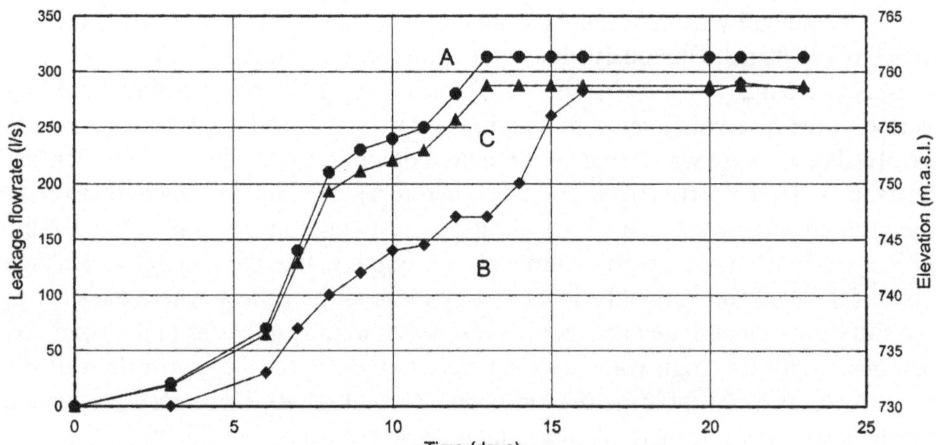

Fig. 2.7. Evolution with time of the water level at the reservoir (line *A*) and the total flow rate of several springs (line *B*) for the reservoir shown in Fig. 2.3. Line *C*corresponds to the flow rate of the same springs computed for the case of very slow filling of the reservoir. The difference between the integrated flow rates of lines *C* and *B* allows the estimation of the minimum capacity of the rock cavities involved in the leakage.

with those obtained under very slow filling conditions, with permanent equilibrium between the level of the reservoir and the flow rate of the springs. The flow rate of each spring can be experimentally measured if the prevailing conditions allow it. Alternatively, it can be computed from the maximum flow rate measured at a given moment, when the water level at the reservoir remains at the maximum elevation for a period of time long enough to achieve a constant value.

For example, in the case of the PUNA reservoir, it was proven that under ascending level conditions, the springs of the right margin gallery started to discharge when the water level at the reservoir reached the elevation of 732 m.a.s.l. The total flow rate reached a maximum value of 287 l/s when the maximum water level of 761.3 m.a.s.l. was reached for a period of time long enough to reach equilibrium (line A of Fig. 2.7). Therefore, assuming the existence of a proportionality between the flow rate of the springs Q and the water elevation at the reservoir h, the value of Q for a given value of h can be analytically calculated by using the equation of a straight line defined by the points $Q = 0$ for $h = 730$ m and $Q = 287$ l/s for $h = 761.3$ m. The equation obtained for this straight line is the following:

$$Q = 9.169 \cdot h - 6693 \tag{2.6}$$

Using this equation, the values of Q corresponding to the water levels experimentally obtained in the reservoir have been calculated. Curve C of Fig. 2.7 has been obtained in this way. The volume of water filling cavities of the rock formation is given by the difference between the surfaces defined by the curves C and B, which can be obtained by using a graphic method. The value obtained in this case was close to 70,000 m^3. Obviously, this value represents a minimum value of the cavities, because many of these cavities are totally or partially filled during the filling period of the reservoir, even if this period is very short. The geological formation involved in this case was formed by fractured dolomites, the porosity of which can be estimated to be between 2% and 5%. Therefore, the volume of rock involved in the groundwater flow that emerges at the springs would range from 1.4×10^6 m^3 to 3.5×10^6 m^3. Such a high volume of rock suggests that the seepage or leakage zone at the reservoir cannot be located very close to the gallery where the springs are situated. As mentioned above, the presence of mining galleries in the zone was responsible for this high volume of cavities, as well as for the high values of the water losses (several cubic metres per second, as total value). Tracer tests performed inside the reservoir using rhodamine and uranine as tracers (see Chapters 7 and 10) allowed the identification of the area with the largest infiltration capacity.

The previous example illustrates the usefulness of these kinds of studies. The determination of the response velocity of the springs as a function of the variation of the water level at the reservoir allows the estimation of the storage capacity of the groundwater system related to the springs. These data also provide information on whether the area of infiltration at the bottom of the reservoir is located close to the dam site. The presence of cavities occupying a large volume requires that the infiltration area be located at a long distance from the dam, and inversely, the opposite deduction can be made.

2.3 PIEZOMETRIC STUDIES

Studies based on the measurement of piezometric levels in the boreholes existing in the vicinity of the dam or in the zone where the reservoir water is emerging are common in leakage investigations. These studies include the following aspects:

1) Determination of the potentiometric surface by plotting maps with piezometric contour lines that inform on the direction of the groundwater flowpaths existing in the study area.
2) Study of the relationship between the changes of water level at the reservoir and the piezometric levels in boreholes.
3) Changes of the piezometric levels in boreholes produced during their drilling. These changes inform the possible hydraulic connection existing between boreholes and the reservoir or between a borehole and the springs, depending on the case.

However, it has to be pointed out that the results obtained in this type of study are not always interpreted in a correct way, in spite of their apparent simplicity. This is because piezometric levels, when used as a single tool, do not always provide unequivocal information, that is, the same set of data can lead to different interpretations, depending on the local hydraulic conditions. This is especially true when dealing with fractured, consolidated rocks, in which groundwater flows may present large variations in very short distances.

For example, a piezometric level higher than the prevailing water level at the reservoir does not mean necessarily that the groundwater flow intercepted by the borehole is completely independent of the reservoir. This higher piezometric level can be derived from the ingress inside the borehole of groundwater flowing, for example, from a perched aquifer, producing a vertical ascending flow. But such a flow can be mixed with a second groundwater flow infiltrated in the reservoir.

In the same way, a piezometric level lower than the level of the reservoir does not necessarily mean the existence of such hydraulic connection and, of course, does not mean that the borehole is intercepting a groundwater flow feeding the downstream springs. On the contrary, the lower piezometric level with respect to the reservoir usually is a consequence of the low permeability of the materials intercepted by the borehole. However, it can also be explained by the proximity of the borehole to the springs where the reservoir water is emerging.

Even the fact that a borehole shows a quick response to the water level variations at the reservoir does not mean necessarily that it is intercepting the groundwater flow feeding the springs. The hydraulic head of the reservoir can be transmitted to the borehole in a short period of time, even through materials of a relatively low permeability and, therefore, with no or negligible groundwater flow.

The previous statements lead to the conclusion that the measurement of piezometric levels in boreholes is always an important tool for the investigation of the stated seepage or leakage problem but should never be used as the single tool. Other complementary tools that can help in the interpretation of piezometric data are the following:

1) Measurement of conductivity and temperature profiles of the water column in boreholes. These profiles permit the identification of groundwater really derived from the reservoir.
2) Measurement of groundwater flows existing in boreholes by means of the techniques described in Chapter 8.
3) Chemical and isotopic analyses performed in water samples collected from boreholes. These analyses will provide information on the origin of groundwater found in certain boreholes, as well as on the eventual hydraulic connections between boreholes and between boreholes and springs.

Usually, when a borehole is drilled in the vicinity of the reservoir, the piezometric water level is located at lower elevation than the water level at the reservoir if the materials intercepted by the borehole show low permeability. Inversely, when the borehole intercepts a permeable stratum hydraulically connected to the reservoir, the piezometric level of the borehole rises immediately until a value close to the water level of the reservoir is reached, unless the borehole position is very close to the zone where reservoir water is emerging. This situation was observed in the case of the Chacuey reservoir, Dominican Republic, as presented in Fig. 2.8. Seven boreholes were drilled at the top of the dam in order to localise the permeable formation responsible for the leakage existing downstream from the dam. Boreholes 2 and 3 were exclusively intercepting impermeable materials of the dam body and its foundation. For this reason, their piezometric levels were much lower than the

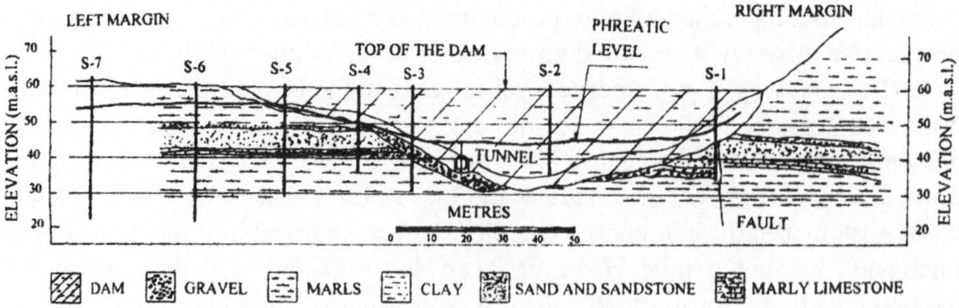

Fig. 2.8. Profile of the phreatic level along the axis of the Chacuey dam in the Dominican Republic. The highest piezometric levels reflect the preferential pathway of the water escaping from the reservoir.

water level at the reservoir, which was situated at an elevation of 56 m.a.s.l. On the contrary, boreholes 1, 4, 5, 6 and 7 were intercepting a permeable stratum of alluvial materials and, accordingly, their piezometric levels were close to the reservoir level. Obviously, the leakage at the reservoir was taking place through this permeable stratum. This point was confirmed by conducting interconnection experiments carried out by injecting tracers in these boreholes.

In the same way, the changes of piezometric level occurring in the borehole during its drilling constitute very valuable information for the identification of permeable strata or fractured stretches that are responsible for the leakage. As mentioned above, when a borehole intercepts exclusively low permeability materials, its piezometric level usually is lower than the level of the reservoir. However, when a permeable stratum or fractured stretch connected to the reservoir is intercepted, the piezometric level immediately rises until a value equal to or very close to the level of the reservoir is reached. This was the case observed at the Sabaneta reservoir, Dominican Republic, where the dam site was founded on limestone formations (see Chapter 12). The drilling of one of the boreholes was stopped at a depth of 60 m. Flow tests were performed inside the borehole using the technique based on the labelling of the whole water column with common salt, according to the technique described in Chapter 8. The flow velocity existing under these conditions was negligible. After that, the drilling continued, and once a depth of 90 m was reached, the piezometric level of the borehole rose by about 10 m because a fractured stratum was intercepted. As a consequence of this piezometric level rise, a vertical ascending flow was generated inside the borehole. The low permeability strata located at depths below 60 m started transmitting water due to the increase of the hydraulic head.

REFERENCES

Hutchinson, G.E., 1975. *Treatise on limnology*, Vol. 1, John Wiley and Sons, New York.

Merlivat, L., 1970. L'ètude quantitative de bilans de lacs à l'aide des concentrations en deuterium et oxygène-18 à l'eau. In *Isotope Hydrology 1970*, International Atomic Energy Agency, Vienna, pp. 89–107.

Nir, A., 1973. Tracer relations in mixed lakes in non-steady state. *J. Hydrol.*, 19, 33–41.

Zuber, A., 1983. On the environmental isotope method for determining the water balance components of some lakes. *J. Hydrol.*, 61, 409–427.

CHAPTER 3

Permeability tests in boreholes

3.1 GENERALITIES

In this chapter, several techniques frequently used for the determination of the permeability of aquifers through tests performed in small diameter boreholes are described. The application of these techniques is especially important during the feasibility studies performed prior to dam construction. Equally important might be the contributions of these tests during the dam construction, which are applied to improve the effectiveness of the impermeabilisation works on the dam foundations or in the abutments at the dam site.

Most dam failures have been attributed, at least partially, to insufficiently detailed information on the hydraulic characteristics of the geological formations outcropping at the dam site. It has to be pointed out that as a general rule the hydraulic parameters of most aquifers present important heterogeneities that are noticeable especially at local scale. The ignorance or underestimation of these heterogeneities has been in many cases the origin of serious accidents. The most important hydraulic parameter that allows the quantification of the aquifer heterogeneity is the hydraulic conductivity or permeability. Numerous studies have shown that this parameter can present important variations with depth. Several authors have described appropriate techniques for the determination of permeability profiles in boreholes (e.g., Cedergren, 1967; Custodio and Llamas, 1983; Halevy et al., 1991; Molz et al., 1988; Taylor et al., 1990). The techniques used by the authors of this book, which are based on the use of artificial tracers, are described in Chapter 8.

Most of the techniques used for the determination of permeability are based on the injection of small water volumes inside the borehole. Usually, these methods can be considered of low precision, providing values of the local permeability of the formation at the precise point where the borehole is located. In some cases, the permeability tests are performed during the drilling of the borehole, while in other cases, these tests are performed in boreholes already drilled. In the latter case, the characteristics of the borehole and the casing play an important role in the precision of the results. Particularly important are the characteristics of the casing (especially its own permeability) and the fitting of this casing to the surrounding terrain. Obviously, the casing has to be perforated in the stretch where the tests are to be performed to ensure the free circulation of groundwater through the borehole. It is very important to ensure that the fraction of perforated surface of the casing is sufficiently high to ensure that the permeability of the rock formation to be measured is not affected by the permeability of the casing. In other words, the permeability of the casing has to be much higher than the permeability of the rock formation, for example, 10 times higher. The permeability of the casing can be approximately calculated by means of the method described in Annex 4. A fraction of perforated surface between 1% and 5% will be sufficient in most cases.

A second cause of error with these methods, when performed in already developed boreholes, is related to the eventual water flow that circulates along the contact between the casing and the terrain. Usually, boreholes are drilled with a diameter slightly higher than the diameter of the casing. If cavities between the casing and the terrain are present, groundwater may vertically circulate along these cavities, even when packers are used to stop vertical flows inside the borehole. The errors made in the permeability determination can be very important in such cases. The same argument is valid when the space between the casing and the terrain is filled up with a gravel pack because the permeability of this gravel filling is usually much higher than the permeability of the terrain.

The methods for the determination of the permeability based on pumping tests usually provide values of this parameter more reliable than those obtained by means of the method described in this chapter. Nevertheless, only very rarely are pumping tests performed in studies related to the construction of a dam or for the investigation of leaks from dams. As a general rule, pumping tests conducted in low permeability formations are not advisable, due to the low values of the radius of influence of the pumping. On the other hand, pumping tests may not be the appropriate tools in cases of very heterogeneous geological formations, for example, in karstified limestone formations, because the determined permeability values may not be sufficiently representative of the whole area. Areas of very high permeability can be present at short distances from other areas where the permeability is almost zero.

Some of the methods described in the following sections are to be used in a saturated media, where laminar flow exists. The equations used for the calculation of the permeability, which are based on the Darcy's law, are valid only in these cases. The need for a saturated medium represents, in practise, a serious limitation for the application of the permeability tests in studies carried out prior to the dam construction. In many instances, rock formations located at high elevations in the dam site, which will later be impounded, are not saturated before and during the dam construction. Therefore, some of the methods cannot be applied at this stage. The previous statement mainly applies to the flow-measuring techniques described in Chapter 8.

On the other hand, it must be pointed out that in many cases, the determination of precise values of permeability is not absolutely necessary in dam leakage studies.

Very frequently, the main objective of the tests is the localisation of permeable strata or fractures through which the impounded water can escape from the reservoir and flow.

Due to the scope and objectives of this book, the techniques referred to in this chapter are not described in a very detailed way. The practical aspects that can assist in the investigation of leakage from dam and lakes have been stressed. For further information, the reader is referred to the literature specialised in this topic.

3.2 GILG-GAVARD TESTS

3.2.1 Constant level method

3.2.1.1 *Traditional method*

In the traditional Gilg-Gavard method, a constant flow rate of water is injected in a borehole so that a constant piezometric level is kept during a certain period of time. Under these conditions, the following equation is fulfilled (Custodio and Llamas, 1983):

$$k = \alpha \frac{q}{\pi \cdot d \cdot \Delta h} \tag{3.1}$$

where:

k = permeability, in m/s,
q = injected flow rate of constant value, in m^3/s,
d = borehole diameter, in m,
Δh = constant hydraulic head, that is, height of water level measured from the initial piezometric level, in m,
α = calibration factor.

The value of the calibration factor α is equal to 1 for a point piezometer, that is, for a piezometer with a small screen length (for example, less than 2 or 3 m). When the point piezometer condition is not fulfilled, the following equation can be used (Vidal, 1962):

$$k = \frac{q}{A \cdot \Delta h} \tag{3.2}$$

where A is a coefficient that depends on the filter length Δh_f, and the borehole diameter. Its value can be obtained by means of the following empirical equation:

$$A = a(1.032\Delta h_f + 30d) \tag{3.3}$$

where:

$a = 1$ if $\Delta h_f > 6$ m and
$a = 0.481 + 0.178\Delta h_f - 0.014\Delta h_f^2$ if $\Delta h_f < 6$ m.

A simple way to perform this test is by filling up the borehole to the mouth and measuring the water flow rate required to keep this level constant. However, the utilisation of this technique might not be possible when dealing with formations of very high permeability or with boreholes where the piezometric level is very deep.

Very frequently, this method is performed during the drilling of the borehole. The drilling is temporarily stopped, and water is injected through the drilling pipeline to the bottom of the borehole, where the infiltration chamber is located. Usually, this infiltration chamber has a length between 1 and 5 m. Under these conditions, a difficult aspect of the technique might be the precise determination of the piezometric level valid for the moment when the test is being performed. If the permeability of the terrain is low, this determination may require a long period of time, until the piezometric water level attains its equilibrium value.

3.2.1.2 *Method based on the measurement of the induced descending vertical flow*

This method has been developed by the authors and is routinely applied when the circumstances demand it. The method is performed in boreholes already developed, which are fitted with a metallic or PVC casing perforated in the whole length or in a certain stretch. The method is based on the measurement of the vertical descending flow induced in the borehole by the injection of water. The variation with depth of the magnitude of the vertical flow reflects the flow rates admitted by the different strata or permeable horizons intercepted by the borehole.

The method is illustrated in Fig. 3.1. Water is injected in the borehole using an appropriate pump or by gravity with a flow rate that permits the borehole to be kept completely full of water. Preferably, the upper stretch that corresponds to the unsaturated zone or a zone of no interest is fitted with an unperforated casing to avoid water losses through it. Under these circumstances, the injected flow rate is distributed through the different permeable strata (or fractures, in the case of consolidated impermeable rocks) intercepted by the borehole with a rate proportional to their respective transmissibilities.

The vertical descending flow can be measured using a technique that employs an artificial tracer, which is described in a detailed form in Chapter 8. A saturated solution of common salt can be used as a tracer. The tracer concentration is then measured at different depths along the water column using an electrical conductivity meter. A plastic tube of an inner diameter of about 8 mm is introduced in the borehole, acting as the injection line. The probe of the electrical conductivity meter is then fixed to this injection line using, for example, an appropriate adhesive tape, in such a way that the probe remains at a known distance below the lower end of this tube, for example, at a distance of 2 m. A thin iron or brass cap provided with several small lateral holes is adapted to this lower end. The task of this cap is to allow the tracer injection to be made horizontally in the borehole, avoiding any initial vertical motion of the tracer cloud induced by the injection flow. A second task of this cap is to keep the tracer losses small when the measuring line is on rest.

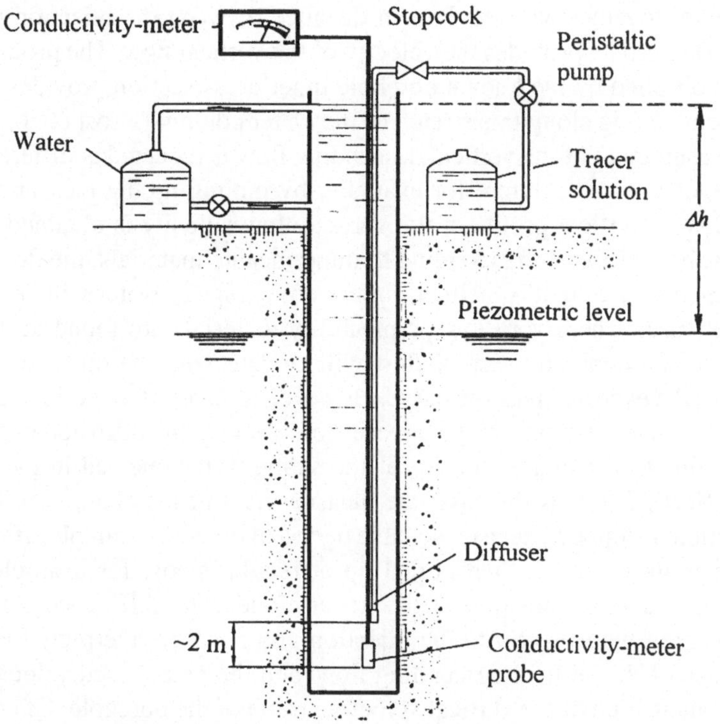

Fig. 3.1. Gilg-Gavard method modified by the authors for the determination of permeability based upon the determination of the velocity of the induced vertical descending flow created by water injection in a borehole.

Later, the plastic tube is first filled up with a saturated solution of common salt using a peristaltic pump. Immediately after the injection of the tracer solution, additional water is introduced again to keep the borehole filled. Once the borehole has been filled up again and the injection flow rate has been adjusted to the minimum value needed to keep the borehole full, the measuring line (injection tube plus conductivity meter probe) is introduced in the borehole until the deepest point of the stretch to be investigated is reached. The upper end of the injection line is kept permanently connected to the tracer container to avoid air entering inside the plastic tube. The system is kept on rest for several minutes to allow the distribution or washing out of any tracer fraction that has eventually escaped through the lower end of the injection line.

For the measurement of the vertical flow prevailing at the stretch between the injection point (cap of the injection line) and the conductivity meter probe, a tracer pulse is then injected by switching on the peristaltic pump for a short period of time, for example, 5 s. If a vertical flow exists, the tracer will arrive to the position of the measuring probe after a certain period of time, resulting in a bell-shaped passage curve, which is recorded. The peak or even better, the centre of gravity of this curve, together with the known distance between the injection point and the measuring probe, provides the velocity of the vertical flow. The product of this velocity multiplied by the known borehole inner cross section provides the mean flow rate circulating along this stretch of the water column (Drost et al., 1968).

The measurement of the vertical descending flow is repeated at different depths of the borehole water column, for example, by pulling up the measuring line in steps of 2 m. A vertical profile of the vertical flow velocity is obtained this way. Obviously, in stretches corresponding to impermeable materials, the flow velocity will be kept constant; that is, neither ingress nor escape of water will be observed.

In other cases where very low permeability materials are found at the lowest stretch of the borehole, no vertical flow will be detected. The injected tracer will not produce the expected passage curve. However, after a certain period of time, the conductivity meter response may increase slowly due to the diffusion of the tracer, especially due to the higher density of the saline water injected in the borehole (density effect). If this is the case, the measurement of the changes in the tracer concentration is stopped after a reasonable period of time, for example, after 30 min, and the measuring line is then pulled up at regular steps, for example, 20 cm, to determine the exact position of the tracer cloud. Any difference between the current position with respect to the injection point can be interpreted as being a consequence of the vertical descending flow, and the determined velocity can be used as an indication of the vertical velocity profile of the borehole. Of course, the above-mentioned density effect will be negligible if a radioactive tracer such as ^{131}I is used instead (see Chapter 7).

Water losses produced through a permeable stretch of the water column are calculated in terms of flow rate units, using the velocity values determined at both ends of the investigated stretch, and from the inner cross section of the borehole. The computed value corresponds to the parameter q in equation (3.1). Therefore, the method does not require the precise determination of the flow rate injected from the surface to keep a constant level in the borehole.

Obviously, the vertical descending flow induced by the water injection system can be measured using other techniques, such as those described in Sections 3.5.2 and 8.6.2, taking into account their own limitations. For example, the use of a current meter fitted with a propeller requires a relatively high flow velocity inside the borehole (above ~2 cm/s), while the method described in the previous paragraphs can be applied with lower groundwater flows, such as those found in low permeability formations.

The velocity of the vertical descending flow can be very slow when dealing with low permeability formations or when the hydraulic head Δh, created during the injection, is small due to the shallow character of the static piezometric or phreatic level. In such cases, a simplified method can be used for the measurement of the vertical flow. A pulse of tracer is injected at a point located, for example, 50 cm below the level of the water inside the borehole. The vertical migration of the tracer cloud is determined by the periodic measurement of conductivity profiles using a conductivity meter whose probe is vertically displaced along the borehole. The position of the tracer cloud as a function of time indicates the flow velocity for different stretches of the water column. The figures determined with this method permit the drawing of profiles representing the magnitude of the vertical descending flow. This technique is described in a more detailed way in Chapter 8.

The method provides the average permeability of the different strata or permeable stretches intercepted by the borehole. This permeability value multiplied by the thickness of the corresponding stratum or permeable stretch gives the respective transmissibility. Obviously, the value determined by this method is valid only for the point where the borehole is located. More representative information on the aquifer permeability will be obtained if the method can be applied in several boreholes.

Case studies Figures 3.2, 3.3 and 3.4 show the results obtained by the authors using the technique described above in three different case studies. In the first one, iodine-131 was used as a tracer due to the high salinity of the water derived

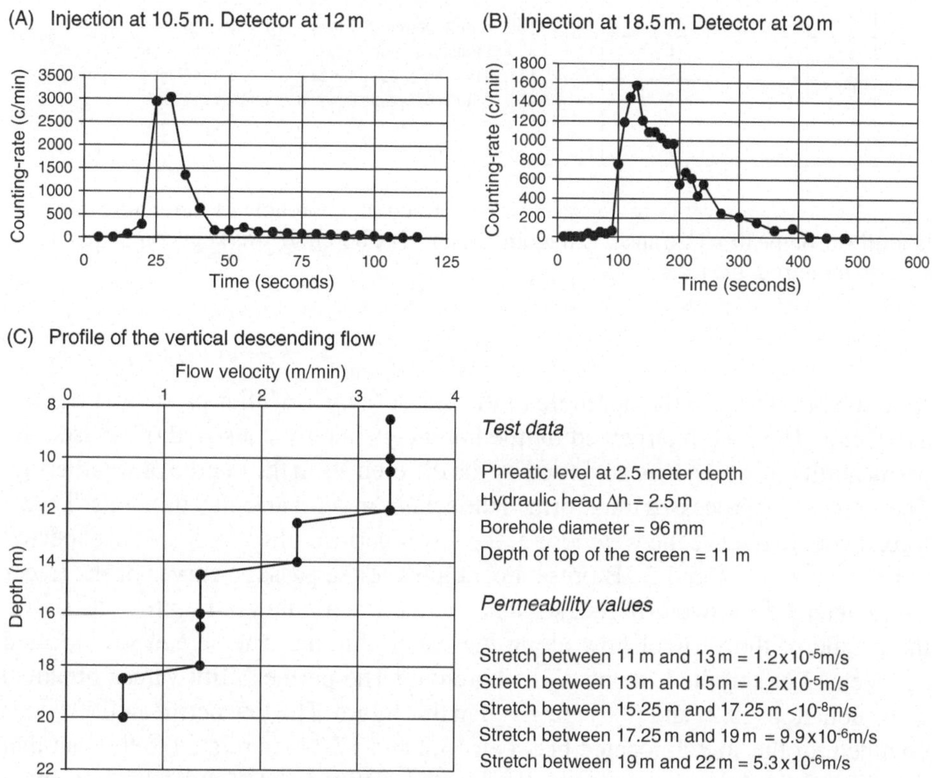

Fig. 3.2. Constant level Gilg-Gavard test performed in a borehole drilled close to the sea shore, Cartagena, Spain.

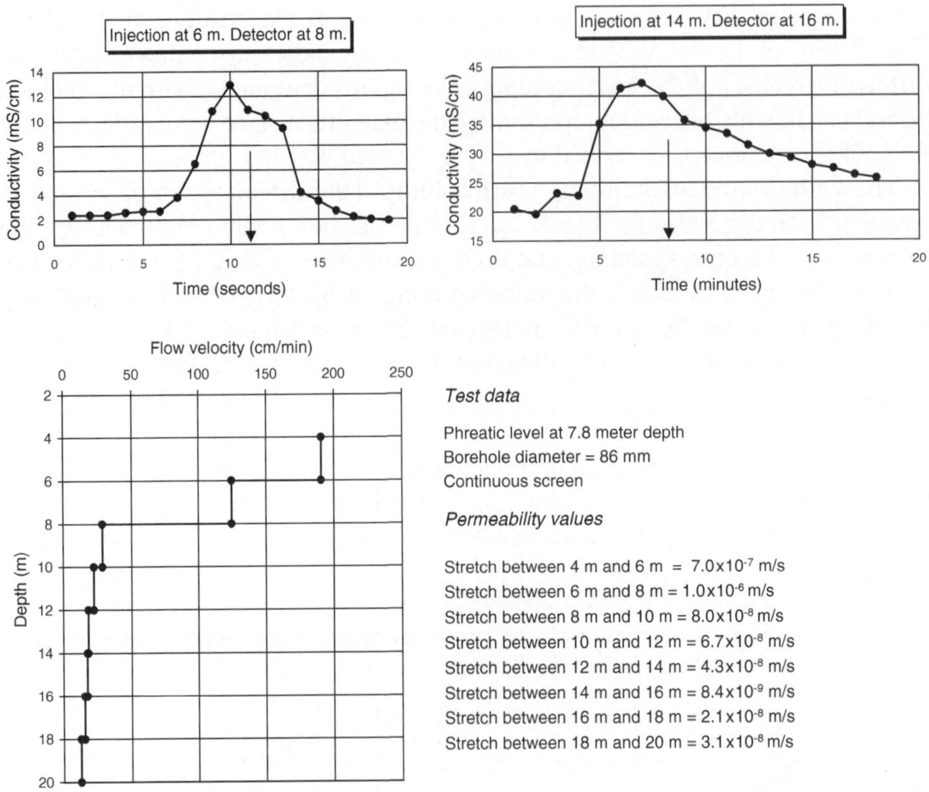

Fig. 3.3. Constant level Gilg-Gavard test performed in a borehole of the industrial waste landfill site of the firm Cetransa, Valladolid, Spain. The borehole was kept completely filled with water during the test.

from the proximity of the borehole to the sea. A Geiger-Müller probe was used as a detector. The rock intercepted by the boreholes mainly consisted of schists, the permeability of which was governed by the differences in the degree of weathering. The weathering processes transformed the schists into silt and clay minerals. Therefore, the more intense the weathering, the lower permeability values were expected to be. Figures 3.2A and 3.2B present examples of the passage curves of the tracer cloud obtained for two different stretches of the water column. Figure 3.2C shows the profile of the vertical flow along the water column. The screen was located between 11 m and the bottom of the borehole. The permeability values obtained using equations (3.2) and (3.3) are given in this figure. The low permeability values obtained for the stretch located between 15.2 and 17.2 m is related to the fact that the vertical flow was kept practically constant between 14.4 and 18 m. Constant flow, under these conditions, reflects the absence of both water ingress or water loss and, therefore, low permeability formations.

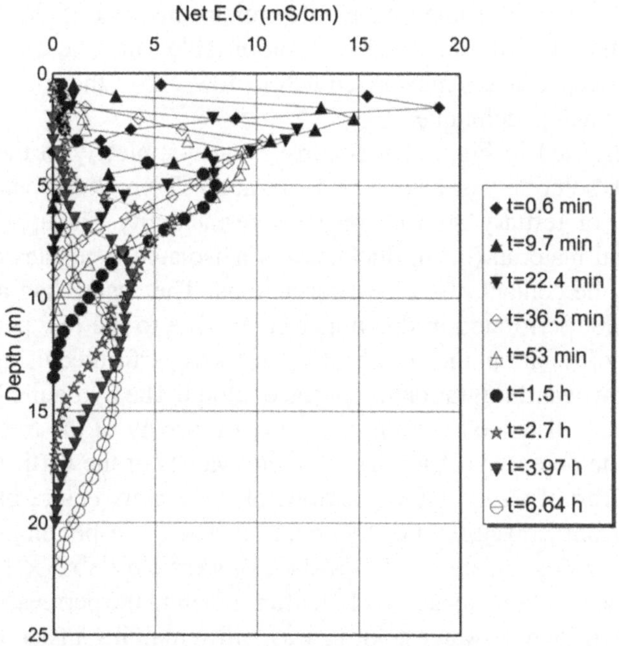

Stretch (m)	Velocity (cm/min)	Stretch length (m)	Hydraulic head (m)	Flow rate m^3/s	A	Permeability (m/s)	Valid stretch (m)
02.21 to 03.17	7.56	19.30	12.35	3.56E-06	21.72	1.32E-08	2.7 to 22
03.17 to 03.77	3.64	18.53	12.73	1.72E-06	20.92	6.44E-09	3.5 to 22
03.77 to 04.68	2.46	17.80	13.10	1.16E-06	20.17	4.39E-09	4.2 to 22
04.68 to 05.71	2.78	16.80	13.60	1.31E-06	19.14	5.03E-09	5.2 to 22
05.71 to 08.07	3.28	15.11	14.44	1.55E-06	17.39	6.15E-09	6.9 to 22
08.07 to 10.51	3.20	12.71	15.66	1.51E-06	14.92	6.46E-09	9.3 to 22
10.51 to 13.65	1.97	9.93	17.00	9.28E-07	12.05	4.53E-09	12 to 22

Fig. 3.4. Constant level Gilg-Gavard test performed in a borehole located at the site of a future industrial waste landfill near Medina del Campo, Spain. The borehole was fully drilled in the unsaturated zone.

In the case presented in Fig. 3.3, a saturated solution of common salt was used as a tracer. The borehole intercepted colluvial materials to a depth of about 8 m, and low-permeability marls from this depth to the bottom. The phreatic level at the time of the experiment was located at a depth of 7.8 m. The results show, as expected, that the high permeability values were exclusively restricted to the upper layer of about 2 m, which corresponds to the colluvial materials. The permeability of these materials was evaluated to be about 10^{-6} m/s. Most of the injected water was escaping through this horizon. The vertical flow induced by the water injection was negligible in the layers formed by marls. Taking into account the possible errors of the measuring technique, it was stated that the permeability of the marl deposits was around 10^{-8} m/s. The main cause of error in this case is derived from the density effect affecting the saturated solution of table salt. The true vertical flow velocity in the deeper layers must be somehow lower than the values determined using the cited tracing technique.

The test illustrated in Fig. 3.4 was carried out completely in the unsaturated zone. Table salt solution was also used as a tracer. The rock formation consisted of detritic rocks of Tertiary age with very low permeability. The upper layer of the pervious alluvial material (6.5 m thickness) was isolated by means of a concrete seal located in the contact of the two formations. The casing had a diameter of 6 cm and was not perforated in this upper layer. Due to the low permeability of the sedimentary formation, the permeability test was performed using the method based on the downward migration of the tracer cloud. The hydraulic head Δh was taken as the mean distance between the centre of gravity of the tracer cloud and the bottom of the borehole (25 m). The velocity values of the vertical descending flow obtained from the successive positions of the gravity centre of the passage curves ranged from 7.56 cm/min to 1.97 cm/min. The corresponding permeability values, obtained using equations (3.2) and (3.3), were close to 5×10^{-9} m/s. The water level in the borehole, measured 18 h after finishing the permeability test, was 13.5 m, that is, 11.5 m of water column was still remaining inside the borehole. This fact supports the low permeability values obtained for the rock formation.

3.2.2 Variable level method

In the variable level method, a large volume of water, sufficient to generate an important rising of the water level in the borehole or, if possible, to fill it up completely, is injected. Later, the recovery of the original water level is measured as a function of time. The average permeability of the saturated formation crossed by the borehole can be computed by the following empirical equation (Vidal, 1962):

$$k = \frac{1.308\, d^2 \Delta h}{A \cdot h_m \cdot \Delta t} \tag{3.4}$$

where:

k = permeability, in cm/s,
d = inner diameter of borehole, in m,
Δh = drop of water level, expressed in m, produced during the time period of time Δt, in min,
h_m = average hydraulic head computed as the difference between the average water level observed during the period of time Δt and the initial piezometric level (static level), in m.

The value of the coefficient A is then calculated as described in Section 3.2.1.1. The method requires the borehole casing be perforated only in the stretch of the saturated zone where the permeability is to be determined. Nevertheless, the method can also be very interesting as a qualitative tool for the identification of permeable areas or strata, even in the unsaturated zone, as is shown in the following case studies.

Case studies This method is used by the authors as a routine tool in dam leakage and, in general, in geotechnical studies. Figures 3.5, 3.6 and 3.7 present examples of case studies managed by the authors in the zone of the industrial waste landfill of the company Cetransa, located in Valladolid, Spain (Plata and Araguás, 2000). In the three cases, the unsaturated zone was not hydraulically isolated; that is, the casing was perforated in its whole length. Therefore, part of the injected water escaped from the borehole through this zone, and as a consequence, the permeability could not be computed using equation (3.4).

In the case of Fig. 3.5, the test was made in a 30-m deep borehole with an inner diameter equal to 5 cm. The initial piezometric level was close to a depth of 10 m. The intercepted rock formation consisted of filling material in the first upper layer (1.5 m thickness), colluvial material up to a depth of about 12 m, and marls to the bottom of the borehole. The results have been plotted in two different time scales. In the first case (Fig. 3.5A), a rapid drop of the water level in the upper layer is observed. A drastic change of the slope of the curve was observed at a depth of about 6 m. Figure 3.5B shows a second change of slope at a depth

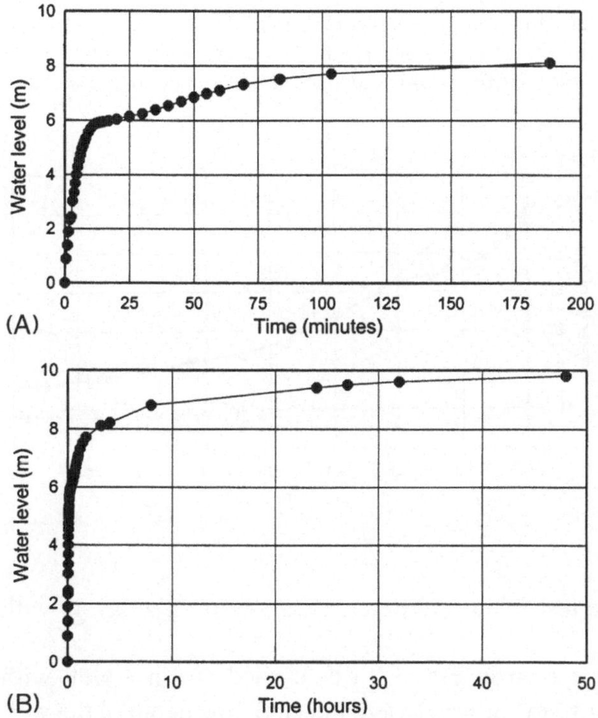

Fig. 3.5. Variable-level Gilg-Gavard test performed in a borehole of the industrial waste landfill site of the firm Cetransa, Valladolid, Spain. The borehole was initially filled with water to the mouth. The test showed the existence of three strata with different values of permeability.

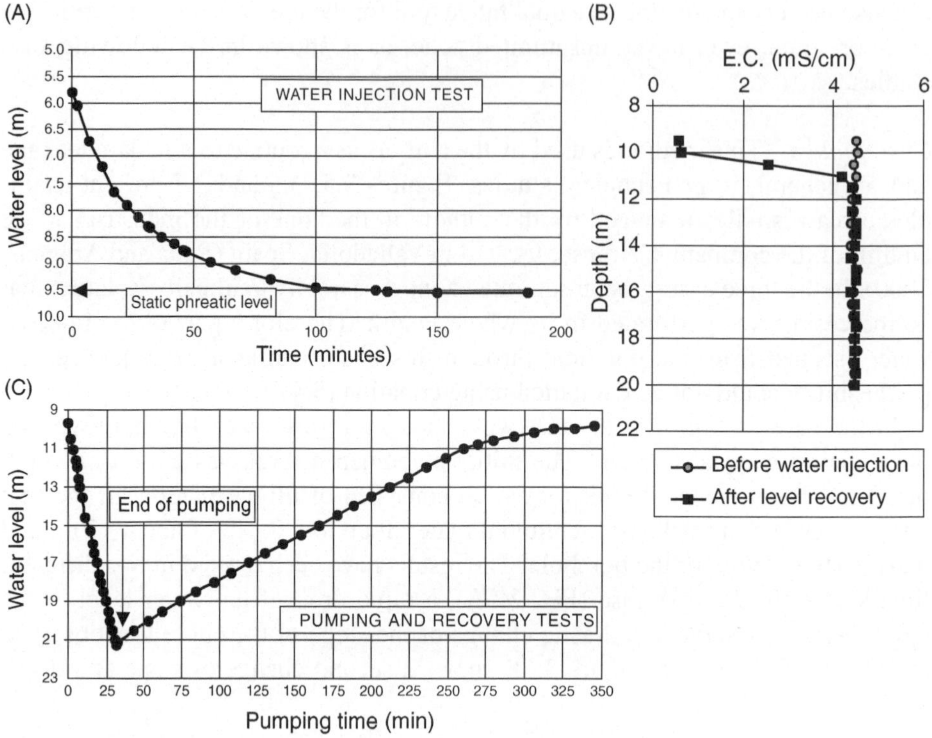

Fig. 3.6. Variable level Gilg-Gavard and pumping tests performed in a 1-m diameter borehole of the industrial waste landfill site of Cetransa, Spain. A: water injection test (water level withdrawn as a function of time). B: EC profiles measured before and after the water injection. C: Pumping test.

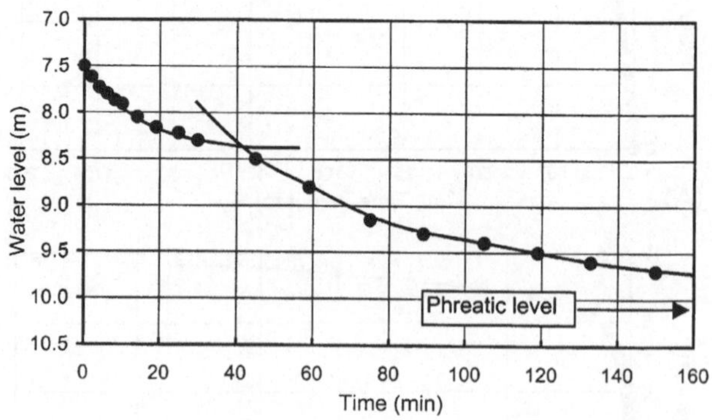

Fig. 3.7. Variable level Gilg-Gavard test performed in a 1-m diameter borehole of the industrial waste landfill of Cetransa, Valladolid, Spain. A bimodal withdrawn curve was obtained, representing two different strata.

of about 8.5 m. Therefore, the test allowed the identification of three layers with different permeability values.

The test whose results are shown in Fig. 3.6 was carried out in a well with 1 m inner diameter, into which 10 m^3 of water were injected. The depth of the well was 22 m, and the phreatic level was located at 9.6 m. Before the injection of water, the whole water column of the well was labelled with a solution of common salt. Twenty-five kg of common salt dissolved in 150 l of water were homogeneously distributed along the water column. Figure 3.6A shows the descent of the water level as a function of time. The level restitution curve is completely normal. The initial phreatic level was reached about 2 h after completing the injection of water. Figure 3.6B shows the conductivity profiles obtained before and after the water injection. It is clear that all of the injected water escaped from the borehole through a layer located above a depth of 11.5 m. Below this depth, no significant amount of the injected water penetrated due to the presence of the low permeability horizons of marls. Later, a pumping test was carried out in this well. The piezometric level was lowered to a depth of about 21 m, and the recovery of the piezometric level was measured. Figure 3.6C shows the results obtained both for the level drop and the level recovery after pumping. The straight lines obtained for both processes prove that the only horizon storing a significant amount of water is located between 9.6 m and approximately 11.5 m. Below this depth, no significant contribution of water was perceptible. Once groundwater reached the level of 11.5 m, the slope of the level recovery curve started to decrease due to the presence of pervious material above this depth.

The third example (Fig. 3.7) corresponds to a well of similar characteristics as the one previously described. The restitution curve of the initial or permanent piezometric level shows a bimodal shape, which is derived from the existence of two layers with different permeability values. The upper layer corresponds to a filling material of a trench that had been previously excavated to settle the drainage pipelines of the leachates produced by the landfill. The deeper one corresponds to the colluvial material of the zone.

3.3 LUGEON TEST

Lugeon tests are usually performed in formations of consolidated rocks where groundwater is transmitted through fractures or fissures intercepted by a well or by a borehole. The fundamentals of the method consist of the measurement of the water volume V that can be injected in a certain stretch of the borehole of length Δh during a time interval Δt and under an injection pressure p. The permeability k is obtained by means of the following equation:

$$k = \frac{V}{p \cdot \Delta h \cdot \Delta t} \tag{3.5}$$

The water admission of the borehole is usually expressed in an unit called a lugeon, in honour of the author who developed the method. This unit is defined as a water admission of 1 l/min per m in a given stretch with a pressure of 10 kg/m^2. The water pressure corresponds to the one induced on the piezometric level, that is, the sum of the water column thickness existing between the surface and this level and the pressure given by the injection pump.

Usually, water is injected using a piston pump, and water admission, given in flow rate units, is measured for different pressures, for example, 0 (only the water column is considered), 1, 2, 5, 7.5, 10 and 12.5 bars. The volume of injected water is determined using a flow gauge or as the difference of level at the tank from which water is taken.

According to Cambefort (1962), if water admission is proportional to pressure, one lugeon is equivalent to a permeability value of about 10^{-7} m/s to 2×10^{-7} m/s. The test is more appropriate in cases of consolidated rocks of low permeability. In granular media, water injection under such pressures would most probably create artificial cavities that would adulterate the expected result of the test.

3.4 LEFRANC TEST

Lefranc tests are frequently carried out to determine the permeability of granular rocks (e.g., alluvial materials, sand, gravel, sandstone, etc.) that require the installation of a casing to avoid internal collapse. The test is performed during the drilling of the borehole. The drilling is temporarily stopped, and water is injected through a pipeline introduced in the borehole at a point close to the bottom, as illustrated in Fig. 3.8A. Water infiltrates the lower stretch of the tested interval (several tens of cm), which is the only stretch without casing (infiltration chamber). The water surplus returns to the surface through the annulus of the borehole, that is, through the circular space between the injection pipeline and the casing.

The infiltration chamber can have a more or less spherical shape (diameter similar to the height) or a cylindrical one. In the first case, the permeability is calculated by the following expression:

$$k = \frac{V}{2\pi \cdot d_s \cdot p_i \cdot \Delta t} \tag{3.6}$$

where:

k = permeability, in m/s,
V = volume of water in m^3 injected during the time Δt, in s,
d_s = diameter of the spherical injection chamber, in m,
p_i = pressure of water injection, in m.

When the injection chamber has a clearly cylindrical shape, that is, height much higher than the diameter, the following equation is used:

$$k = \frac{V}{B \cdot p_i \cdot \Delta t} \tag{3.7}$$

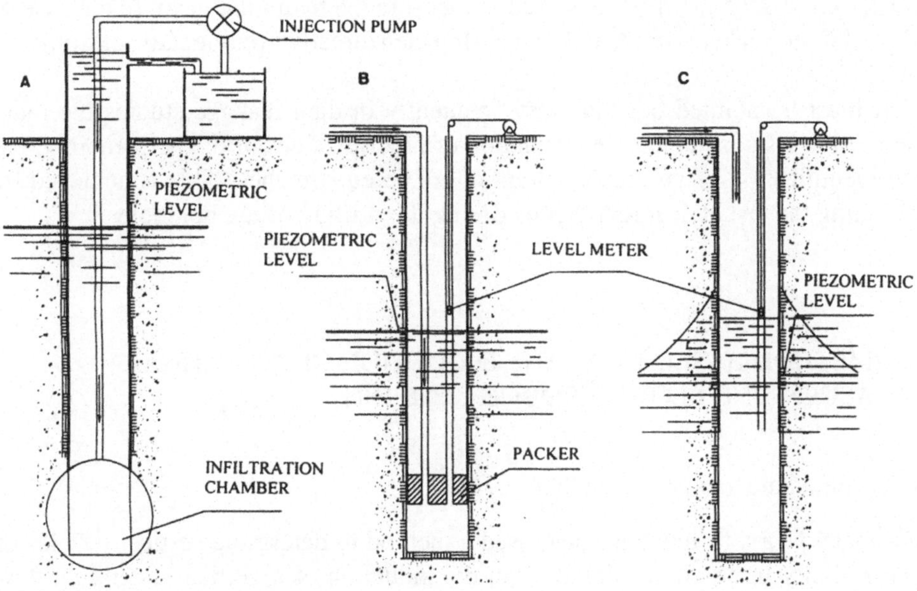

Fig. 3.8. Illustration of the Lefranc test. Graph A corresponds to the traditional method. B and C correspond to alternative methods suggested by Sahún and Sahuquillo (1969).

with *B* being given, after Castany (1971) and Schneebeli (1966), by the expression:

$$B = \frac{2\pi \cdot d_c \sqrt{(h_c/d_c)^2 - 1}}{ln(h_c/d_c + \sqrt{(h_c/d_c)^2 - 1})} \tag{3.8}$$

where:

d_c = diameter of the cylindrical injection chamber,
h_c = height of the cylindrical injection chamber.

The method requires the fulfilment in the injection chamber of the Darcy's law and, therefore, the existence of a laminar flow inside the injection chamber. It is achieved if the injection pressure is kept low enough. According to Castany (1971), this pressure should be limited to some centimetres for permeability values close to 1000 m/day, to some decimetres for permeability values close to 100 m/day, and to some metres for permeability values close to 10 m/day. Due to these requirements, the technique illustrated in Fig. 3.8A cannot be used when the piezometric level is situated at a depth higher than the above-mentioned pressures or hydraulic heads. In these cases, the methods illustrated in Figs. 3.8B and 3.8C, suggested by Sahún and Sahuquillo (1969), can be used. The pressure during injection corresponds then to the hydraulic head over the piezometric level, which is directly measured using a water level meter. In the case of Fig. 3.8B, a packer is used to define the height of the injection chamber, and water is injected through an appropriate pipeline fitted to it. A second pipeline is used to measure the hydraulic head. In the case of Fig. 3.8C, no packer is used, and water is injected directly over the static piezometric level.

It must be pointed out that very frequently in dam leakage studies the main task of the tests is not the determination of precise values of the permeability, but the identification of permeable strata or fractured stretches. This is achieved by repeating the test at different depths during the drilling of the borehole.

3.5 METHOD BASED UPON THE INDUCTION OF A VERTICAL ASCENDING FLOW THROUGH PUMPING

3.5.1 Fundamentals of the method

For many years, technicians have used a method to determine vertical profiles of permeability in boreholes based upon the induction of a vertical ascending flow by pumping water from a point located near to the top of the water column. The variation of this vertical flow along the water column provides information on the contribution to the pumping flow rate of the different strata intercepted by the borehole and, therefore, on their respective transmissibility values (Molz et al., 1994; Taylor et al., 1990). The vertical flow inside the borehole is measured using a method appropriate to the velocity range generated by pumping. Obviously, the method requires the casing to be perforated in the stretch between the piezometric level and the bottom.

Pumping is done using a medium-size pump whose diameter is small enough to permit access to the lower end where the vertical flow is measured. A piezometric level depression of several metres would be enough in most cases. Assuming Δh represents this lowering of the piezometric level, Q the pumping flow rate, T the overall transmissibility of the intercepted formation, r the inner radius of the pumping well or borehole and R the influence radius of the pumping, the Tiem equation establishes the following relationship between these parameters for a free recharged aquifer and a fully penetrating well:

$$\Delta h = \frac{Q}{2\pi \cdot t} ln \frac{R}{r} \quad \text{and, therefore:} \quad T = \frac{Q}{\Delta h} \frac{1}{2\pi} ln \frac{R}{r} \tag{3.9}$$

The ratio R/r varies, in most of the cases, between 10^2 and 10^4. Therefore, the expression $(1/2\pi)ln(R/r)$ varies between the values 0.73 and 1.46. According to these values, the transmissibility can be approximately calculated by means of the following equation:

$$T \approx (1.1 \pm 0.4)Q/\Delta h \tag{3.10}$$

In the case that several strata contribute to the extracted groundwater, for example, when dealing with a multilayer aquifer, it can be stated:

$$T = T_1 + T_2 + T_3 + \cdots \tag{3.11}$$

where T_1, T_2, T_3, etc. are the respective transmissibility values of these strata.

However, taking into account that the relative contribution to pumping of the different strata is proportional to their respective transmissibility values, the following equation is also valid:

$$Q = Q_1 + Q_2 + Q_3 + \cdots \tag{3.12}$$

The values of Q_1, Q_2, Q_3, etc. are then obtained from the change of the vertical flow velocity along the water column.

Figure 3.9 shows the results of a test performed in Austria by Rank (pers. comm.) in a multilayer aquifer. It was found that nine permeable strata were contributing to the observed pumping rate. The most important stratum was located at a depth between 12 and 14 m. This kind of test can provide very valuable information on the hydrological characteristics of the alluvial materials eventually present in the dam site.

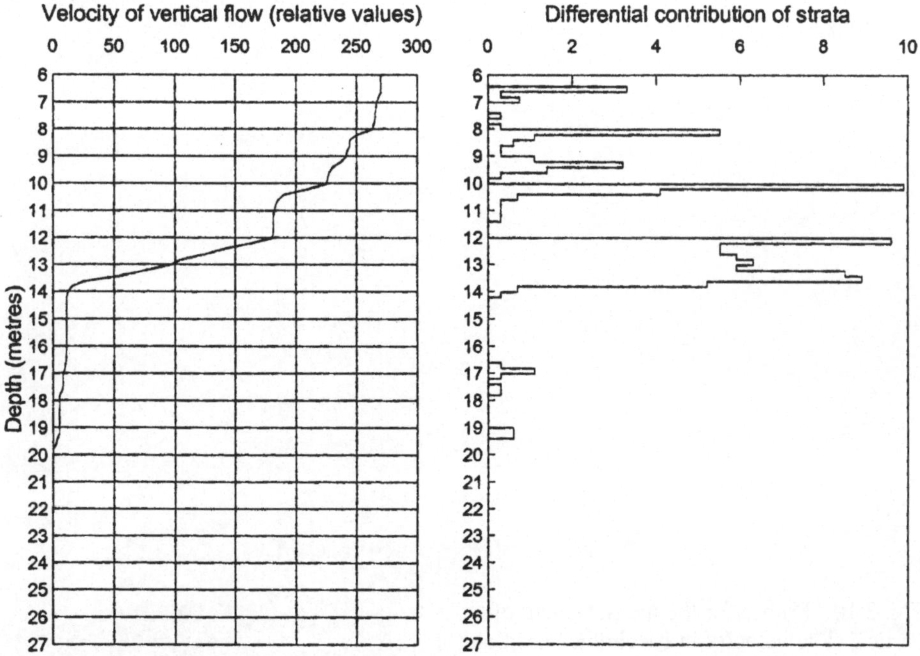

Fig. 3.9. Permeability test performed by Rank (pers. comm.) based upon pumping water from the borehole. A vertical ascending flow is created this way.

3.5.2 Gauges for measurement of vertical flows (flow meter)

Apart from the techniques described in Chapter 8 for the measurement of vertical flows based on the use of artificial tracers, three types of flow meters are known for the same purpose. These are the following:

1) Current meters based on the use of a propeller (e.g., Hufschmied, 1986; Leblanc et al., 1991). The flow velocity is obtained from the turns per time unit of the propeller, according to its calibration. In recently developed current meters, the propeller turns are measured using an optical method that reduces the friction with fixed parts of the device and, therefore, increases sensitivity (see Fig. 3.10).
2) Devices based on the transmission of a heat pulse (Hess, 1986). The heat pulse is produced through the activation of an electrical filament or metallic grid for a short period of time (some seconds). The heat pulse is transmitted upward or downward, depending on the flow direction, and is detected by a thermistor placed at a known distance from the heat source. The quotient of this distance and the travel time of the heat pulse provides a figure of the vertical flow velocity (see Fig. 3.10).
3) Electromagnetic devices based on Faraday's law of induction. This law states that the voltage produced by a conductor moving at right angles through a

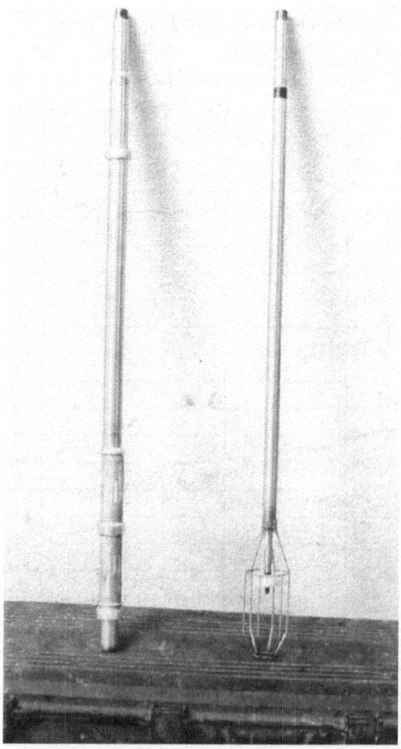

Fig. 3.10. Probes for the measurement of vertical flow velocity in boreholes based on the use of a propeller (right) and the transmission of a heat pulse (left) manufactured by the US company Mount Sopris Instrument Co., Inc.

magnetic field is directly proportional to the velocity. Flowing water then becomes the conductor due to the presence of dissolved ions. A coil generates the magnetic field, and the voltage signal is measured by means of two electrodes that are connected to the central unit of the gauge (Molz and Young, 1993). Accuracy of 2% can be obtained for flow velocities higher than 1 cm/s. The resolution is about ±2 mm/s, and the threshold flow velocity is 1 mm/s, which is determined by electronic noise. The induced voltage potential (or electromagnetic force) E is given by the following equation:

$$E = \int_0^\infty \nu \cdot B \cdot dL \tag{3.13}$$

where ν is the velocity of the flow through the electrodes, B is the strength of the applied magnetic field supplied by the coil and L is the distance from the centre of the coil. The electric current induced by the voltage potential can be measured directly and converted into values of flow velocity using laboratory calibration factors.

The propeller current meters are the most popular. Their main disadvantage is that they do not provide satisfactory results with flow velocities lower than about 2 cm/s. Heat transmission devices do not work properly in cases of high velocities with a turbulent flow regime. The electromagnetic devices can operate both with low and high velocities and are stronger due to the absence of mobile parts.

The three measuring devices mentioned above operate better when they are adapted to a sort of packer that forces groundwater to circulate exclusively through the chamber where the velocity sensor is located. However, this makes the measurement more cumbersome.

Apart from these devices, vertical flows can also be measured using the techniques described in Chapter 8, which are based on the use of artificial tracers. The application of these techniques is rather simple. However, their main disadvantage is the poor definition derived from the fact that the flow velocity is measured in a stretch of the water column and not in a concrete point. This represents a limitation for the precise identification and localisation of permeable strata or fractured zones of small thickness.

3.6 TEST FOR THE DETERMINATION OF THE EFFECTIVE POROSITY

The method for the determination of the effective porosity is illustrated in Fig. 3.11. The method is linked to a pumping test. Water is pumped from a well, and a certain amount of tracer is injected in boreholes located at a known distance. The pumping flow rate Q is known. The effective porosity is obtained from the time needed for the tracer to arrive at the pumping well. If the aquifer is homogeneous, this time is equivalent to the time needed for all groundwater stored in a cylinder of radius r,

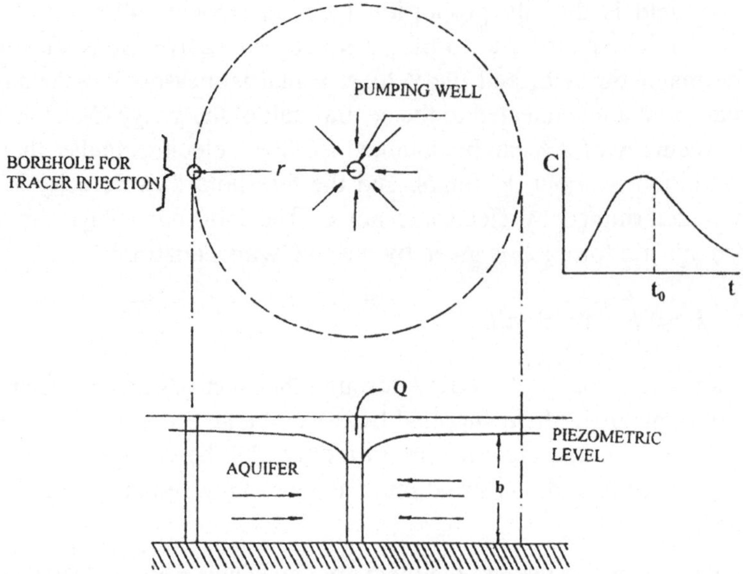

Fig. 3.11. Illustration of the method for the determination of the effective porosity of an aquifer based on pumping water from a borehole and tracer injection in a neighbouring borehole.

equal to the distance between the injection and pumping wells, and height b, equal to the thickness of the aquifer, to be pumped out. Calling m the effective porosity, this water volume is given by the expression $\pi r^2 b \cdot m$. On the other hand, if Q represents the pumping flow rate, the volume of groundwater pumped out from the time of the injection until the arrival time of the tracer, t_a, is given by the product $Q \cdot t_a$. Therefore, the following equation is valid:

$$\pi r^2 b \cdot m = Q \cdot t_a \quad \text{and, therefore:} \quad m = \frac{Q \cdot t_a}{\pi r^2 b} \tag{3.14}$$

Obviously, to obtain a reliable value for the effective porosity, the following conditions have to be fulfilled:

1) Radial symmetry of pumping, which implies: (i) isotropic aquifer, and (ii) groundwater velocity induced by pumping much higher than the natural flow.
2) The empty volume linked to the water level depression produced by pumping should be small compared with the volume of the cylinder. However, if this is not the case, the volume given by the expression $\pi r^2 b \cdot m$ can be corrected accordingly.
3) The distance r must be much higher than the radius of the injection and pumping wells.
4) The tracer adsorption in the aquifer should be negligible.

5) Fully penetrating well. If this is not the case, the volume of aquifer involved in the pumping can be corrected accordingly.

Usually, the tracer is injected after a steady state of the pumping regime has been achieved. The arrival or transit time of the tracer equals the time required for the centre of gravity of the recovery or breakthrough curve, which is measured at the pumping well. A discussion of the most appropriate tracers for this test appears in Chapter 7.

This method provides in almost all cases valuable information on the location and hydraulic characteristics of the permeable strata or fractured stretches intercepted by the well, even when the previous conditions are not fulfilled and, therefore, no valid value of the effective porosity is obtained. The different permeable strata intercepted by the wells can be easily identified through the repetition of the tracer injection tests at different depths or, even better, through the labelling of the whole water column using the technique described in Chapter 8. The relative contribution of these strata to the total volume of groundwater extracted, and, therefore, their respective transmissibility and permeability, can be obtained in this way. The information will be of better quality if the tracer injection can be repeated in different wells located around the pumping well.

The described technique has been repeatedly applied by the senior author for the investigation of the hydraulic characteristics of rock formations at dam sites. One of the studies was carried out in the Contreras reservoir, Spain. The injection and pumping wells were drilled to a depth of 180 m, and the distance between them was only 20 m. The intercepted formations consisted of marl and limestone. The tracer was injected at the depth corresponding to the marls, but never arrived to the pumping well. On the contrary, when the tracer was injected in the limestone stretches, the cloud arrived in the monitoring well after about 30 min from the injection. The tests provided precise information on the location of the permeable strata intercepted by the wells.

3.7 ESTIMATION OF THE TRANSMISSIBILITY FROM THE LEVEL RECOVERY AFTER PUMPING

In case of a homogeneous and isotropic aquifer, the transmissibility T can be estimated from the level recovery at a full penetrating well after pumping during a period of time τ. Let us call Q the pumping rate, t the time elapsed from the end of pumping, h_P the water level attained at the end of pumping, and h_R the water level measured at time t. The residual water level withdrawn at time t is given by the expression $W_r = h_P - h_R$. The Jacob equation stated that this residual water level is given by the following equation:

$$W_r = \frac{Q}{4\pi T} ln \frac{t+\tau}{t} \tag{3.15}$$

If we call $(W_r)_{1/10}$ the residual water level withdrawn measured for a value of the expression $(t + \tau)/t$ equal to 10, the previous equation gives:

$$T = 0.183 \frac{Q}{(W_r)_{1/10}} \tag{3.16}$$

The value of the term $Q/4\pi T$ is calculated as the slope of the straight line obtained when the residual water level W_r is plotted versus the expression $ln[(t + \tau)/t]$. Therefore, the transmissibility can be computed from the pumping flow rate Q. If the term $(W_r)_{1/10}$ is given in m and Q in m^3/s, T is expressed in m^2/s. The condition $(r/b) < 0.1$ has to be fulfilled, r being the radius of the pumping well and b the thickness of the aquifer.

REFERENCES

Cambefort, H., 1962. *Perforaciones y sondeos*. Editorial Omega S.A., Barcelona.

Castany, G., 1971. *Tratado práctico de las aguas subterráneas*. Editorial Omega S.A., Barcelona.

Cedergren, H.R., 1967. *Seepage drainage and flow nets*. John Wiley and Sons, New York.

Custodio, E. and Llamas, M.R., 1983. *Hidrología subterránea*, 2nd ed. Editorial Omega S.A., Barcelona.

Drost, W., Klotz, D., Koch, A., Moser, H. and Rauert, W., 1968. *Point dilution methods of investigating ground water flows by means of radioisotopes*. Report GSF-R 15, Institut für Radiohydrometrie, Gesellschaft für Strahlenforschung m.b.H., Neuherberg, Munich, Germany.

Halevy, J.L., Hanson, B., Enfield, C. and Glass, J., 1991. Evaluating the effectiveness of groundwater extraction systems. *Ground Water Monit. R.*, XI, 119–124.

Hess, A.E., 1986. Identifying hydraulically conductive fractures with a slow-velocity borehole flowmeter. *Can. Geotech. J.*, 23, 69–78.

Hufschmied, P., 1986. Estimation of three-dimensional statistically anisotropic hydraulic conductivity field by means of a single well pumping test combined with flowmeter measurements. *Hydrogeologie*, 2, 163–174.

Leblanc, D.R., Garabedian, S.R., Hess, L.M., Gelhar, L.W., Quadri, R.D., Stollenwork, K.G. and Wood, W.W., 1991. Large-scale gradient tracer test in sand and gravel, Cape Cod, Massachusetts, 1. Experimental design and observed tracer movement. *Water Resour. Res.*, 27, 895–910.

Molz, F.J., Güven, O., Melville, J.G., Nohrstedt, J.S. and Overholtzer, J.K., 1988. Forced-gradient tracer tests and inferred hydraulic conductivity distributions at the Mobile site. *Ground Water*, 26, 570–579.

Molz, F.J. and Young, S.C., 1993. Development and application of borehole flowmeters for environmental assessment. *Log Analyst*, 3, 13–23.

Molz, F.J., Boman, G.K., Young, S.C. and Waldrop, W.R., 1994. Borehole flowmeters: field application and data analysis. *J. Hydrol.*, 163, 347–371.

Plata, A. and Araguas, L., 2000. *Estudio hidrogeológico del entorno del vertedero de resíduos industriales de CETRANSA, Valladolid*. CEDEX, Madrid (in Spanish).

Sahun, P. and Sahuquillo, A., 1969. Medida de la permeabilidad en terrenos rocosos. Coloquio Nacional de Mecánica de Rocas, Madrid, *Bol. Serv. Geol. Obras Públicas*, 33, 1–45 (in Spanish).

Schneebeli, G., 1966. *Hydraulique souterraine*. Ed. Eyrolles, Paris.

Taylor, K., Wheatcraft, S.W., Hess, J., Hayworth, J.S. and Molz, F.J., 1990. Evaluation of methods for determining the vertical distribution of hydraulic conductivity. *Ground Water*, 27, 88–98.

Vidal Pardal, M., 1962. Estudio acerca de la permeabilidad del terreno. *Inform. Est. Serv. Geol. Obras Públicas*, 15, 1–66.

CHAPTER 4

Geophysical techniques

4.1 GENERALITIES

The use of geophysical techniques for the investigation of leakage from dams and lakes is certainly very scarce. This is due to the following reasons:

1) The number of known geophysical techniques is very high, but many technicians engaged in this field are not familiar with the real possibilities they offer. This is partially due to the rapid development during the last few years of most of these techniques; these rapid changes have been mainly derived from the use of integrated electronic circuits. The availability of such units substantially improves data collection, processing and analysis with very high precision and velocity.
2) Both the performance of field tests and the interpretation of the results require the collaboration of specialised and well-experienced technicians. Furthermore, the interpretation of the results is, in many instances, linked to subjective appraisals, and incorrect interpretations are not uncommon. These failures have generated mistrust in the techniques in many cases.
3) Usually, geophysical techniques alone do not provide definitive solutions to dam leakage problems. The results obtained need to be confirmed by other independent techniques.

In spite of the previous statements, some geophysical methods can play an important role in the studies carried out prior to and during the dam construction. Valuable information can be obtained on the hydraulic characteristics of the rock formations that will contain the future reservoir. Concretely, these techniques can provide information on the existence, depth and dimensions of permeable formations, on the presence of fractures and other cavities in consolidated rocks and, in general, on the possible existence of structural accidents that may affect the waterproofing characteristics of the geological formations at the dam site. Consequently, the results obtained with these techniques might be of great help during the planning of grouting or other waterproofing works at the dam site.

The above-mentioned possibilities justify the decision to devote a full chapter to the geophysical techniques in this book. However, due to the specialised scope of the book, it is not possible to make a detailed description of these techniques. The most relevant features of the techniques will be stressed for the mentioned applications. A large number of publications, reports and journals on this subject are available to the interested reader. A selection of the literature is presented at the end of this chapter (e.g., Allaud and Martin, 1977; Repsold, 1989; Serra, 1987; Vogelsang, 1995).

Classically, geophysical prospecting techniques can be classified in two main groups: (i) surface methods, and (ii) borehole or downhole methods (well-logging). Both categories include numerous specific techniques, and one specific technique may have a number of variants. Surface prospecting methods are especially interesting and more attractive for the investigation of dam leakage because, in principle, they offer the possibility of detecting groundwater flows without the need for drilling expensive boreholes.

It must be said that geophysical prospecting methods provide better results when used in an integrated way, that is, when different geophysical techniques are applied for the investigation of the same problem or the same geological formation. A difficulty for this integrated use arises from the fact that, very often, the working groups dealing with this evaluation do not have access to all techniques that may provide valuable information on the hydraulic properties of the relevant geological units.

4.2 SURFACE PROSPECTING GEOLOGICAL METHODS

4.2.1 Self-potential

4.2.1.1 *Theoretical considerations*

The self-potential technique aims at the localisation of groundwater flows through the surface measurement of potentials generated by flowing waters. This method has been used by different authors for several decades in connection to dam leakage studies (e.g., Al-Saigh et al., 1994; Bogolovski and Ogilvy, 1970a, b; Corwin, 1988a; Cuéllar et al., 1996; Haines, 1978; Ogilvy et al., 1969; Schiavone and Quarto, 1984). However, in spite of the experience gained by some groups, the technique is still rather unknown.

It has been well established that water flows circulating through some conduits in the terrain can generate along their pathway differences of electrical potential, called *streaming potential*, which can be measured from the surface. This potential is called electrokinetic potential and is produced as a result of the interaction of flowing water with solid material of the terrain (Corwin and Hoover, 1979). Groundwater acts as a conductor, transporting electrical charges from one point to another. The phenomenon is attributed to the fact that a potential difference, the so-called *z-potential*, exists at the interface between the two media, water and rock. The electrokinetic potential V generated by flowing water circulating through a capillary tube can be expressed by the following equation:

$$V = C\Delta p, \quad \text{with } C = \frac{\rho\varepsilon\xi}{4\pi\eta} \tag{4.1}$$

where:

Δp = pressure drop along the water flow,
ρ = liquid resistivity,
ε = water dielectric constant,
ξ = electrokinetic potential,
η = water viscosity,
C = electrokinetic coupling coefficient.

In porous media, the equation giving the electrokinetic potential is more complicated than the previous one due, among other factors, to the influence of the grain size distribution, tortuosity and surface of contact between water and rock. For practical purposes, the coefficient C cannot be determined.

Equation (4.1) shows that the generated self-potential is proportional to the pressure drop along the water flow, that is, to the difference of hydraulic head, which increases with the flow velocity. Fitterman (1978) and Corwin and Hoover (1979) postulated that the existence of a pressure gradient along the water flow is not a sufficient condition to ensure the existence of a measurable difference of electrical potential at the surface. According to these authors, it is also necessary for the pressure gradient to be parallel to the separation surface of two zones of the terrain, with different values of the electrokinetic coupling coefficient C. The polarity of the potential anomaly depends on the sign of the difference $C_1 - C_2$, and its magnitude depends on the absolute value of such difference. Bogolovski and Ogilvy (1970a) postulated that the electrical potential increases along the flow direction and that the differences of potential are proportional to the velocity and flow rate of groundwater flow.

Schiavone and Quarto (1984) have described the different types of anomalies that can be obtained for different geometrical patterns of groundwater flow, e.g., flow along geological faults, vertical ascending flow, vertical descending flow, etc.

4.2.1.2 *Experimental procedure*

The measurement of self-potential is performed by placing on the ground measuring electrodes distributed in a convenient array over the surface at the zone where it is assumed that water escaping from the reservoir is flowing. When the technique was developed, the electrodes were made of metallic rods, generally of copper, which were driven into the soil. To improve the electrical contact with the soil, the zone around the electrodes was kept wet through irrigation. The main problem with these electrodes is their polarisation. This polarisation can create local potential anomalies that are frequently much higher than the anomalies derived from groundwater flow.

Nowadays, unpolarisable electrodes are used, which are formed by a copper rod introduced inside a saturated solution of copper sulphate (Corwin [1988b] has used electrodes of the type silver/silver chloride). The solution of copper sulphate is contained in a ceramic pot with half-permeable walls. For the construction of these pots, a ceramic material is used whose behaviour is similar to the pitchers that allow drinking water to be kept cool in the summer time. These electrodes are buried in the soil and are interconnected and linked to the volt meter by means of an electrical conductor. The infiltrated solution throughout the walls of the pot ensures a good electrical connection with the ground. The measurements are performed using a high impedance millivolt meter. The values of the generated self-potential are usually lower than 200 mV.

The electrodes are placed in the soil with different geometrical arrangements. The simplest and most frequently used arrangement consists of placing the electrodes along a straight line, more or less perpendicular to the supposed direction of groundwater flow, as shown in Fig. 4.1A. The distance between the electrodes can vary between 5 and 20 m, depending on the magnitude and extent of the anomaly.

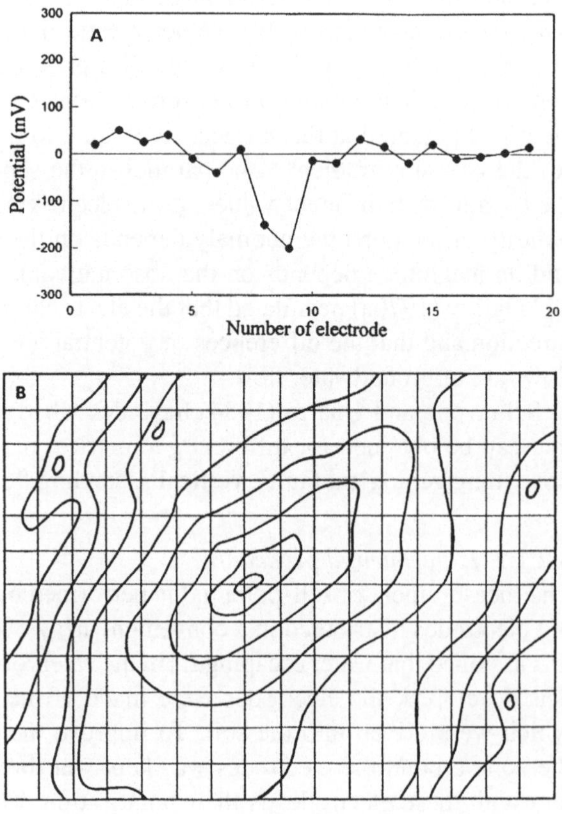

Fig. 4.1. Typical arrangement of electrodes for the localisation of groundwater flows in areas close to dams using the self-potential method. A = linear arrangement; B = reticular arrangement.

The potential differences are measured with respect to a reference electrode that is placed in a zone not affected by the flow. The results obtained with this technique may reveal potential anomalies such as the one shown in Fig. 4.1A, which can be interpreted as being a consequence of the prevailing groundwater flow.

A more complicated array of the electrodes consists of deploying them in a grid, such as the one shown in Fig. 4.1B, that is, an electrode is placed at each junction. The measured values are plotted then in the form of a contour map with lines of equal potential, the distribution of which reflects the possible existence of anomalies derived from the groundwater flow prevailing in the zone.

In the opinion of the authors, the few studies reported in the literature using the described technique for dam leakage investigations are affected by two important limitations:

1) The results obtained during the investigation were never contrasted in an irrefutable way using other independent and more direct techniques. In other words, the true existence of the assumed groundwater flows was not unequivocally proven, as indicated by this technique.
2) The technique has been applied as an isolated tool. The combined and integrated use of other geophysical techniques, such as electrical resistivity, seismic refraction, microgravity and penetrating radar, could have provided, in most cases, additional information supporting the self-potential results.

To evaluate the real possibilities offered by this technique, it is absolutely necessary to carry out additional experimental studies in areas where the existence of groundwater flow is evident or well proven, as happens, for example, in areas close to the discharge points of karstic springs. It must be taken into account that self-potential anomalies can be derived from other causes besides the existence of groundwater flows, such as tellurian currents, some particular biological processes and/or changes of temperature and humidity in the soil.

4.2.2 Electrical resistivity

The electrical resistivity method is one of the most frequently used in geophysical prospecting. The method is based on the measurement of the propagation through the ground of an induced direct electrical current by means of an appropriate network of electrodes. The method measures the resistance to the flow of electricity through the geological horizons. The specific resistance of the soil materials is mainly governed by the presence of water and its salinity. Usually, dry soil materials show resistance values much higher than the typical resistance measured in pure water. The specific resistance of water varies between 0.03 ohm · metre for sea water and about 300 ohm · metres for low salinity water. Contrarily, rocks with low water content show resistivity values of several thousands of ohm · metres.

The method is illustrated in Fig. 4.2. The electrical current is applied to the ground by installing *current electrodes* in the soil surface (electrodes *A* and *B* of

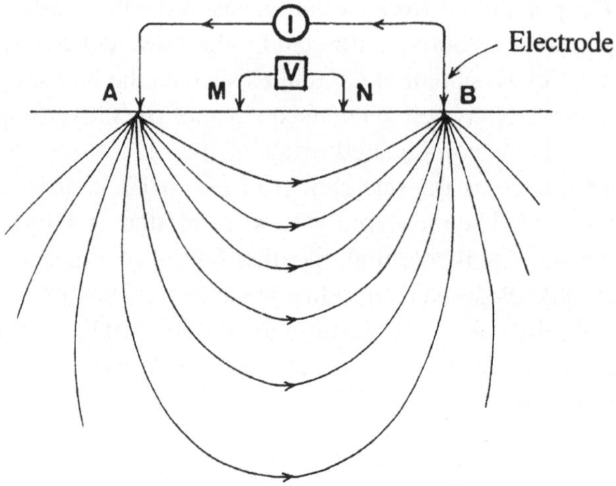

Fig. 4.2. Measurement of resistivity using a four-electrode arrangement and generated pattern of flow current. *A* and *B* are current electrodes, *M* and *N* are potential electrodes. The resistivity is given by the ratio V/I.

Fig. 4.2). Therefore, direct electrical contact with the ground surface is necessary. This current is measured by a volt meter located between the two remaining electrodes, called *potential electrodes* (electrodes M and N). Different arrangements of the electrodes can be used depending on the measurement requirements. The mean depth of the investigated rock formation is directly related to the distance between the current electrodes.

Resistivity tests frequently have been used for the localisation of leaks from dams and lakes. The magnitude of the anomalies observed in the resistivity profiles is controlled by the characteristics of the geological formations and the existence of groundwater flow. Usually, fractured zones produce low resistivity anomalies because they act as active pathways or because the presence of clay or other weathering products fills cavities.

Anomalies derived from the presence of karstic channels can be of high or low resistivity. Channels filled with groundwater or with clayey materials also present low resistivity anomalies. The contrary is valid for empty channels.

Generally, for the determination or estimation of the depth of fractured horizons that generate resistivity anomalies, the measurement of several horizontal profiles of resistivity with different electrode distances is necessary.

4.2.3 Seismic waves refraction

The geophysical prospecting method known as seismic refraction is based on the transmission through different soil strata of the vibrating waves generated by a point explosion produced at the surface. A powerful explosive, such as dynamite rubber, is used for this purpose. However, when the identification of very shallow strata is required, a hammer specially developed for this purpose is frequently used. The soil surface is strongly beaten with this hammer at a selected point. In both cases, the refracted waves are measured by a number of detectors, called seismometers or geophones, that record the soil vibrations. These geophones are usually placed in a linear array and connected to a multichannel measuring system that records the seismograph corresponding to the different geophones (waveform versus time).

The time interval elapsed from the explosion or blow to the arrival of the refracted waves at the geophones is measured. The transmission velocity of these waves depends on the nature of the soil strata. As mentioned, the hammer technique is used exclusively for near-surface investigations. The explosive technique allows the investigation of subsurface layers up to depths of several tens of metres. Properly interpreted, the refraction data allow the estimation of the thickness and depth of unconsolidated materials and the bedrock, including lateral changes of facies.

Typical values of the wave transmission velocity, given in metres per second, are the following:

Vegetal soils, silt, sand and gravel in the unsaturated zone	250–900
Sand and gravel in the saturated zone	1400–1800
Water	1400–1600
Clay	900–2500
Sandstone, shells, marls and limestone, granite and metamorphic rocks	1900–7000

As previously mentioned, the measuring system records the transmission velocity of the vibrating wave. In the case of consolidated rocks, the measured values are compared with the typical values corresponding to the nonfractured rock formations. If a measured value is much lower than the typical one, it is concluded that the quality of the rock is poor due to the presence of fractures, fissures, intense weathering or other processes. High water content, derived from high density of fractures or from the presence of karstic cavities, produces a decrease in the transmission velocity of the vibrating waves.

The technique permits the localisation of faults and other tectonic structures when the average thickness is sufficient to produce a significant delay in the wave transmission. In the same way, this technique may provide quantitative information on the thickness of alluvial, colluvial, or in general, detritic layers deposited over impervious consolidated rocks with high transmission velocity of the vibrating waves.

When studying leaks from dams and lakes, the technique can play a complementary role in the localisation of pathways of flowing groundwater, providing additional arguments to explain the existence of pathways for groundwater flow.

4.2.4 Microgravity

The microgravity test is based in the detection of gravity anomalies from the ground as a result of variations in the density of near-surface formations. The technique has experienced an important development over the last several years. The gravitational field is expressed in terms of acceleration (cm/s^2 or gal). The instruments used for conventional gravity measurements are torsion balances, pendula and gravity meters. All of them can detect gravity anomalies as small as one ten-millionth of the gravitational field. Microgravimeters are calibrated with a precision of microgals, which are sufficient to detect cavities in the subsurface.

Gravity data obtained in the field must be corrected for elevation, rock density, latitude, and influence of surrounding topographic variations. The corrected data are plotted in the form of anomaly maps (Bouguer anomaly), which show similar shapes and forms as those found in topographic contour maps.

Basically, the technique is used for the localisation of shallow karstic cavities. A second important application of the technique is the evaluation of the effectiveness of grouting walls constructed for stopping or decreasing water flows. This effectiveness is usually determined by gravity measurements performed before and after the construction of the grouting wall. Bedrock valleys buried by unconsolidated materials can usually be evidenced with this technique, especially if groundwater flow is involved.

4.2.5 Penetrating radar

The geophysical technique known as penetrating radar is very valuable for the localisation of fractures and cavities affecting shallow formations (depth less than 15 m) (Ballard, 1983; Olhoef, 1984, 1988). The method is based on the subsurface penetration of the radio- and microwave portion of the electromagnetic spectrum. The technique has been used in the past for detecting land mines and subsurface tunnels.

The penetrating radar method uses a small antenna to radiate high-frequency radio waves into the subsurface and a receiving antenna to record variations in the reflected return signal. Dragging the antennae along the ground surface generates a continuous profile of the return signal. The method, when applicable, provides a vertical resolution much higher then the results provided by all other geophysical prospecting methods. However, the depth of penetration is usually very low (between 1 and 15 m) and is strongly limited by the presence of water, soils with high electrical conductivity and fine-grained materials. However, if such materials are not present, the penetrating radar technique can detect cavities with very high vertical and horizontal resolution.

The technique can provide interesting information on the presence below the soil surface of materials with very high concentrations of crystallisation water, as well as on the possible existence of cavities, trench boundaries and highly fractured formations, which can be of great importance in dam leakage studies. The presence of fillings and recently reworked materials can also be detected and evaluated. Apart from that, the technique has been increasingly used to estimate the depth to the water table and to identify soil horizons and lithological contacts, faults, bedding joints and planes in rocks.

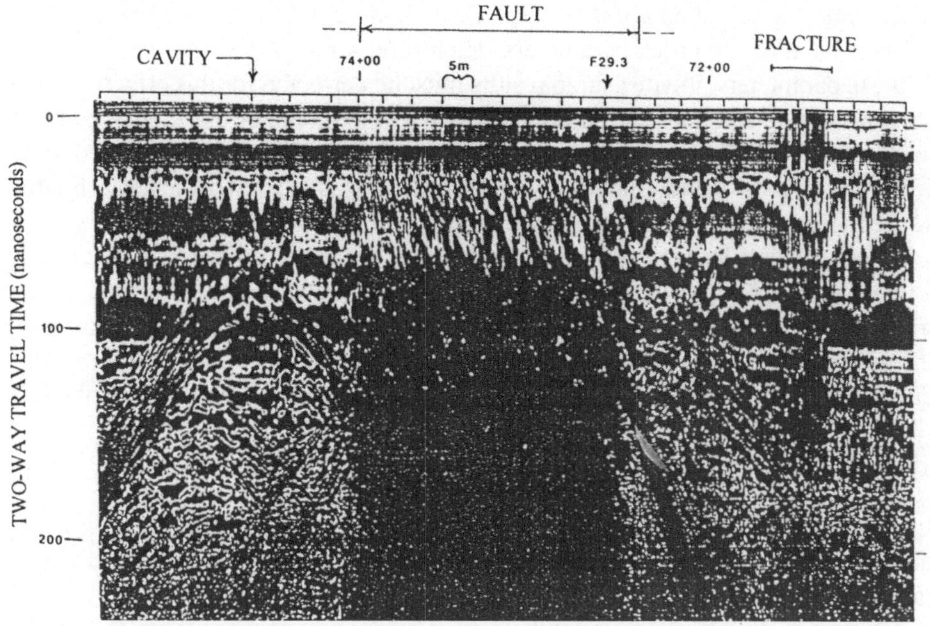

Fig. 4.3. Penetrating radar record obtained by Llopis and Butler (1988) at the Beaver dam, Arkansas, USA.

Figure 4.3 shows the results of a profile obtained by Llopis and Butler (1988) at the Beaver dam in Arkansas (USA). The profile was obtained in a faulty area of a graben located at the dam site. The fault is clearly reflected in the record by a vertical dark band, where no return signal is observed due to the presence of a band with a high water content.

4.3 GEOPHYSICAL WELL-LOGGING

4.3.1 Generalities

Geophysical well-logging methods have been used for over 50 years in the field of oil exploration and in the search for groundwater. The number of available techniques is very large, and their use can provide very valuable information for the study of diverse hydrological problems. Most of these methods have been drastically improved during the last few years. Other methods of measurement, such as single and multiple borehole methods for determining groundwater flow, are described in Chapter 8.

For hydrological applications, measuring probes with a diameter of 35 to 50 mm are predominantly used. The ratio of the borehole diameter to the probe diameter determines the so-called *borehole effect*, which more or less distorts the determination of the unknown rock parameters. Therefore, for a quantitative determination of these parameters, the measured results must be corrected for this effect. To keep the borehole effect small, the slimmest possible hole should be used. For the above-mentioned probe diameter, borehole diameters of 50 to 100 mm will be appropriate. However, in most cases, quantitative information is not required; instead, relative vertical changes of the measured parameter are sufficient.

About 40 well-logging techniques have been described so far. Very frequently, two or more techniques are used in a combined way because some logs require the measurement of other logs for interpretation. Due to the scope of this book, only the methods that, in opinion of the authors, are considered more useful for studies dealing with the investigation of leaks from dams and reservoirs are described.

4.3.2 Natural gamma logging

In the natural gamma well-logging method, the gamma radiation emitted by the radioactive materials contained in the surrounding rocks is measured. These radioactive materials are basically potassium-40 and the members of the natural radioactive series of uranium and thorium. The concentration of these radioactive components strongly varies from one type of rock to another. The isotope ^{40}K is a natural component of the element potassium and, therefore, is present in all rocks containing this element. The natural abundance of this isotope amounts to 0.0118%. Rocks with high potassium content are those formed by all micaceous minerals, such as muscovite, biotite, illite and feldspar.

Table 4.1 shows the range of concentrations of radioactive materials in the most important sedimentary rocks and other rocks, expressed as radium equivalent

Table 4.1. Radioactivity content of some sedimentary rocks expressed in radium equivalents.

Type of rock	Radium equivalents (10^{-12} g of radium per gram of rock)
Anhydrite	0.5
Brown coal	1.0
Halite	2.0
Dolomite	0.5 to 10
Limestone	0.5 to 12
Sandstone	1 to 15
Clayey sandstone	2 to 20
Clayey limestone	2 to 20
Carbonated lutite and shale	3 to 25
Noncarbonated lutite and shale	4 to 30
Potassic feldspars	10 to 45
Bottom sea clay	10 to 60

per gram. This unit is defined as the equivalent response given by a scintillation detector supposing that the only radioactive material existing in the measured rock is radium in equilibrium with all members of its family. It can be seen that the highest radioactivity contents correspond to clayey rocks and shells, while the lowest values are found in anhydrite, coal and halite. On the other hand, the high variability of the radioactivity levels observed for most of the rocks shown in Table 4.1 must be noticed. This fact produces, in some cases, a serious difficulty in the interpretation of the results. Nevertheless, this difficulty can be accounted for if the lithological composition of the formations intercepted by some wells or boreholes is known. The gamma ray signal obtained for each type of rock can be taken as an intrinsic characteristic in the study area. Natural gamma profiles obtained in areas where this information is not available can be used for the identification of the intercepted rock.

Usually, the probes used for natural gamma profiling consist of a scintillation detector fitted with a crystal of sodium iodide activated with thallium (NaI(Tl)). The interaction of the gamma rays with this crystal produces light flashes (scintillations) that are converted by a photomultiplier tube and its associated electronic circuit into electrical pulses. The number of pulses per time unit (counting rate) is proportional to the intensity of the gamma ray flux emitted by the surrounding rock formation in the zone where the detector is placed. Due to the gamma ray attenuation in the rock, only a thickness between 20 and 40 cm is effective for gamma counting.

On the other hand, the counting rate measured by the gamma ray probe depends on the following factors:

1) The content of radioactive material in the surrounding rock formation.
2) The size of the scintillation crystal and the setting of the associated electronic measuring system.
3) The borehole diameter.
4) The type and thickness of the casing.
5) The position of the probe in the borehole (centred or not centred).

A quantitative and reproducible conclusion on the radioactivity of the rock under investigation cannot be easily derived from the counting rate. To do that, a borehole correction and a previous calibration of the gamma ray probe are necessary. However, investigators are more interested in achieving qualitative or half-quantitative gamma ray profiles than in obtaining absolute values of radioactivity. Frequently, the gamma ray intensity (I) produced by a layer or stratum is expressed as the gamma ray index (GRI) given by the expression:

$$GRI = \frac{I_L - I_{min}}{I_{max} - I_{min}} \tag{4.2}$$

where:

I_L = gamma radiation intensity produced by the layer concerned,
I_{max} = maximum gamma radiation intensity observed in the study area, usually produced by pure shale or pure clay,

I_{min} = minimum gamma radiation intensity observed in the study area, usually produced by clean sand or sandstone.

An equation similar to the previous one is used for the determination of the shale content in Tertiary and even older rocks. The measurement is based on the high radioactivity content of shales, due to the fact that their origin is the consolidation of clayey materials.

An important advantage of this technique is that it may be applied in all types of wells and boreholes, regardless of their diameter and casing. Of course, the presence of casing produces a certain attenuation of the gamma ray flux, but usually this does not have much influence on the possibilities offered by the technique because, most frequently, the interest is the measurement of relative values of this gamma ray flux more than the absolute values. For this reason, natural gamma profiling is one of the techniques most frequently used. Usually, probes are constructed in such a way that both the gamma ray detector and the self-potential and resistivity units are combined.

The most important applications of the gamma ray profiling technique are the following:

1) Investigation of stratigraphic correlations, mainly oriented toward the determination of the continuity of pervious and impervious strata.
2) Identification of rock types on the basis of their radioactivity, although this can be difficult to achieve when this technique is applied as the only well-logging method, due to the usual large variability (see Table 4.1). Different types of rocks can provide similar readings of gamma ray intensity, and inversely, the same type of rock can present very different values of intensity. Nevertheless, once the gamma ray intensities produced by the different types of rocks existing in the study area have been established, the identification of these rocks in other places can be a relatively easy task.
3) Recognition of clay-rich formations in sedimentary sequences, due to their higher natural gamma activity. This is due to the high concentrations of potassium, uranium and thorium isotopes absorbed by clay lattices during deposition.
4) Discrimination of overburden from nonweathered rocks. In areas where hard rocks are overlain by weathered material, the gamma ray counting rate obtained for the overburden can be much lower than the value measured in the nonweathered rocks, due to the preferential leaching of minerals containing potassium and uranium.

Figure 4.4 shows a natural gamma log obtained by the senior author in an alluvial aquifer of Vietnam. As expected, a high intensity of gamma radiation was observed in the clayey horizons and low intensity in the sandy ones.

When the technique is used in consolidated rocks, the fractured or karstified stretches may show a gamma radiation intensity higher than the typical value found

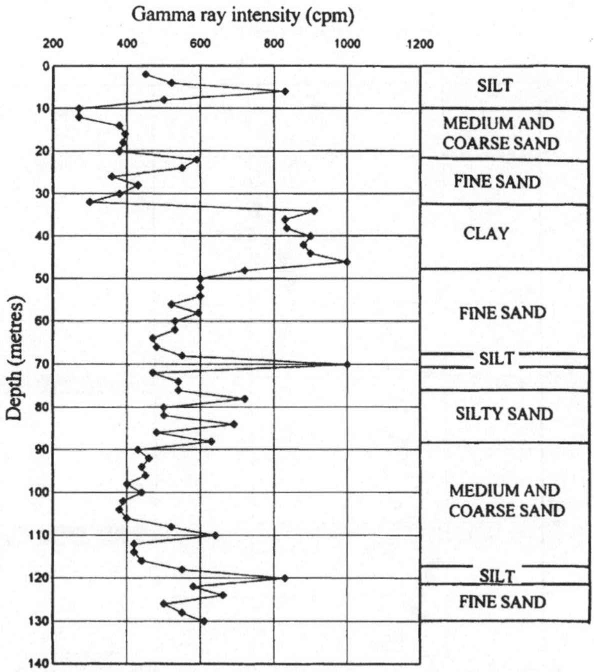

Fig. 4.4. Natural gamma radiation log obtained by the senior author in a borehole of an alluvial aquifer located in an area close to Ho Chi Minh City, Vietnam.

for the matrix rock if the fractures or cavities are filled with clay, or lower if these are filled with water. In the latter case, water acts as a shielding for the gamma rays emitted by the hard rock, and at the same time, water does not contribute significantly to the gamma counting due to its low content of radioactive components. That was clearly observed in the gamma log obtained by the senior author in a borehole located at the Sabaneta dam site (Dominican Republic), which intercepted limestone rocks exclusively (Fig. 4.5). The fractured stretches generate in the gamma ray log peaks of low intensity due to the presence of water. The same fractured stretches were also identified through the measurement of the groundwater flow velocity using the tracer techniques described in Chapter 8. Similar results were obtained in two other boreholes in the same dam site.

During dam construction, the natural gamma technique can be of high relevance to study the permeability of alluvial strata at the dam site. Similarly, this method, together with other geophysical techniques, may assist in the identification of fractured stretches in compact rocks both in the saturated zone and in the part of the unsaturated zone that will be impounded.

Nowadays, the use of scintillation detectors for the identification of radionuclides through gamma ray spectrometry is well established. The method is based on the fact that the height of the electrical pulses produced by these detectors is proportional to the fraction of energy released by the gamma rays in the NaI(Tl) crystal. Most of the gamma rays are incompletely absorbed in the crystal and produce

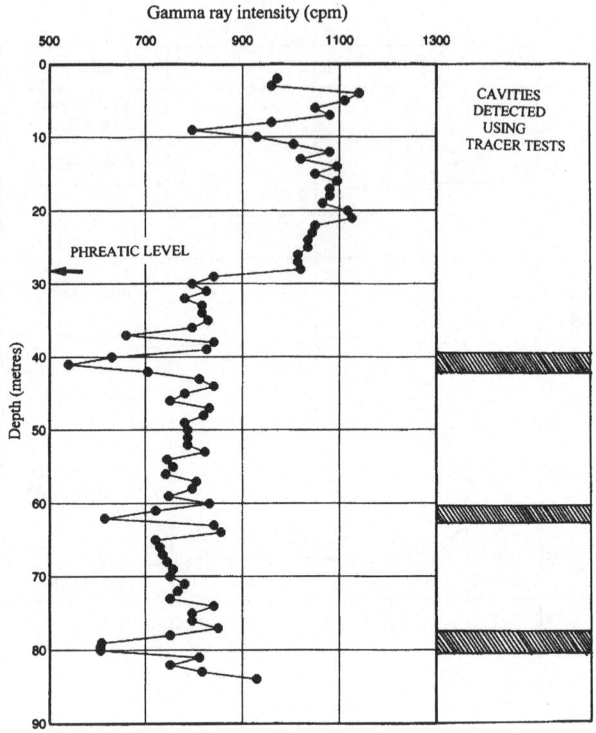

Fig. 4.5. Natural gamma radiation log obtained by the senior author in a borehole located at the dam site of Sabaneta reservoir (Dominican Republic). The log shows a good correlation between the gamma ray peaks and the location of the fractures detected by means of tracer techniques.

pulses of relatively low height. This is known as the Compton effect. However, some of the gamma rays are completely absorbed (photoelectric effect) and produce pulses of the maximum possible height, whose value is governed by the energy of the gamma rays. Such pulses form the so-called photopeak. The position of this photopeak in the gamma ray spectrum allows the determination of the gamma ray energy, facilitating the identification of the radionuclide present in the sample.

As already mentioned, most rock formations usually contain many different radioactive components, which emit different types of gamma rays. Therefore, the gamma ray spectra are usually very complex and show many photopeaks. However, due to the low resolution of the scintillation detectors with NaI(Tl) crystals, the spectra provided by them do not allow a clear identification of the gamma rays emitted by the rock formation. The different photopeaks overlap each other.

In spite of this difficulty, the U.S. company Mount Sopris Instrument Co., Inc. has recently developed a spectral gamma probe for the measurement of gamma rays in the energy interval between 100 KeV and 3 MeV (see Chapter 7 for unit definition), with a resolution of 12 KeV if a 256-channel pulse analyser is used and 6 KeV resolution using a 512-channel pulse analyser. The detector uses both CsI(Tl) and NaI(Tl) crystals. The incorporated software allows the selection of up to 10 energy windows from the spectrum to be examined during logging. The

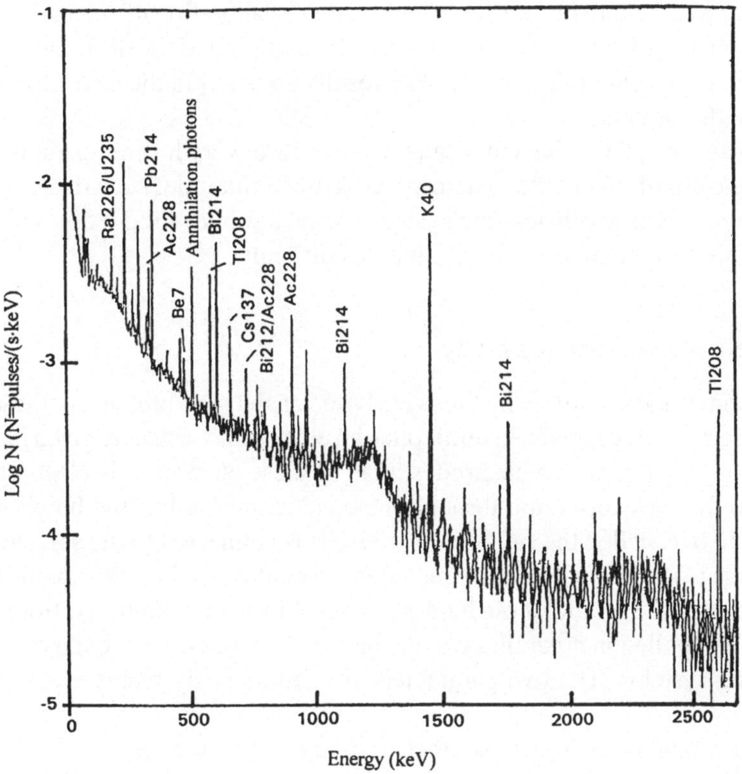

Fig. 4.6. Gamma ray spectrum obtained by Rank (1985) for a sediment sample collected at the Aschach reservoir in the Danube River, Austria, using a germanium–lithium detector. Specific activities of different radioisotopes expressed in Bq/l: ^{7}Be $= 11 \pm 4$; ^{40}K $= 597 \pm 36$; ^{137}Cs $= 7.4 \pm 0.8$; ^{226}Ra $= 56.2 \pm 4.2$; ^{232}Th $= 46.5 \pm 3.4$. Concentration of elements expressed in mg/kg: Potassium $= 19400 \pm 1200$; Thorium $= 11.2 \pm 0.6$; Uranium $= 4.6 \pm 0.3$.

contribution of ^{40}K and consequently the clay content can easily be evaluated due to the high gamma radiation energy emitted by this radionuclide. This technique represents an important improvement in the characterisation of the rock formations intercepted by the borehole.

On the other hand, it is well known that the identification of the different gamma rays emitted by different rock formations is possible using solid-state detectors of the type germanium-lithium (Ge-Li). These detectors are characterised by a very high resolution and provide gamma ray spectra of the type shown in Fig. 4.6, which was obtained by Rank (1985) using sediment collected at the reservoir Aschach in the Danube River (Austria). In this spectrum, the photopeaks corresponding to the gamma rays emitted by the radionuclides ^{40}K, ^{137}Cs, ^{7}Be, as well as those emitted by different members of the uranium and thorium radioactive series, are clearly identified. The determination of the area of the different photopeaks, given by the respective total number of pulses, allows, after a calibration of the measuring system, the quantitative determination of the concentration of the different radioactive components present in the sample. The results obtained in the case of Fig. 4.6 are shown in the caption.

Unfortunately, these detectors can only operate at very low temperatures, which require the use of cryogenic systems. Due to this limitation, such detectors cannot be incorporated into well-logging systems. Maybe similar solid-state detectors will be developed in the future to overcome this difficulty.

4.3.3 Self-potential and resistivity

Self-potential and resistivity methods are based on the measurement of these parameters (electric voltage, given in millivolts and electrical resistance, given in ohm · m, respectively) between two electrodes. One of these electrodes is electrically connected to the rock formation through the conducting fluid of the borehole, while the second one, called the *reference electrode*, is connected to the soil surface at a point close to the borehole. These methods are mainly used for the identification of permeable strata in granular media, for example, in alluvial aquifers. Both methods are usually applied in a parallel way using double array probes that combine both measuring systems. The two parameters are continuously recorded as a function of depth.

Self-potential is not a rock parameter but rather an effect derived from the borehole itself as a result of the interaction of borehole water, pore water and rock formation (see Fig. 4.7). The measured resistivity is not the true rock resistivity but rather

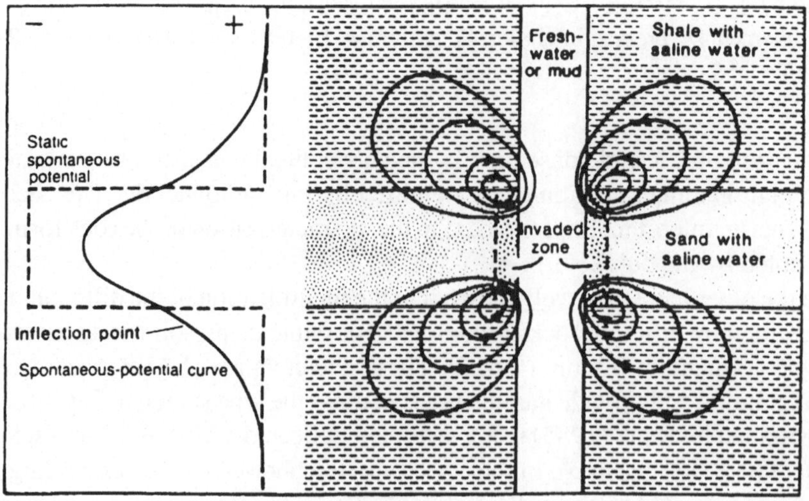

Fig. 4.7. The flow of current at the contacts between strata and the resulting self-potential profile (after Boulding, 1993).

the so-called apparent resistivity because the measurement is always performed in a borehole filled with mud or water, the influence of which cannot be ignored.

Technicians using these logging methods usually assume that the technique can only be applied in boreholes without casing, filled with drilling mud to avoid the collapse of the rock materials. This limitation requires good coordination between the drilling team and the logging technicians so that the logging can be performed before the casing is installed. The measured profiles are used to identify the most favourable stretches for the installation of screens, which is extremely important if the borehole will be used for groundwater withdrawal. Sandy permeable strata are usually characterised by negative values of the self-potential associated with a high resistivity. The contrary is valid for clayey and silty strata.

However, Araguás et al. (1998) have demonstrated in a recent study that the method can also provide valuable results in cases of boreholes equipped with a perforated PVC casing when water of high salinity (electrical conductivity higher than about 10 mS/cm) is present. The study was carried out in the Aznalcóllar dam of Spain, which was constructed to store mine tailing materials. The self-potential and single-point resistance (SPR) logs obtained in a number of boreholes allowed the identification of geophysical anomalies related to the dam collapse in April 1998, which was produced by the sliding of its foundation. An example of the logs obtained is shown in Fig. 4.8. The anomaly observed at a depth between 24 and 27 m was also observed in other boreholes. Temperature and electrical conductivity anomalies were also observed at the same depth in the respective profiles. The same anomalies were also reflected by the single-well tests performed for the measurement of groundwater flow, using the techniques described in Chapter 8.

Concerning dam construction, these logging methods can be highly valuable for the identification of permeable strata in alluvial formations eventually outcropping at the dam site. The thickness of permeable and impermeable strata can be precisely evaluated.

4.3.4 Electromagnetic induction method

The electromagnetic induction method was firstly described by Schlumberger in 1948. In this method, the inverse of the resistivity of the rock formation, that is, the electrical conductivity, is measured. Resistivity is expressed in ohm · metre, with 1 ohm · metre being equal to 1000 mmhos/metre. To obtain the true conductivity (or resistivity) of the rock, the use of correction or calibration diagrams specifically valid for the measuring system being used is necessary. The method does not work properly when dealing with rock formations of very high resistivity.

The measuring probe is fitted with a transmission coil, which generates a low frequency electromagnetic field (ca. 20 kHz), and one or two receiving coils. The electromagnetic field induces eddy currents around the borehole, the strengths of which are directly proportional to the conductivity of the rock. The magnetic fields produced by these eddy currents induce voltages in the detector coils of the probe,

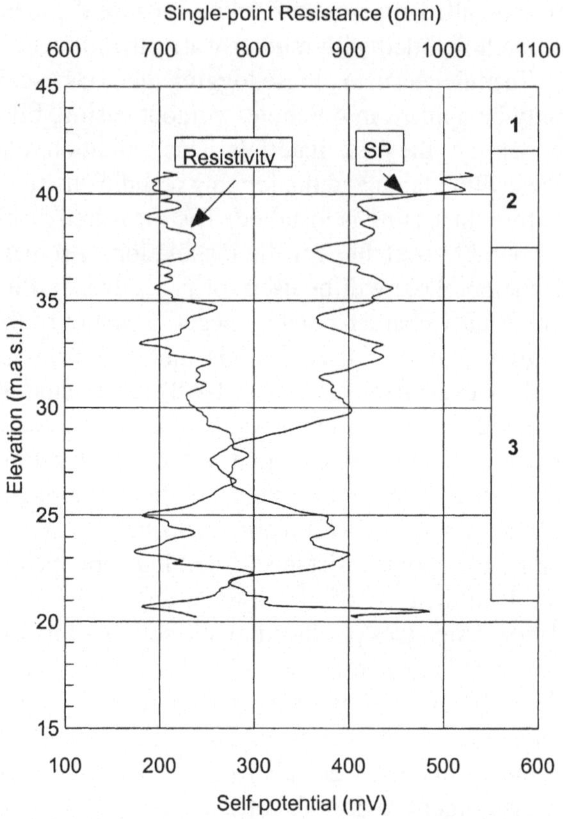

Fig. 4.8. Self-potential and single-point resistance (SPR) logs obtained by Araguás et al. (1998) in a borehole located at the Aznalcóllar dam, Spain, after installing a perforated plastic casing. The borehole contained high salinity water. Lithological column: 1) breakwater; 2) alluvial; 3) "blue marls" formation.

the amplitudes of which are approximately proportional to the conductivity of the rock. The induction log thus measures a conductivity signal that is mostly converted into a resistivity signal by means of a reciprocator.

An important advantage of this technique is derived from the fact that the presence of a conducting medium such as groundwater to electrically connect the probe and the rock formation is not necessary, as in the case of the previously described resistivity technique. This factor allows the logging to be carried out in dry boreholes without casing or fitted with a casing consisting of a plastic tube. In other words, the measuring system can be used both in the saturated and unsaturated zones. This renders the technique very interesting for studies prior to or during the dam construction. The technique can also provide valuable information for the localisation of groundwater flows derived from the leaks occurring at the reservoir.

4.3.5 Gamma–gamma logging method (γ–γ)

4.3.5.1 *Theoretical considerations*

The (γ–γ) method provides information on the bulk density of the rock formation expressed in g/cm^3. The measuring probes are fitted with a radioactive source

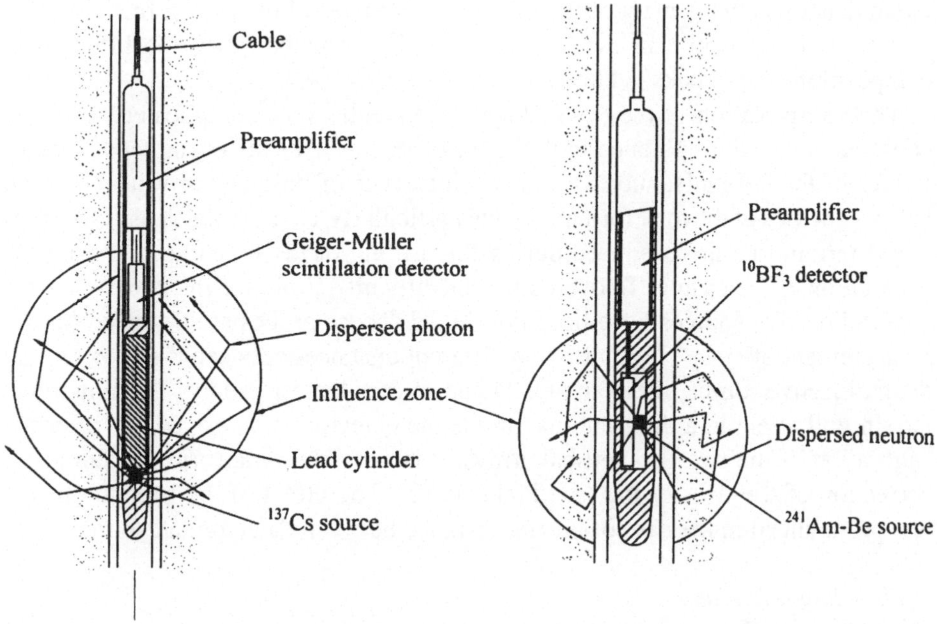

Fig. 4.9. Illustration of the gamma–gamma and neutron–neutron probes.

of an appropriate gamma emitter, usually caesium-137, and a Geiger-Müller, or better, a scintillation detector with an NaI(Tl) crystal (Fig. 4.9). A lead shielding is placed between the radioactive source and the detector to stop the gamma rays from impinging directly on the detector. The gamma rays are then forced to travel through the terrain, where they are partially dispersed by the Compton effect and absorbed by the photoelectric effect. Obviously, water stored inside the borehole, when applied in the saturated zone, and the borehole casing also contributes to the dispersion and absorption processes. If these contributions are kept constant, the variations of the intensity of the gamma ray beam that reaches the detector can be related to changes in the density of the rock formation. The net counting rate of the detector tends to be approximately inversely proportional to this density.

The radius of the influence zone depends on the distance between the radioactive source and the detector, as well as on the density of the rock formation, which ranges from 1.4 to 2.9 g/cm^3. For the commercially available measuring systems, this radius usually varies between 15 and 35 cm.

When the measurements are made in the saturated zone, the porosity of the rock formation can be computed using the following equation:

$$Porosity = \frac{grain\ density - bulk\ density}{grain\ density - water\ density} \tag{4.3}$$

Grain density is the density of the individual particles of the rock formation. This parameter is measured in the laboratory or is estimated when the lithological composition of the rocks is known.

The interpretation of the $(\gamma-\gamma)$ logs is frequently very complicated. This is a consequence of the influence in the detector response of the constructive irregularities of the borehole, such as casing joints, type of fluid (air or water), natural heterogeneities of the rock formation, and particularly, cavities and density changes of the terrain that are produced during the drilling of the borehole or during the installation of the casing. To overcome this difficulty, probes using two exchangeable radioactive sources have been developed. These probes are designed for the measurement of $(\gamma-\gamma)$ logs with two different distances between the detector and the radioactive source, for example, 24 and 45 cm. Any irregularity existing in the terrain will affect in a different magnitude the detector response, so that after the calibration of the two geometrical arrays, it is possible to perform the automatic correction of the density log (see Plata, 1972). The effect of the irregularities is less pronounced in the case of greater distance between detector and source.

4.3.5.2 *Suggested uses*

The $(\gamma-\gamma)$ well-logging method is recommended for routine investigation of density changes produced in earth-filled dams. Very frequently, such changes are associated with sitting-down processes affecting the dam, which can take place during the first filling or even afterwards. In a similar way, sitting-down of the boundary granular rock formations during the filling and emptying of the reservoir can be studied using this technique. These surveys are particularly interesting in the case of dams constructed in zones with high seismic activity. Differential sitting-down processes can produce permeable paths through which the impounded water can escape (pipe formation). Contact surfaces between concrete structures and granular materials are especially sensitive to this differential sitting-down, and therefore to the development of leaks, because the concrete structure can act as a roof for the pipe formation.

The density measurements are to be performed in boreholes drilled along the dam body or in the adjacent rock formations. Even boreholes drilled for inclinometric measurements can be used, provided that they are not cemented. The interpretation of the results will be, in principle, easy due to the fact that the $(\gamma-\gamma)$ logs will always be repeated in the same boreholes. Borehole irregularities will not have any influence on this interpretation because the changes of density will be determined from the comparison of logs measured in successive field campaigns.

Particularly interesting is the use of this technique in reservoirs intended to impound acidic waters, for example, mine tailings and wastewaters from mining plants. If the dam materials contain a significant fraction of calcareous rocks, these can be strongly attacked by acidic waters and a drastic change in the density can be produced over time. The chemical attack may result in important accidents due to the formation of preferential infiltration pathways. This process was probably a contributing factor in the disaster produced in 1998 at the Aznalcóllar dam in southern Spain.

4.3.6 Neutron–neutron (n–n) logging

4.3.6.1 *Theoretical considerations*

The (n–n) logging technique provides information on the humidity of the terrain. The measuring probe uses fast neutrons from a radioactive source prepared by mixing an alpha particle emitter (usually, americium-241 or plutonium-239) and beryllium powder. The neutrons are generated by the following nuclear reaction:

$$^{10}\mathrm{Be} + \alpha \rightarrow {}^{12}\mathrm{C} + \mathrm{n}$$

Fast neutrons (average energy of 4.5 MeV) are slowed down in the rock formation by successive collisions with the atoms existing in their influence zone (Fig. 4.9). Due to the fact that the mass of the hydrogen atoms is very similar to that of neutrons, the slowing-down power of this element is much higher than that of the rest of the elements. This means that the energy loss undergone by neutrons is also higher when the collision occurs with a hydrogen atom. After several collisions with hydrogen atoms, neutrons lose most of their energy (velocity) and are transformed into thermal neutrons (energy about 0.025 eV). On average, 18 collisions are necessary for the thermal energy level to be achieved. With silicon or chlorine atoms, about 261 and 316 collisions are needed, respectively.

The detection of the thermal neutrons is accomplished by means of a proportional detector that contains boron tri-fluoride (BF_3) enriched in ^{10}B as a filling gas. The thermal neutrons are detected through the following nuclear reaction:

$$^{10}\mathrm{B} + \mathrm{n} \rightarrow {}^{7}\mathrm{Li} + \alpha$$

The generated α particle produces the ionisation of the filling gas of the proportional detector, which, in turn, produces an electrical pulse that is detected by the electronic system.

The previous nuclear reaction occurs only with thermal neutrons, and as a consequence, the detector is unable to detect the fast neutron emitted by the radioactive source. The number of pulses per time unit (counting rate) given by the detector is proportional to the hydrogen content in the influence volume, and therefore, to the water content. This is the case because most rocks generally do not contain minerals made up of hydrogen. Obviously, the measuring system has to be calibrated using appropriate samples as standards to relate the counting rate with the water content or the humidity of the sample.

The influence volume has a more or less spherical shape, and its radius is approximately given by the following equation:

$$r(\mathrm{cm}) = 15(1/H_v)^{1/3} \tag{4.4}$$

where, H_v is the volumetric humidity given in g/cm^3. In the case of pure water, H_v is 1 and r is equal to 15 cm.

When the measurement is performed in the saturated zone, the method directly provides a value for the porosity of the formation because it is assumed that all pores are filled with water. In the case of the unsaturated zone, the volumetric humidity is obtained. The difference between the bulk density, evaluated by the $(\gamma-\gamma)$ log as described in previous sections, and the volumetric humidity, reflects the dry density of the terrain. This parameter is of critical importance in most geotechnical investigations.

As in the case of $(\gamma-\gamma)$ logging, the most important errors derived from the use of this technique are related to the presence of irregularities within the volume of influence. Further information on the technique can be found in Plata (1972).

4.3.6.2 *Suggested uses*

The suggested uses of the (n–n) technique in dam leakage studies are those described for the $(\gamma-\gamma)$ logging technique in Section 4.3.5.2. In most cases, both techniques are applied in a parallel way. The requirements for borehole construction are the same for both techniques. The measurements in the saturated zone are only possible in boreholes with a small inner diameter. This diameter must be only slightly greater than the diameter of the measuring probe in order to keep small the influence of the water stored in the borehole.

When the technique is applied in boreholes drilled in consolidated rocks, both with and without casing, the logs obtained within the saturated zone provide information on the size of the intercepted fractures and cavities. Obviously, the largest counting rate of the measuring system is obtained when the probe is surrounded in all directions by a water volume of thickness equal to or greater than about 15 cm. The magnitude of the largest counting rate can be easily obtained by introducing the probe inside a water-filled container of appropriate size. The lowest counting rate will be obtained when the probe is located in a nonfractured stretch of the rock formation. Intermediate values are generally related to the size of the fractures and cavities.

4.3.7 Sonic velocity log

Well-logging methods based on changes in the sonic velocity in the rock have seen great development in recent years. The method is more appropriate for the identification of fractured stretches in consolidated rock formations. The measuring probes are fitted with a transmitter of elastic waves of a certain frequency and one or two receivers. The measuring system determines the time required for the longitudinal waves to travel the distance between the transmitter and receiver or between two receivers. This time depends on the type of rock and its porosity and, consequently, on the presence of fractures or cavities in the rock formation and

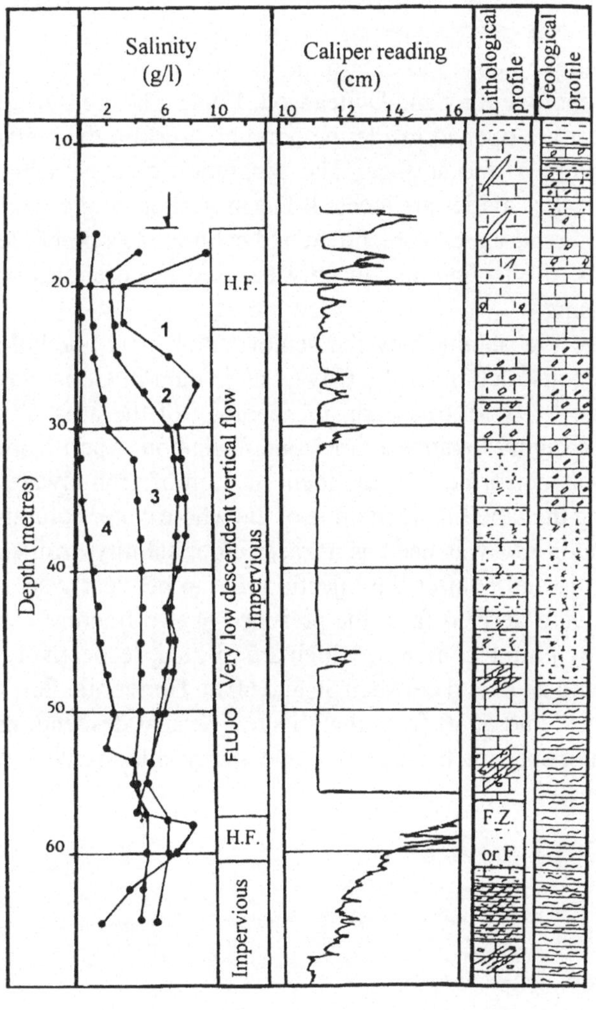

Fig. 4.10. Caliper and tracer dilution test logs obtained by Plata and González (1989) in a borehole located near the planned Cuyaguateje reservoir, Pinar del Rio, Cuba. The logs show a good correspondence for the identification of the position of the fractures using both techniques. The values of t_1, t_2, etc. indicate the time elapsed after tracer injection. Common salt was used as a tracer. F.Z. or F. = fractured zone or fault; H.F. = horizontal flow. $t_1 = 0$; $t_2 = 68$ min; $t_3 = 130$ min; $t_4 = 17$ hours.

on their size. The transmission velocity of the wave is higher in compacted rocks than in water (test in the saturated zone) or in air (test in the unsaturated zone). If the matrix velocity and the velocity of the pore fluid are known, it is possible to evaluate the rock porosity from the recorded sonic travel time.

The results of the method are satisfactory in boreholes without casing or in those equipped with a PVC casing. If an iron casing is installed, sonic waves travel through this material at a much higher velocity than through the rock formation. However, even in this case, the measuring system can distinguish and separate waves travelling through the casing and those travelling through the rock formation. The method becomes less appropriate when dealing with unconsolidated rocks that present lower sonic transmission velocities.

4.3.8 Caliper

The caliper well-logging technique is used to determine the size of fractures and cavities intercepted by the borehole (Paillet and Duncanson, 1994). The measuring probe has two sprung arms that are opened inside the borehole to skim the walls when travelling from the bottom to the surface. The separation of these arms increases in the positions where the cavities are located. The movement of the arms is converted into an electrical signal, which is recorded and calibrated as borehole diameter. For obvious reasons, the technique can only be used in boreholes of consolidated rocks without casings.

Figure 4.10 shows a caliper log obtained by the senior author in a borehole located at the area where the construction of the Cuyaguateje dam in Cuba was planned (Plata and González, 1989). In this case, the geology of the area was controlled by the presence of a highly karstified limestone formation. Apart from the caliper log, a tracer test was performed for the identification of groundwater flows inside the borehole, using the labelling technique of the whole water column described in Chapter 8. Common salt was used as a tracer. Four salinity profiles were measured at times t_1, t_2, t_3 and t_4 after the injection. It can be clearly seen that the detected water flows were derived from the presence of two fractured or karstified stretches identified by both techniques, which are located, respectively, at depth intervals between 15 and 21 m and between 56 and 60 m. Horizontal flows were identified at both stretches, and apart from that, a slow vertical descending flow was also present, with water entering through the upper permeable stretch and going out through the lower one.

REFERENCES

Allaud, L.A. and Martin, M.H., 1977. *Schlumberger, the history of a technique*. John Wiley and Sons, New York.

Al-Saigh, N.H., Mohammed, Z.S. and Dahham, M.S., 1994. Detection of water leakage from dams by self-potencial method. *Eng. Geol.*, 37, 115–121.

Araguás, L., Plata, A. and Díaz, M.F., 1998. *Estudio hidrológico de la zona afectada por la rotura de la presa minera de Aznalcóllar, Sevilla*. Internal Rep., CEDEX, Ministry of Public Works, Madrid (in Spanish).

Ballard, R.F., 1983. *Cavity detection and delineation research, Rep. 5: Electromagnetic (radar) techniques applied to cavity detection*. Tech. Rep. GL-83-1, US Army Engr. Waterways Exp. Station, Vicksburg, MS.

Bogoslovski, V.A. and Ogilvy, A.A., 1970a. Application of geophysical methods for studying the technical status of earth-dams. *Geophys. Prosp. Suppl.*, 18(4), 758–773.

Bogoslovski, V.A. and Ogilvy, A.A., 1970b. Natural potential anomalies as a quantitative index of the rate of seepage from water reservoirs. *Geophys. Prosp.*, 18, 261–268.

Boulding, J.R., 1993. *Subsurface characterization and monitoring techniques: A desk reference guide. Vol. 1, Solids and ground water*. EPA/625/R-93/003a.

Corwin, R.F., 1988a. *Geotechnical applications of the self-potential methods. Rep. 3: Development of self-potential interpretation techniques for seepage detection*. Tech. Rep. REMR-GT-6, US Army Engr. Waterways Exp. Station, Vicksburg, MS.

Corwin, R.F., 1988b. *Data quality for engineering self-potential surveys*. In *Proc. Int. Symp. on Detection of Subsurface Flow Phenomena*, Karlsruhe, Germany.

Corwin, R.F. and Hoover, D.B., 1979. The self-potential method in geothermal exploration. *Geophysics*, 44, 226–245.

Cuellar, V., Plata, A., Ruiz, J.M. and Cuadra, A., 1996. Estudio de filtraciones en la presa de El Tejo mediante técnicas geofísicas. *V Jornadas Españolas de Presas, Valencia, Comité Español de Grandes Presas, Madrid* (in Spanish).

Fitterman, D.V., 1979. Calculations of self-potential anomalies associated with dilatant regions in a layered earth-dam. *J. Geophys. Res.*, 83(B12), 5923–5928.

Haines, B.M., 1978. The detection of water leakage from dams using streaming potentials. In *Society of Professional Well Log Analyst Annual Logging Symposium, 19th, El Paso, Texas 1978, Transactions*, p. K1–K13.

Llopis, J.L. and Butler, D.K., 1988. *Geophysical investigation in support of Beaver Dam Comprehensive Seepage Investigation*. Tech. Rep. GL-88-6, US Army Engr. Waterways Exp. Station, Vicksburg, MS.

Olhoef, G.R., 1984. Applications and limitations of ground penetrating radar. *54th Ann. Intern. Meeting Soc. Explor. Geoph., Expanded Abstracts*, 147–148.

Olhoef, G.R., 1988. Selected bibliography on ground penetrating radar. In *Proc. Symp. Appl. Geoph. Engr. Environ. Probl. Soc. of Engr. and Mineral Explor. Geoph.*, 462–520.

Ogilvy, A.A., Ayed, M.A. and Bogoslovski, V.A., 1969. Geophysical studies of water leakages from reservoirs. *Geophys. Prosp.*, 17, 36–62.

Paillet, F. and Duncanson, R., 1994. Comparison of drilling reports and detailed geophysical analysis of groundwater production bedrock wells. *Ground Water*, 32, 200–206.

Plata, A., 1972. *Isótopos en hidrología*. Editorial Alhambra S.A., Madrid (in Spanish).

Plata, A. and González, R., 1989. *Estudios sobre riesgos de fugas en el futuro embalse de Cuyaguateje, Cuba*. Internal report, IAEA, Vienna (in Spanish).

Repsold, H., 1989. *Well logging in groundwater development. Verlag Heinz Heise*.

Rank, D.I., 1985. *Tätigkeitsbereich und unverbindliche Richtpreise der Abteilung Hydrogeologie und Angewandte Geophysik sowie anderer Abteilungen des Geotechnischen Institutes*. Bundesversuchs- und Forschungsanstalt ARSENAL, Vienna (in German).

Schiavone, D. and Quarto, R., 1984. Self-potential prospecting in the study of water movements. *Geoexploration*, 22, 47–58.

Serra, O., 1987. *Fundamentals of well log interpretation*. Elsevier, Amsterdam.

Vogelsang, D., 1995. *Environmental geophysics*. Springer Verlag, Berlin.

CHAPTER 5

Natural tracers of water: (A) Temperature, electrical conductivity and chemical composition

5.1 INTRODUCTION

Chapters 5 and 6 are devoted to the description of the most important natural tracers of water that are commonly used to obtain information on the origin (i.e., whether the emerging water is exclusively formed by the reservoir water) and the localisation of infiltration points in lakes and reservoirs. More concretely, natural tracers of water are used to obtain information on the followings aspects:

1) Determination of the origin of the groundwater emerging at the discharge points located downstream from the dam as well as groundwater flowing through the abutments. As shown in the following chapters through the presentation of case studies, groundwater emerging at these springs does not always entirely correspond to reservoir water. Very frequently, some springs located close to the dam site represent the discharge of adjacent aquifers or a mixture of groundwater recharged by local precipitation and reservoir water. For this reason, when a leakage study is initiated, the first aspect to be tackled should be the confirmation of the origin of the emerging water, that is, whether the origin can be proven to be exclusively related to the reservoir, even in cases where it appears to be obvious. Experience proves that nature frequently reserves unexpected surprises.
2) Investigation of the dynamic behaviour of groundwater flows emerging downstream. These studies should be based on the analysis of certain changes produced as a function of time, specifically the concentration of determined characteristic chemical or isotopic components contained in the reservoir water. Initially, the monitoring work should focus on the study of the trends in the concentration of certain parameters in the reservoir water and in the emerging water, and in an affirmative case, on the determination of the delay with which any concentration pulse is transmitted. This time lag provides information on the transit time of water between the reservoir and the springs, as well as between the reservoir and the boreholes located close to or downstream from the dam. If various springs discharge downstream from the dam, the transit time can be

different for each flow path, and therefore, individual surveys for each spring should be carried out. Obviously, the first aspect to check in such cases is to determine whether groundwater feeding these springs and the reservoir water have the same chemical and isotopic composition.

The following natural tracers will be discussed in this and in the next chapter:

1) Temperature of water.
2) Electrical conductivity.
3) Chemical composition, including the major dissolved ions.
4) Stable isotopes of water: deuterium and oxygen-18.
5) Natural tritium, mainly introduced in the water cycle as a result of the thermonuclear weapons tests performed in previous decades.
6) Carbon-13 and carbon-14 in the total dissolved inorganic carbon, TDIC.

It must be pointed out that apart from the above-mentioned tracers, many other substances incorporated or transported by natural waters through different mechanisms can be used, under favourable conditions, as tracers. Examples of such substances are the following:

1) Urban and industrial sewage as well as agricultural wastewater spilled into a river, which is regulated by a reservoir. These wastewaters may contain specific chemical components, depending on their origin, that can provide clear evidence of the hydraulic connection between the reservoir and the springs or groundwater bodies existing downstream of the dam. For example, in cases of urban sewage, the impounded waters in the reservoir may become labelled with high concentrations of nitrogen compounds, as well as other ions, such as Na^+, K^+ or HCO_3^-. In the case of industrial wastewaters, heavy metals, such as iron, chromium, vanadium, manganese, mercury, lead, etc., as well as organic compounds such as phenols, hydrocarbons, etc., can be detected in the reservoir and in other water bodies recharged from the reservoir. Wastewaters derived from agricultural practices may produce specific labelling with pesticides, copper compounds, nitrates, etc. All of these components may be of great help in the investigation of dam leakage.
2) Suspended sediments. Sediments accumulate in reservoirs during high water periods as a result of soil erosion in the catchment area, especially during periods of heavy storms. These sediments may label a large fraction of the impounded water in the reservoir. The coloured waters can be detected in the springs connected with the reservoir, especially in cases of leaks flowing through fractures in consolidated rocks. Obviously, after a few days, the sediments will be deposited at the bottom of the reservoir and, for this reason, this type of labelling should be considered a pulse. This kind of labelling was observed in the reservoir La Honda in Venezuela, but, unfortunately, it was not possible to carry out an appropriate control on the appearance of the sediments in the springs.

3) Certain minor dissolved ionic species and other trace elements incorporated into the reservoir water, for example, in cases of reservoirs located in seismically active areas. This was the case observed in the Lake Asososca in Nicaragua, which is described in a detailed form in Chapter 12. This lake has a surface of about 0.6 km^2 and is used for water supply to the population of Managua. At a distance of only 2 km is located Lake Managua, with a surface of about 1000 km^2, which is strongly polluted with sewage of urban, industrial and agricultural origin. Due to the residual volcanic activity of the zone, Lake Managua shows a boron concentration of about 2.5 mg/l. However, the boron concentration in Lake Asososca was lower than the detection limit. This observation led to the conclusion that no hydraulic connection existed between the two lakes, in spite of the fact that Lake Asososca had a water level about 4 m lower than the head of Lake Managua.
4) Dissolved gases, such as radon. These gases can also play an important role in establishing the supposed hydraulic connection between the reservoir and downstream springs or groundwater bodies. Radon-226 (^{226}Rn) is a radioactive noble gas that is a member of the natural uranium-238 series. Consequently, radon is present with variable concentrations in the atmosphere and in all aquifers. Being a gas, radon incorporates into surface waters and tends to escape to the atmosphere. Therefore, its concentration in rivers and reservoirs is generally very low. This is not the case for groundwater, however, where radon concentrations can be very high, especially in the case of rocks with high uranium content, for example, granite. Therefore, a simple analysis of radon in the reservoir and spring waters can indicate whether the emerging water in the springs is fed by the reservoir.

Some natural tracers can undergo large variations of concentration with time in the reservoir, for example, seasonal variations. The use of these tracers for the investigation of leakage problems may require prolonged monitoring over time and a defined sampling strategy to establish a possible correlation of concentration between the concentration at the reservoir and at the emerging waters downstream.

5.2 WATER TEMPERATURE

5.2.1 Theoretical considerations

Water temperature is, in many cases, an excellent tracer for the investigation of dam and reservoir leakage. The reason behind this affirmation lies in the thermal stratification that most reservoirs develop, not only in temperate climates, but also in tropical areas. In fact, only very shallow reservoirs (depth less than about 10 m) are characterised by a permanent uniform temperature, which is a consequence of the mixing effect derived from wind action. The cause of the thermal stratification

Table 5.1. Density of pure water for different temperature values.

Temperature (°C)	Density (g/ml)	Temperature (°C)	Density (g/ml)	Temperature (°C)	Density (g/ml)
4	1.000000	13	0.999406	22	0.997801
5	0.999992	14	0.999273	23	0.997569
6	0.999968	15	0.999129	24	0.997327
7	0.999930	16	0.998972	25	0.997075
8	0.999877	17	0.998804	26	0.996814
9	0.999809	18	0.998625	27	0.996544
10	0.999728	19	0.998432	28	0.996264
11	0.999634	20	0.998234	29	0.995976
12	0.999526	21	0.998022	30	0.995678

is the result of heat exchange between the lake surface and the atmosphere. The exchange obviously operates in both directions. In summertime, water of the uppermost layer of the lake is rapidly heated (positive heat flow), and the heat energy absorbed is progressively transmitted to deeper layers. Due to the fact that water density decreases at increasing temperatures (see Table 5.1), the heating of the shallow part of the water column creates a stable water column. On the contrary, in wintertime, shallow water in the uppermost layer of the lake becomes progressively colder (negative heat flow). Due to its higher density, cold water sinks, and the final result is the vertical mixing of the water column. In most cases, water columns of lakes and reservoirs present during the cold season a more or less uniform vertical temperature profile.

The thermal stratification of lakes and reservoirs produces the development of a thermocline, which corresponds to the transition zone between the warm and cold waters. The position of the thermocline migrates toward greater depths during the period of positive heat flow, and it is rapidly destroyed with the arrival of lower temperatures during autumn or winter, that is to say, when the vertical mixing of the reservoir or lake takes place. The water layer located above the thermocline is called the epilimnion and the layer located below is the hypolimnion. At high latitudes, lakes can be permanently stratified. These lakes are called meromictic lakes. Similarly, permanently stratified lakes can be found in tropical areas and at low elevations due to the lack of a sufficiently cold atmosphere, at least throughout a season.

Thermal stratification of the reservoir or lake may seriously affect the physical, chemical and biological properties of the lake water. For example, dissolved oxygen reaches saturation only in the epilimnion, while in the hypolimnion, anaerobic conditions usually prevail. The study and modelling of stratified reservoirs is always a very difficult task due to the disturbing effects derived from the meteorological conditions (solar radiation and wind), from the incoming and outgoing fluxes of

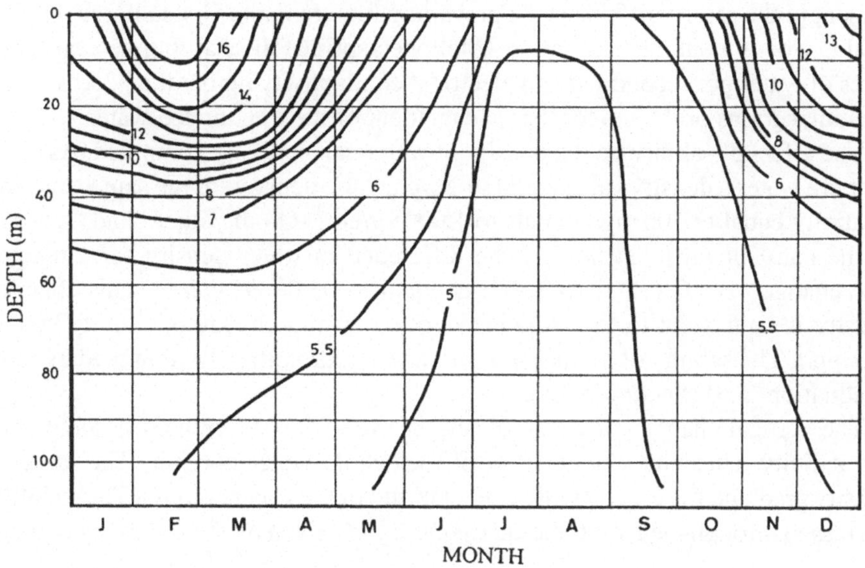

Fig. 5.1. Diagram illustrating the annual thermal evolution of Lake Laja (Chile).

surface waters (with contrasting temperatures) and from the irregular topography of the bottom of the reservoir. Strong winds facilitate the mixing of water in the upper layer of the epilimnion, while their effect in the hypolimnion is generally negligible.

The thermal stratification of most lakes can be expressed in the form of diagrams of the type illustrated in Fig. 5.1. This diagram corresponds to Lake Laja, located in the Andes Cordillera, southern Chile, at an elevation of about 1350 m.a.s.l. In these diagrams, isotherm lines are plotted of the impounded water throughout the whole year. Obviously, due to the normal cyclic character of the local climatic conditions, these diagrams present small yearly variations. In the case shown in Fig. 5.1, the deep water of the lake (below 50 m) is kept permanently on the bottom with a temperature ranging from 5°C to 6°C. Complete mixing of the water column takes place during the months of July and August, that is, during the winter months of the Southern Hemisphere. As shown in Chapter 12, thermal changes affecting the upper layer of lakes and reservoirs are used for the identification of the area(s) of the lake where the infiltration of waters emerging downstream from the dam takes place.

On the other hand, it must be pointed out that apart from the thermal stratification mentioned above, many lakes and reservoirs also develop saline stratification, and consequently, deep waters show higher salinity contents (total dissolved solids expressed in g/l) than waters in the upper layer. The line dividing the two types of waters is called, in this case, a picnocline (in Greek, "piknos" means density). This situation is found when submerged springs discharging high salinity waters are present in the reservoir. This is the prevailing feature in the Guadalhorce reservoir, Spain, where a submerged spring (Meliones spring) discharging several tens of litres of water per second with an electrical conductivity higher than 250 mS/cm is creating serious problems for the management of the impounded water.

The effect of salinity on the density of water can be easily computed. Assuming for pure water a density value equal to 1 g/cm^3, the density for the same water with a salinity equal to 500 mg/l equals to 1.0005 g/cm^3. On the other hand, according to the data shown in Table 5.1, the difference in water density for a temperature change of 5°C (15°C to 20°C) is equal to 0.000895 g/cm^3. This change in density is equivalent to the variation derived from a change of salinity equal to 895 mg/l. The salinity of a water sample can be approximately computed by means of equation (5.3) (Section 5.3.1).

The vertical changes of water density derived from differences of temperature and salinity determine the stability of the whole water column. The larger the density gradient, the greater will be the stability of the water column. The stability is expressed through the Brunt-Väsälä frequency N, given by the following equation:

$$N^2 = \frac{g}{\rho}\frac{d\rho}{dz} \tag{5.1}$$

where ρ is the density and g the gravity acceleration.

According to the previous statements, it is possible to explain that the coldest waters will not always be located at the deepest layers of the reservoir. Inverse conductivity profiles can provide a stable water column due to the density compensation derived from the temperature profile. However, unstable water columns (lower density at deeper layers) are not infrequent. This situation corresponds to transient situations derived from the dynamic behaviour of the incoming water. Obviously, the differences of salinity between the different layers are small.

The vertical diffusion coefficient D_z decreases when N increases, because more stable water columns require additional work against gravity to produce and maintain the vertical turbulent mixing.

Fischer et al. (1979) published a detailed study on the mixing processes occurring in stratified lakes, including the energy exchange between lake and atmosphere. The convective flows occurring in the epilimnion are responsible for the rapid mixing of water inside this layer, as well as for the changes observed at the thermocline. The vertical diffusion coefficients of heat and the concentration of dissolved components can vary over very wide limits, ranging from the values corresponding to molecular diffusion (around 10^{-5} cm^2/s) up to values as high as 10 cm^2/s. The vertical diffusion coefficient determines the heat exchange between different layers of the lake, which is approximately expressed by the well-known Fick's equation:

$$F = -D\frac{dC}{dz} \tag{5.2}$$

where F represents the heat flow per unit of area perpendicular to the axis z given in $cal \cdot cm^{-2} \cdot s^{-1}$, dC/dz is the thermal gradient prevailing in a given moment expressed in $cal \cdot cm^3\, cm^{-1} = cal \cdot cm^{-4}$, and D is the vertical diffusion coefficient, in $cm^2 \cdot s^{-1}$. In the case of dissolved components, the unit mol per litre is used instead of calorie.

The increment of turbulence is proportional to the square of the vertical velocity gradient expressed as $(dv/dz)^2$. On the contrary, the turbulence suppression is proportional to N^2. The turbulence evolution is determined by the Richardson number, which is given by the following equation:

$$Ri = g(d\rho/dz)/(du/dz)^2 \tag{5.3}$$

Turbulence increases when Ri is higher than 0.25 and decreases for lower values. The horizontal mixing of a tracer dissolved in water, both in the epilimnion and in the hypolimnion, is expressed as a function of the turbulent diffusion coefficient D_x^*. The value of this coefficient increases with the size of the tracer cloud. Fischer et al. (1979) have reported for this increment the relationship $D_x^* = \sigma_x^{1.3}$, where σ is the standard deviation of the cloud along the x-axis, that is, along the direction of migration. However, this relationship has to be taken with caution because of the lack of thorough studies on this matter. In fact, the relationships between the parameters D_x^*, D_z^*, N^2 and Ri for different regions of the same lake are not well known.

5.2.2 Obtainable information

The comparison of the temperature of the emerging waters with the vertical temperature profile measured in the reservoir can provide valuable information on the depth at which the emerging waters infiltrate in the reservoir. Under favourable conditions, a single measuring campaign undertaken at a specific time can be sufficient. However, in most cases, a relatively long observation period is required to firmly establish correlations between the thermal changes observed in the reservoir and in the emerging waters. Usually, conductivity profiles are measured using composite probes that can measure both conductivity and temperature parameters (thermoconductivity meters).

Obviously, temperature and conductivity profiles measured in boreholes drilled in the zone where groundwater is emerging will be always of great help for the localisation of preferential pathways of the groundwater flow fed by the reservoir (see Figs. 5.2 and 5.3).

Field campaigns devoted to carrying out systematic measurements of temperature and conductivity in the reservoir, boreholes and springs should start at the initial stage of the leakage investigation because these simple parameters may provide valuable hints for defining the most appropriate techniques to be applied. Some of the case studies described in Chapter 12 illustrate the importance of these simultaneous measurements of temperature and conductivity. The success of the investigations carried out in the Sabaneta reservoir (Dominican Republic) and

Fig. 5.2. Meter-counter used for the measurement of temperature and conductivity profiles in boreholes.

Fig. 5.3. Measuring temperature and conductivity profiles at the Caribbean Island of San Andrés, Colombia.

Lakes Colbún and Laja (Chile) were mainly derived from the information provided by the temperature measurements.

When the transit time of water between the reservoir and boreholes or springs is lower than a few days, the temperature of water does not change rapidly as it flows in the subsurface. This happens because the specific heat capacity of water is much higher than in the case of most natural materials. On the other hand, due to the low values of the thermal conductivity of dry soil materials (about 5×10^{-3} cal $\cdot$ cm$^{-1} \cdot$ s$^{-1} \cdot$ °C^{-1}), the gain or loss of temperature along the subsurface pathway derived from heat exchange with the atmosphere may represent a small fraction of one degree centigrade. On the contrary, the measured temperature variations at the reservoir, boreholes and springs are generally on the order of several degrees centigrade. In other words, the amount of thermal energy transported by water is usually high in comparison with the gains or losses. This is especially true when the flow rates involved in the leaks are high and the transit times are short. Obviously, important changes of water temperature can be produced when water infiltrated in the bottom of the reservoir mixes with groundwater of a different origin.

In most reservoirs, the average temperature of water in the epilimnion shows seasonal variations that can be approximately described as a sinusoidal thermal wave, with low temperature values in wintertime and high values in summertime. This was the case of Lake Laja shown in Fig. 5.1. Therefore, if water escaping from the reservoir derives from the epilimnion and the transit time between the reservoir and a borehole or spring is short, a similar thermal wave is transmitted to the spring, although somewhat smoothed. The smoothing effect is derived from the mixing in the subsurface of reservoir water infiltrated at different times and also from the heat exchange between water and the surrounding rock materials.

The smoothing effect of the thermal wave derived from the mixing of waters with different temperatures can be evaluated using the same equations developed for the migration of conservative tracers in the case of a sinusoidal input function (Oeschger and Gugelmann, 1974). However, the smoothing effect derived from the heat exchange between water and rock is more difficult to assess because the parameters involved in this evaluation are usually unknown, and only estimated values can be used. This smoothing effect is not negligible due to the following reasons:

1) The specific heat of most rocks is not negligible compared with the accepted value for pure water (about 0.2 cal $\cdot$ g$^{-1} \cdot$ °C^{-1} for the most rocks and about 1 cal $\cdot$ g$^{-1} \cdot$ °C^{-1} for water).
2) The mass of rock in contact with water can be much higher than the mass of water.

Therefore, rock formations act as a heat storage system, gaining or losing calorific energy, depending on the sign and magnitude of the temperature gradient between rocks and water. This property can produce a delay between the temperature cycles observed in the reservoir and in the springs, respectively, the magnitude of which may vary from a few days to months. However, if the net heat gain or loss of the whole system is low (negligible heat exchange with the atmosphere), as happens in most cases, the average temperature during the full annual cycle will be the same for the reservoir and the springs. In the case of Lake Laja (see Chapter 12), the seasonal variations of temperature allowed the determination of the average elevation at which the infiltration of the emerging water was taking place. The observed delay was, in this case, greater than two months.

Vertical temperature and conductivity profiles measured in boreholes are very frequently used for the identification of stretches of the water column where groundwater flow is significant. Such stretches are characterised by presenting different values of these parameters in comparison with the rest of the water column. Most frequently, water in low flow zones of the water column shows higher temperature and conductivity values. Obviously, when a vertical ascending or descending flow is present inside the borehole, both temperature and conductivity profiles will show constant values of these parameters along the whole stretch affected by this vertical flow. Therefore, constant values of these two parameters in a long stretch of the water column are usually indicators of some kind of vertical flow. The measurement of these profiles is strongly recommended as a systematic tool to be applied in all boreholes around the area affected by the leaks.

Figure 5.4 shows the temperature and conductivity profiles obtained in three boreholes drilled at the formation supporting the Maguaca dam (Dominican Republic). In the first diagram (Fig. 5.4A), the stretch of lower temperature and conductivity values located between 17 and 21 m corresponds to the groundwater flow (horizontal flow) connecting the reservoir and the springs located downstream from the dam. This flow was detected and measured using the artificial tracer techniques described in Chapter 8. Above 17-m depth and below 21-m depth, the borehole intercepted low permeability strata consisting mostly of marls. In the case

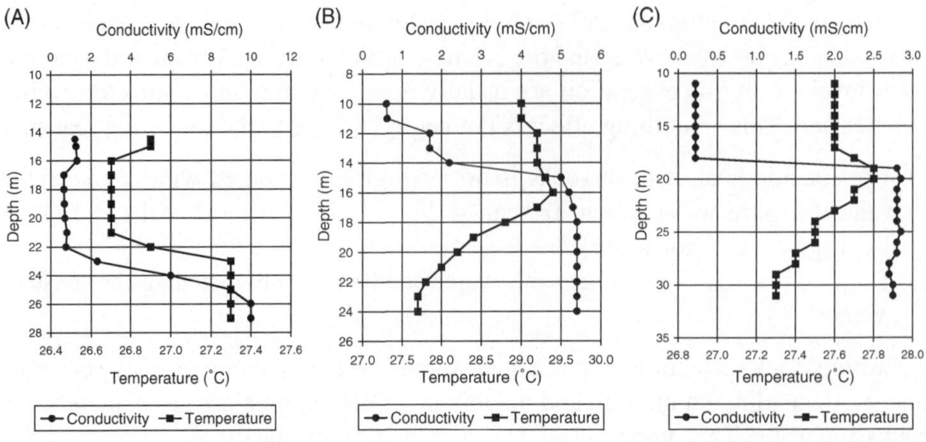

Fig. 5.4. Temperature and conductivity profiles obtained in three boreholes located close to the Maguaca dam (Dominican Republic).

of Fig. 5.4B, the horizontal flow was located at the stretch between the phreatic level and 14-m depth.

Below this depth, the materials intercepted by the borehole were practically impermeable. In the case shown in Fig. 5.4C, the constant temperature and conductivity values observed at the stretch between the phreatic level and 17-m depth were the consequence of a vertical descending flow. Water was entering the borehole at a depth just above the phreatic level. Groundwater was leaving the borehole through a fractured stretch located just above 17-m depth.

Figure 5.5 shows some temperature profiles measured at different times of the year at the Sabaneta reservoir in the Dominican Republic. The ordinates indicate the topographic elevation measured from the water surface. The temperature profiles start at different elevations due to the variations of the reservoir water level prevailing at the time of measurement. The arrows in each profile correspond to the temperature measured at the same time in the springs fed by the reservoir water. These arrows clearly indicate the elevation at which the spring water infiltrated the reservoir. An average value equal to 606 m.a.s.l. was obtained for this elevation. Figure 5.6 shows the isobath line (line of equal depth) corresponding to this elevation, which should intercept the infiltration zone in the reservoir. In Chapter 12, the complete study performed in this reservoir is described in a more detailed manner.

Apart from the previous applications, soil temperature can also provide valuable information on the presence of underground water flow. Due to the fact that

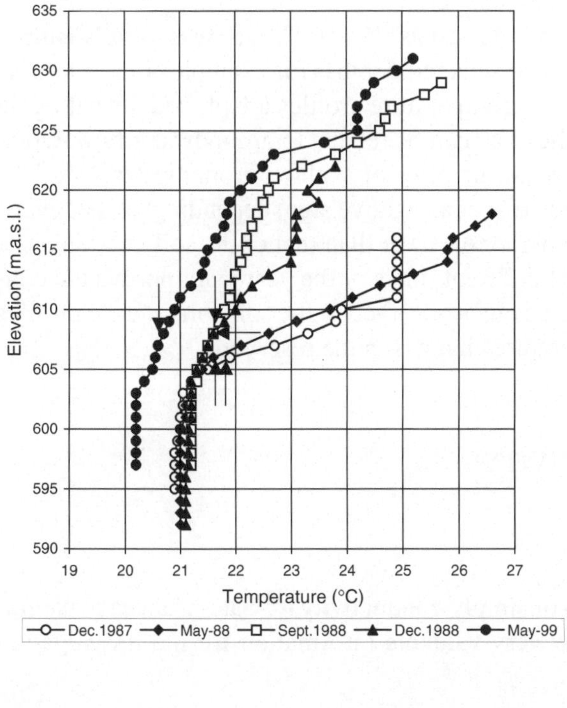

Fig. 5.5. Temperature profiles measured at the Sabaneta reservoir (Dominican Republic) at different times. The arrows show the temperature values prevailing at the downstream springs.

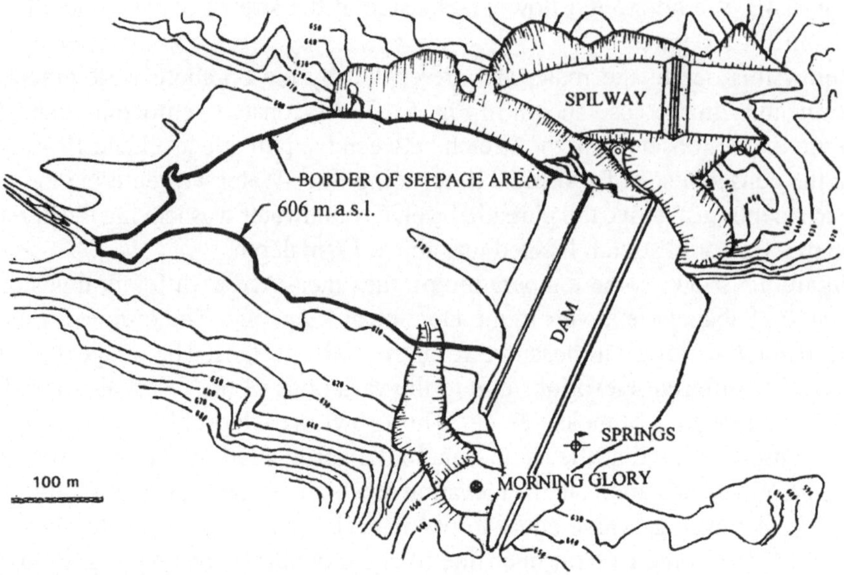

Fig. 5.6. Contour map showing the limit of the infiltration area at the Sabaneta reservoir inferred from the data shown in Fig. 5.5.

water shows a high specific capacity compared to most natural soil materials, temperature changes in water are slow as it migrates downward through the subsurface. Consequently, shallow-earth temperatures can be related to the existence of a groundwater flow (Cartwright and McComas, 1968). The study is usually carried out by measuring soil temperature at selected depths (for example, 1 m or less) at numerous stations of the study area. Temperature profiles are plotted, revealing, in many cases, temperature anomalies that can be related to groundwater flow. Obviously, the method will perform better in cases of shallow groundwater flow. The observed anomaly may show a positive or a negative sign, depending, respectively, on whether water temperature is higher or lower than that of the soil material. The repetition of the measurements at different times of the year can improve the confidence of the results because the changes observed in the soil temperature profiles can be related to seasonal temperature changes in the reservoir.

5.3 ELECTRICAL CONDUCTIVITY

5.3.1 Theoretical considerations

The electrical conductivity (EC) or simply conductivity of water is an easy parameter to measure that can provide very valuable information for the investigation of leaks from lakes and reservoirs. As previously indicated, the measurement of the EC along the water column of the reservoir and in boreholes, as well as in the springs, is strongly recommended as a reconnaissance tool during the first stage of the investigation on leakage from dams and reservoirs.

As indicated by its name, EC represents the capacity of water to conduct electricity. More concretely, EC is defined as the intensity of the electrical current that circulates between two electrodes of 1 cm^2 surface, submerged in water, at a distance of 1 cm, between which a potential difference of 1 volt is applied. EC is the inverse of the resistivity, which is measured in ohm · cm. Therefore, EC is expressed in $(\text{ohm} \cdot \text{cm})^{-1}$ or mhos/cm. Several years ago, the conductivity unit Siemens/cm (S/cm) was defined in honour of the German scientist. One S/cm is equivalent to 1 mho/cm. This unit is very high for the conductivity values usually found in natural waters, and for this reason, the units mS/cm and µS/cm are commonly used, which correspond, respectively, to 10^{-3} S/cm and 10^{-6} S/cm.

The EC of water increases with temperature at an approximate rate of 2% per degree centigrade. Usually, the measuring systems developed for the determination of the water conductivity (thermoconductivity meters) automatically correct the temperature effect and directly report the conductivity referred to a temperature of 20°C or 25°C. However, this correction is only approximate because the correction factor depends, to a certain extent, on the chemical composition of water. If a more precise conductivity value is needed, the measuring system has to be calibrated using an appropriate standard solution. Usually, a solution of known concentration of KCl is used as a standard solution. A solution of 0.01 N KCl (0.01 mol of KCl per litre of solution, i.e., 745.51 mg/l) gives an EC value equal to 1416 µS/cm at 25°C.[1]

However, when very high EC values have to be measured, for example higher than 10 or 20 mS/cm, the linearity of the measuring system has to be checked. Very frequently, conductivity meters do not produce accurate results above 3–5 mS/cm because of the lack of linearity in the response of the electrode. In these cases, the measuring system has to be calibrated using solutions of KCl with concentrations that cover the full range of values to be measured.

Obviously, the EC of water increases with the concentration of the dissolved components. Distilled water has an EC value close to zero. The following empirical equation relates the EC (K) of an aqueous solution at 25°C and the total amount of dissolved solids (TDS):

$$TDS\ (\text{mg/l}) \approx (0.75 \pm 0.12) K\ (\mu\text{S/cm}) \tag{5.4}$$

[1] The ratio between the conductivity K and the concentration C for a KCl solution at 25°C is given by the expression: K (µS/cm) = 1.9C (mg/l). For a temperature of 18°C, the equation is: K (µS/cm) = 1.57C (mg/l).

In the same way, the EC can be related to the number of milliequivalents per litre (meq/l)[2] by the following approximate empirical expression:

$$\textit{Anions}\ (\text{meq/l}) \approx \textit{cations}\ (\text{meq/l}) \approx K/100\ (\mu\text{S/cm}) \qquad (5.5)$$

This equation is only valid for EC values lower than about 2000 μS/cm and can be used to check the overall quality of the chemical analyses performed in the hydrochemistry laboratory.

However, from a practical point of view, when dealing with dam leakage studies more than precise absolute values of the EC, what is really needed is relative variations of the EC in water samples collected at different points (reservoir, boreholes, springs). To achieve this goal, it is important that all measurements be made using the same conductivity meter and referred to the same temperature.

Very frequently, the graphite sensors of the electrode become coated with a layer of a nonconductive substance, such as oil, hydrocarbon or other organic materials. When this happens, the electrical signal given by the conductivity meter is lower than it would be without this coating. Usually, the problem is solved by cleaning both surfaces with an organic solvent, such as ethyl alcohol. If the problem continues, the surfaces of the graphite conductors have to be scraped with fine-grain sandpaper. To check the condition of the coating of the electrode, it is recommended to perform from time to time a calibration of the probe using a standard 0.01 N KCl solution. The coating of the electrode can be prevented if the measuring probe is kept under distilled water when it is not going to be used during long periods of time.

5.3.2 Applications

The surveys of water EC in all water samples are conducted with the aim of obtaining various types of information. Apart from that, the use of saline tracers (for example, a saturated solution of common salt) added to the water offers many possibilities in dam leakage studies. Examples of these possibilities are the determination of the types and magnitude of groundwater flows in boreholes (see Chapter 8), the measurement of small flow rates in open channels or springs (see Annex 2), the investigation of the renewal rate of water in small water bodies, etc. The conductivity meter is then used as a detector to measure in situ the variation of electrical conductivity as a function of time (see Fig. 5.7).

The EC measurements are therefore performed in the water column of the reservoir, in boreholes as well as in springs located downstream from the dam. Concerning the reservoir, the measurement of vertical profiles of EC at the deepest zones is recommended to investigate possible saline stratification. Usually, if

[2] The chemical equivalent is given by the quotient between the atomic or molecular weight and the valence. For example, for Ca^{++} with an atomic weight of 40.078 g and a valence of 2, the chemical equivalent is the half, i.e., 20.039 g (1 meq = 20.039 mg). For the ion $SO_4^{=}$, the molecular weight is $32 + (4 \times 16) = 96$ g, and the chemical equivalent is 48 g.

Fig. 5.7. Measuring conductivity and temperature profiles at the Aznalcóllar dike, SW Spain.

submerged springs discharge high salinity waters, these waters will probably be accumulated in the deepest section of the reservoir due to their higher density. However, as mentioned above, in some cases, the inverse situation can be found, that is, water with slightly higher EC values can be detected in the upper layers of the reservoir. The appearance of such apparently anomalous EC profiles can be explained in three different ways:

1) Transient situations derived from the dynamic behaviour of water flows circulating in the reservoir.
2) As previously mentioned, compensation for water temperature by the change in density due to higher salinity, so that a stable water column is created. For example, low temperature water of higher density in the deepest sections may generate stable water columns with water of slightly higher salinity located in the upper layers.
3) The contribution to the EC of water that for the same mass concentration of trivalent ions is higher than for bivalent and univalent ions. Therefore, waters with lower density can present a slightly higher EC if the contribution of trivalent ions is also higher.

Obviously, the EC of the reservoir waters can show great variations with time and space. Very frequently, reservoirs show cyclic or seasonal variations of EC, with lower salinity water during the rainy periods and higher during the dry ones. The main reason for that is the origin of the water. During rainy periods, most of the water entering the reservoir is provided by the surface runoff occurring in the catchment area, and consequently, the salinity is lower. On the contrary, the relative contribution of groundwater discharges into the river will be higher during the dry periods (base flow conditions), and consequently, the salinity of the entering water can be much higher. These seasonal variations in the EC can be of interest in the investigation of hydraulic connections between the reservoir and the boreholes or springs located downstream. If monitoring is maintained for sufficient time, the hydraulic connection becomes evident and the flow velocity between reservoir and observation points can be determined. This is achieved by means of the correlation between the EC peaks observed in the reservoir and in the boreholes or springs.

Electrical conductivity profiles obtained in boreholes located at the vicinity of the reservoir frequently show very large variations (see Fig. 5.4). Generally, groundwater from boreholes with null or very low flow velocity shows higher EC values than waters from boreholes with rapid groundwater flow. There are two main reasons for that. One factor controlling the salinity of groundwater is the residence time of water. Under slow groundwater motion, the residence time of water in the aquifer reaches time intervals on the order of years, allowing a more intense dissolution of rock minerals and water–rock interaction processes. The second reason for the increment is that in most cases the salinity of the reservoir water is lower than that of local groundwater, and generally, rapid subsurface flow is typical for most leaks.

When important changes in the EC are detected between water samples collected in the reservoir, boreholes and/or springs, the next step should be to perform full chemical analyses of major components (see Section 5.4.2) to identify the ions responsible for such increment. Valuable information on the reason for such increment can be provided by chemical analyses. Several case studies are described below.

5.3.3 Case studies

Maguaca reservoir in the Dominican Republic As mentioned above, when dealing with the study of leaks, EC and temperature of water should jointly be measured periodically in samples representing the reservoir, the relevant boreholes, and the downstream springs. Figure 5.8 shows the results obtained by the senior author at the Maguaca reservoir in the Dominican Republic. The EC of the reservoir water was 180 μS/cm at the surface and 160 μS/cm at the bottom. However, groundwater monitored in boreholes and springs showed values ranging from 210 μS/cm to 2970 μS/cm. The arrows marked in the figure indicate the location of preferential pathways of the water coming from the reservoir, as detected by means of tracer experiments using ^{131}I as a tracer. This tracer was injected at different points of the reservoir. This preferential pathway basically coincides with the zones

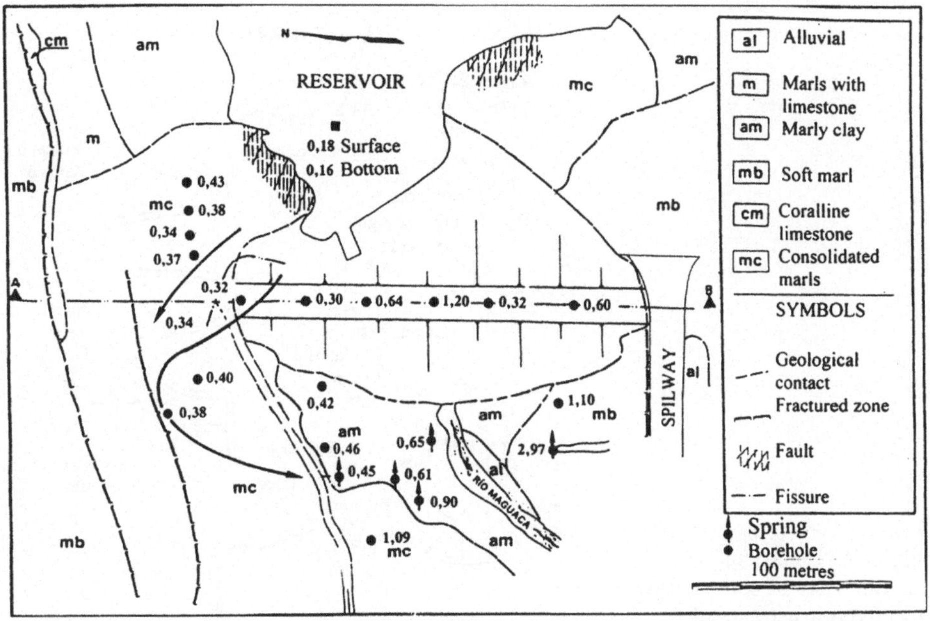

Fig. 5.8. Conductivity values observed in different springs located downstream of the Maguaca dam (Dominican Republic).

where groundwater was characterised by lower EC values. After each ^{131}I injection, radioactivity profiles were measured periodically in the existing boreholes using a scintillation detector. The measured profiles allowed the localisation of the strata through which the water infiltrating in the reservoir was flowing. The results confirmed that the conductivity of the water provided satisfactory information on the relative values of the residence time of water. The higher this residence time was, the higher was the EC. Of course, this may not be valid in cases where the rock formations do not contain important quantities of soluble components. In the case of the Maguaca reservoir, the rock consisted of marls with important fractions of halite and gypsum.

Lake Laja, Chile Figure 5.9 shows the case of the Lake Laja in Chile, which is described in a more detailed way in Chapter 12. A large number of springs exist in the zone where water coming from the lake is emerging. The total flow rate of these springs ranges from 24 m^3/s to 40 m^3/s, depending on the water level in the lake. Although most of the emerging waters are supplied by the lake, a significant fraction of groundwater represents the discharge of the so-called intermediate basin. This component is mainly fed by melting of snow accumulated during the wintertime at higher elevations of the neighbouring Antuco volcano. During the periods of snowmelt, a small lake is formed at such high elevations, which becomes rapidly

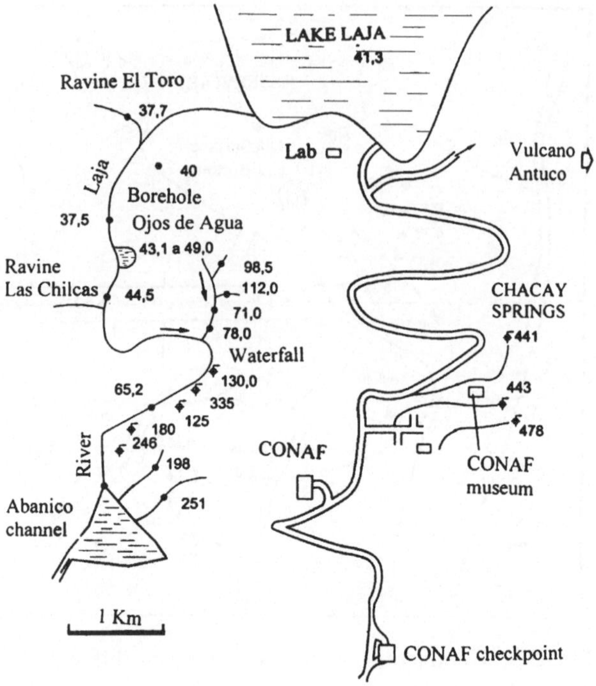

Fig. 5.9. Electrical conductivity of some springs located downstream of Lake Laja (Chile). Values expressed in μS/cm.

dry because of the infiltration through the permeable lava cover as soon as the snowmelt period ends. Part of the water forming this small lake emerges at three springs (Chacay springs) located in the intermediate basin at an elevation higher than the lake and, therefore, with no contribution from Lake Laja. The average EC of these springs is about 450 μS/cm. The remaining water emerges through many springs located at lower elevations along the left side of the River Laja. This water component is mixed before emerging with water fed by the lake, with an EC value of 41.3 μS/cm. As a consequence, groundwater emerging at these springs shows intermediate conductivity values. The fraction f of the contribution of the lake to the flow rate measured at each spring can, therefore, be calculated by means of the following expression:

$$41.3f + 450(1 - f) = K_{SPRING} \tag{5.6}$$

This simple method allowed the quantitative evaluation of the contribution of the intermediate basin, as shown in Chapter 12.

Reservoir La Honda, Venezuela Sometimes, EC data alone provide a strong argument to disregard the possibility that water emerging in some springs represents a leak fed by the reservoir. This was the case at the La Honda reservoir in Venezuela. The left abutment of the earth-filled dam is embedded in detritus eroded from the surrounding mountains, presenting a relatively high permeability. To reduce or eliminate the water losses through this formation, a continuous concrete wall 80 m deep and about 500 m long was constructed. Behind this wall, two drainage galleries were also constructed with the aim of collecting and evacuating groundwater eventually infiltrated through the border of the wall. The result was the appearance of more than 40 springs inside the galleries. The first aspect addressed in the study was the question of whether all these emerging waters really represented reservoir waters. The alternative origin for these waters was the infiltration of surface waters fed by local precipitation through lateral permeable materials. The EC values measured in the gallery springs ranged from 26 μS/cm to about 150 μS/cm, whereas the reservoir water showed an almost constant EC value equal to 54 μS/cm. Therefore, it became clear that spring waters with electrical conductivity lower than the value corresponding to the reservoir could not be fed by it. These low conductivity waters were recharged by recent local precipitation, and for this reason, their salinity was very low.

Beninar reservoir, Spain Sometimes, conductivity measurements reserve unexpected surprises, as in the case of the Beninar reservoir in Almería, Spain. Most of the area flooded by the reservoir consists of highly fractured and karstified dolomites. A large water volume infiltrates through these materials and emerges at the Marbella spring, located about 4 km downstream. This is a perennial spring that was present before the dam was constructed, although with a flow rate ranging approximately from 50 l/s to 400 l/s. The construction of the reservoir led to an important increase in the flow rate. Values ranging from 2 to 3 m^3/s are typical nowadays. Surprisingly, the EC of the spring water, with the reservoir level close to its maximum value, was about 2500 μS/cm, whereas the EC of the reservoir water was only about 500 μS/cm. The salinisation of groundwater is mainly produced by gypsum dissolution. Most of the area of infiltration in the reservoir is located in a well-defined gypsum karstified formation. A second surprising result was obtained when several tracer tests were carried out with the aim of determining the transit time of water between the reservoir and the Marbella spring. The injected artificial tritium in the reservoir was detected in the spring waters only a few days after the injection (Pérez Zabaleta et al., 1990). The transit time varied according to the reservoir water level. The amount of gypsum being dissolved per unit of time is very high, clearly indicating that the leak is being responsible for the formation of enormous cavities in the evaporitic rocks.

PUNA reservoir in Africa In the case of the PUNA reservoir in Africa, the combined use of EC and temperature data provided very valuable information. The study was carried out by the senior author (see Section 2.2.2, Chapter 2). One draining gallery was constructed on either side of the dam, where about 160 draining tubes were installed. Under full reservoir conditions, water was emerging through almost all draining tubes. Figure 5.10A shows the profiles of these two parameters measured in the reservoir. Both profiles reveal the existence of thermal and

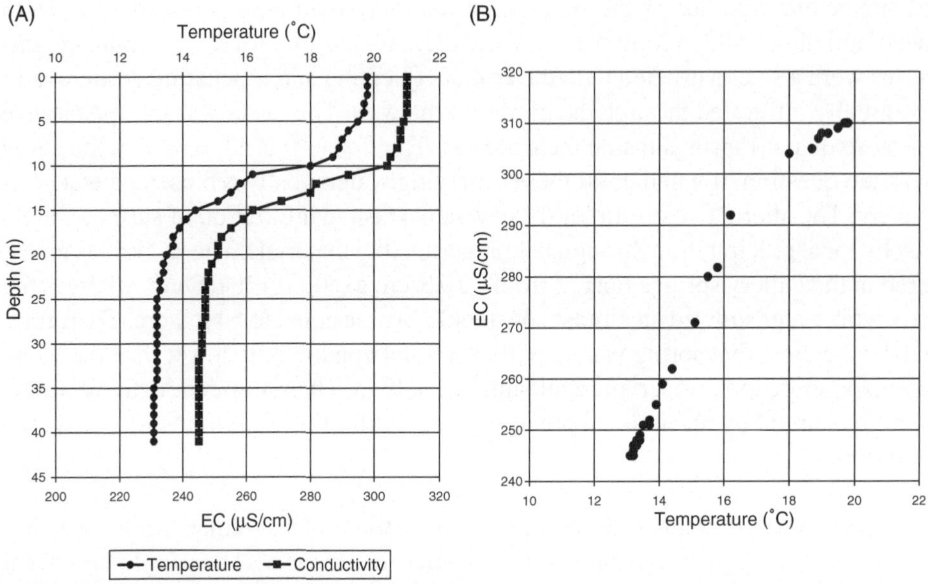

Fig. 5.10. A) Temperature and conductivity profiles obtained in the PUNA reservoir. B) Relationship between these two parameters.

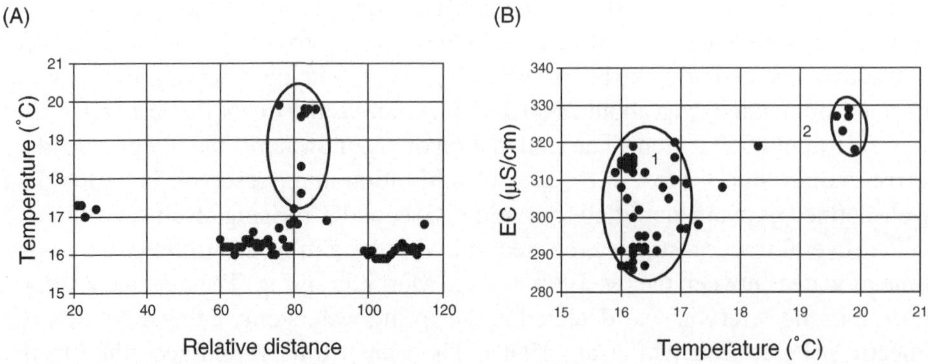

Fig. 5.11. A) Distribution of temperature in several springs located along the right side gallery of the PUNA reservoir. B) Relationship between the temperature and conductivity values measured at the same gallery.

saline stratification, with both the thermocline and picnocline located at depths between 10 and 15 m. Figure 5.10B shows the correlation between temperature and electrical conductivity.

Figures 5.11A and 5.11B show the temperature profiles measured at the drains of the right and the left side galleries, respectively. Figures 5.12A and 5.12B show the relationship between temperature and conductivity, respectively, for the drains

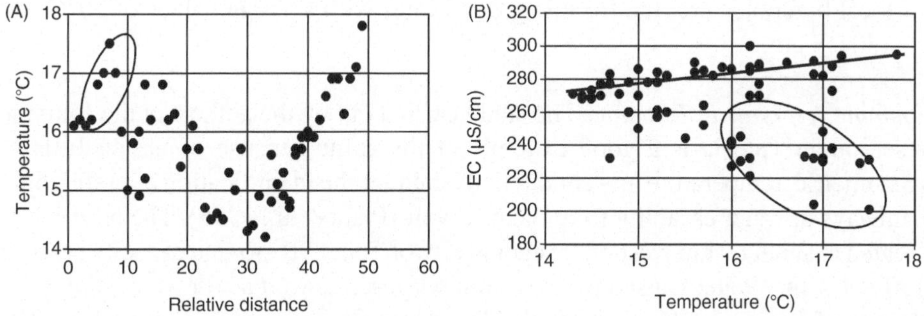

Fig. 5.12. A) Distribution of temperature in several springs located along the left side gallery of the PUNA reservoir. B) Relationship between the temperature and conductivity values measured at the same gallery.

located along the same galleries. These diagrams were interpreted as follows:

1) Most of the water emerging at the right side gallery showed temperature values between 16°C and 17°C. The weighted average temperature measured at the outlet of the gallery, where all the discharges were mixed, was 16.2°C. This temperature value was intermediate between the values measured at the epilimnion and the hypolimnion of the reservoir.
2) The high temperature values found at draining tubes 75 to 85 (left-side gallery) indicate that these tubes were draining water infiltrated in the epilimnion (depth less than 10 m), demonstrating that there is a clear hydraulic connection between the upper layer of the reservoir and this zone of the gallery. These water samples are identified with ellipses in Figs. 5.11A and 5.11B (ellipse number 2 in the case of the latter figure).
3) Waters with an average temperature of 16.2°C in the same gallery showed EC values ranging from 285 to 320 µS/cm (ellipse 1 in Fig. 5.11B). The EC changes are interpreted as being produced by the dissolution of materials injected into the rock formation, on the occasion of the grouting works performed in the previous months. These waters could not be related to locally recharged groundwater nor related to the reservoir because their temperature was higher than 18°C (average value of 20°C).
4) Water samples collected in the left side gallery showed lower temperature values in the central zone, close to drainage tube number 30. This is because this flow line infiltrates in the hypolimnion (coldest water) but laterally mixes with different fractions of epilimnion water.
5) The water samples from the first 10 drain tubes of the same gallery, marked with an ellipse in Fig. 5.12A, showed high temperature values (between 16°C and 17.5°C) associated with an EC lower than the typical value observed in the reservoir water (ellipse of Fig. 5.12B). This is explained as being a mixture of reservoir water with aquifer water. Measurements carried out at different

local boreholes provide for this groundwater with average values of 20°C and 150 μS/cm.

Joumine reservoir in Tunisia The study carried out by the authors at the Joumine reservoir in Tunisia is a good example of the combined use of the analysis of piezometric, temperature and conductivity data for the identification of preferential pathways of water escaping from the reservoir (Plata et al., 1998). The reservoir is located at about 60 km west of the capital city of Tunis. Its storage capacity is about 130×10^6 m^3. Water is used for irrigation purposes as well as for water supply for the city of Tunis. The dam is an earth-filled dam with impervious clay nucleus. The length of the dam is about 600 m, and its maximum height reaches 90 m. The two abutments of the dam are embedded in outcrops of Cretaceous limestone, separated by a thin marl layer. The Cretaceous limestone is underlain by Triassic formations.

Shortly after the first filling of the reservoir, which took place in 1987, an important leak was detected at the dam toe. The emerging flow rate reached a value close to 1 m^3/s at the maximum level in the reservoir. When the water level in the reservoir reached a certain elevation, an important sinkhole was detected in a limestone block outcropping in the left abutment at a short distance from the uptake tower (see Fig. 5.13). Most of the water emerging downstream infiltrated through this sinkhole. The technicians involved in the leakage study decided to undertake the impermeabilisation of the sinkhole by filling it with about 2,000 tons of mine tailings from a blast furnace. The result of the mine tailing injection was that the escaping flow rate was lowered down to about 200 l/s at the maximum level of the reservoir. At the time the study was conducted, water emerged in two drains located at the dam toe (drains D1 and D2 in Fig. 5.13).

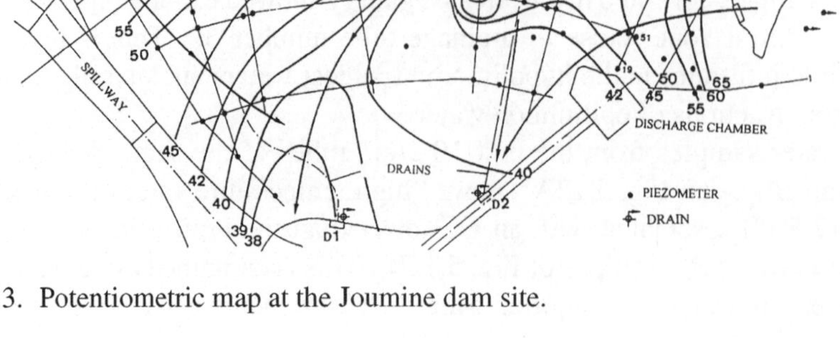

Fig. 5.13. Potentiometric map at the Joumine dam site.

The investigation of the leakage problem was made by the authors and other collaborators within the framework of an international workshop on dam leakage studies organised in Tunis by the International Atomic Energy Agency (Vienna) and the National Centre of Nuclear Science and Technology of Tunisia. Figures 5.13, 5.14 and 5.15 show, respectively, the results obtained from the measurements of piezometric levels, conductivity and temperature profiles in all accessible piezometers. For simplicity, the individual values of these parameters obtained at each borehole have been omitted. The flow rates measured in drains D1 and D2 by means of the tracer technique described in Annex 2 were, respectively, 45 l/s and 58 l/s.

The piezometric contour lines shown in Fig. 5.13 indicate that the subsurface waters emerging at both drains are related to the infiltration occurring at a zone of the reservoir close to the uptake tower, that is, from the zone where the sinkhole was initially found. The piezometric high observed in the zone of piezometer 57 is explained by the infiltration of local precipitation into the two affected piezometers. Local precipitation remained stored inside the casing of these piezometers due to the partial clogging of their screens, which are located at the bottom. The piezometric level outside this casing is normal. The low conductivity of the waters of these piezometers confirms the previous statement. The arrows marked in the figure show the direction of the preferential pathway of the water escaping from the reservoir.

Figure 5.14 shows the distribution of the EC values in the subsurface waters. The EC of the reservoir water was 0.73 mS/cm. The EC value of the water emerging at drain D1 was 0.76 mS/cm, that is, very similar to the value found in the reservoir. On the contrary, the EC of drain D2 was 2.34 mS/cm. It was not clear if the higher salinity of this water was due to the dissolution of soil material from the

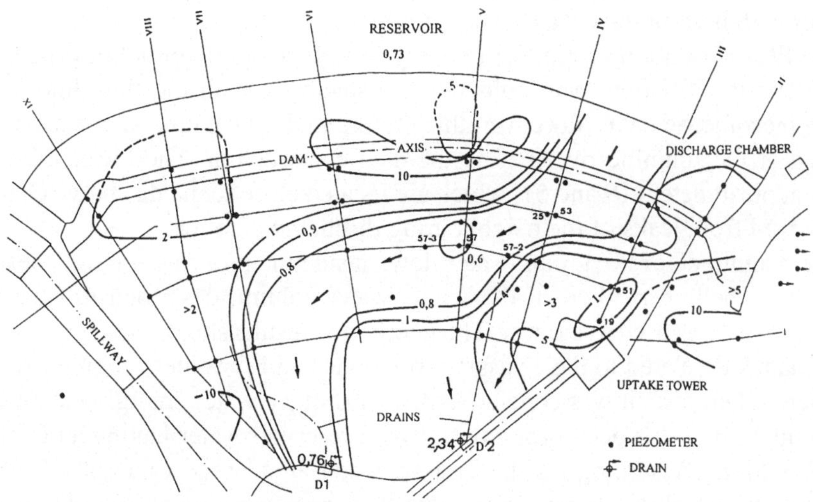

Fig. 5.14. Distribution of conductivity values at the Joumine dam site. Values are expressed in mS/cm at 25°C.

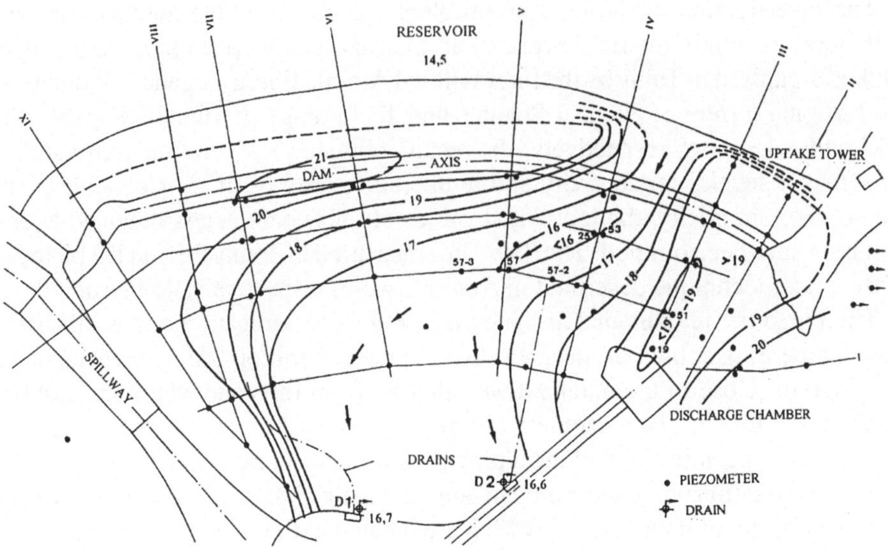

Fig. 5.15. Distribution of temperature values at the Joumine dam site. Values are expressed in °C.

dam site or if it was a consequence of a contribution of water discharging from the underlying Triassic formation. The studies conducted by the team demonstrated that the latter hypothesis was the valid one. Important groundwater flows were detected in some boreholes where the conductivity of water was higher than 4 mS/cm, using the tracer technique described in Chapter 8. Moreover, the analysis of historic hydrogeological data of the study area showed that before the dam construction, springs with high salinity water were present at this site.

The EC distribution of Fig. 5.14 clearly reflects the preferential flow path from the reservoir. This flow path coincides satisfactorily with the flow line defined by the piezometric data. Moreover, this preferential flow path coincides also with the bed of the Joumine river before the dam construction. The lower EC values found at piezometers 19 and 51, which are located close to the discharge chamber, are derived from leaks of the neighbouring discharge pipeline.

The magnitude of the groundwater flows inside the piezometers was measured using the labelling method of the whole water column described in Chapter 8. Common salt was used as a tracer. The interpreted results showed that groundwater at the dam site related to the Triassic formations had EC values ranging from 4 to 5 mS/cm. Therefore, it was concluded that subsurface water emerging at drain D2 was a mixture of this deep groundwater with reservoir water having an EC equal to 0.73 mS/cm. Assuming for the Triassic water an average value of 4.5 mS/cm, the contribution of this component to the flow rate emerging at drain D2 is given by the expression: $(2.34 - 0.73)/(4.5 - 0.73) = 0.43$, that is to say, 43%. This percent corresponds to a flow rate of $0.43 \times 45 = 19.4$ l/s.

Figure 5.15 shows the distribution of water temperature in the study area. The temperature of the reservoir water was 14.5°C, measured at the surface. Due to the fact that the study was performed during the cold season (December), no significant thermal stratification existed in the reservoir. Therefore, this temperature can be considered representative for the whole water column of the reservoir. On the contrary, the temperature values measured at the piezometers fluctuated between 15.8°C and 21.6°C. The piezometers located along the preferential pathway of the water escaping from the reservoir showed temperature values ranging from 15.8°C to 16.7°C. An increasing gradient of temperature in the direction of the flow was also observed at these piezometers. Outside this preferential pathway, the temperature values were higher due to the presence of stagnant water or of water not related to the reservoir. The lowest temperature value was obtained for piezometer 52, indicating that this borehole was preferentially connected to the reservoir. The difference in temperature between the reservoir and the piezometers located at the preferential flow area can be explained by heat exchange with soil materials. These materials could still show a higher temperature due to the heat stored during the previous months, when the temperature of the reservoir water was much higher.

As expected, a rapid groundwater flow velocity was found in all piezometers located along the preferential pathway, as measured using the labelling method of the whole water column described in Chapter 8. The tracer injected in these boreholes (saturated solution of common salt) escaped almost instantaneously.

Finally, it must be mentioned that two interconnection experiments were performed. Small quantities of uranine and rhodamine WT were injected in piezometers 57.2 and 57.3, respectively. Approximately 95% of the injected uranine was recovered in drain D2 and 5% in drain D1. Therefore, the uranine recovery was about 100%. The transit time of uranine from borehole 57.3 to drain D2 was about 4 hours and about 20 hours for drain D1. On the contrary, most of the rhodamine WT was recovered at drain D1 with a transit time of about 10 hours. The percentage of recovery was much lower than in the case of uranine, and the tracer breakthrough curve showed a long tail. This was interpreted as a consequence of the interaction of rhodamine WT with the soil material (see Chapter 7). This interaction was much lower in the case of uranine.

5.4 CHEMICAL COMPOSITION OF WATER

5.4.1 General considerations

In the previous section, the possibilities offered by the measurement of the electrical conductivity of water for the investigation of leakage from dams have been discussed. However, in most of cases, the results obtained can be improved if, apart from this parameter, the major chemical components of water are also analysed. In some cases, some minor components can also provide valuable information.

As a general rule, the performance of chemical analyses of major components at the more representative waters of the study area is recommended at the initial stage of the investigation, as well as when important variations of the leakage regime, also affecting the EC of water, are observed. The selection of the water points to be sampled for such analyses can be made on the basis of EC values measured in the field. These measurements are used to classify the different water points in families or chemical facies showing similar values of total salinity so that the number of samples collected for complete chemical analysis can be drastically reduced. Very frequently, spring and/or borehole waters have the same origin and, therefore, the same chemical composition.

Obviously, the chemical composition of water will provide additional information on its origin and geochemical evolution, facilitating the differentiation between reservoir and aquifer waters. Similarly, chemical data is used to identify and evaluate mixtures of both types of waters. Complementary tools for these studies are the stable isotopes of water, deuterium and oxygen-18, as well as natural tritium. These isotopic natural tracers are discussed in the next chapter.

Even when it seems evident that the emerging water is coming entirely from the reservoir, the comparison of the chemical composition of the reservoir water and the emerging spring water can provide valuable information on the lithological composition of the rock formation through which the emerging water is circulating. On the other hand, the seasonal as well as sporadic changes in the chemical composition of water in the reservoir can be used for the investigation of the dynamic characteristics of the groundwater flows feeding the springs. These dynamic characteristics include the determination of transit times and the evaluation of the storage capacity of the rock formation involved in such flows. For example, significant changes in the chemical composition can be derived from the arrival to the reservoir of sewage or wastewaters discharged in a river feeding the reservoir.

Obviously, due to the scope of this book, it is not possible to present here a detailed synopsis on the chemical composition of natural waters. Therefore, the following sections will include only some practical indications related to the topic of this book. For more detailed information on the principles of water chemistry applied to hydrogeological studies, the reader is referred to the specialised publications on this matter (e.g., Appelo and Postma, 1996; Custodio and Llamas, 1976; Wedepohl, 1969, 1976). The reason for including this topic in the book is to help those readers from the engineering field who might not be familiar with the terminology used in hydrochemistry, especially those aspects related to its application in hydrological investigations.

5.4.2 Major chemical components of water

Usually, the major dissolved components of water are the following:

Anions: Cl^-, $SO_4^=$, NO_3^-, HCO_3^- and $CO_3^=$.
Cations: Na^+, K^+, Ca^{++} and Mg^{++}.

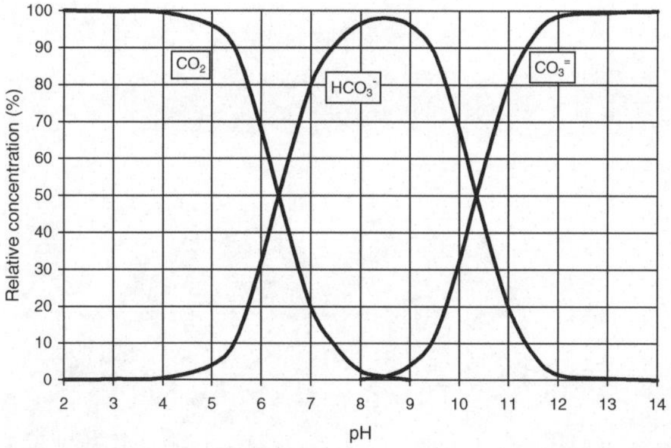

Fig. 5.16. Variation of the concentration of ion HCO_3^- in water as a function of pH.

The carbonate anion $CO_3^=$ is present in natural waters in significant concentrations only if the pH is higher than 8.3 and becomes the predominant carbon component for pH values higher than 10.2. This is illustrated in Fig. 5.16. The bell shape line of this figure gives the percent of ion HCO_3^- present in a water sample as a function of the pH. The carbonic acid (or CO_2) predominates in the left region of the diagram (i.e., at low pH values), while in the right region the major carbon component is $CO_3^=$. For the most frequent pH values found in natural waters (between 6.5 and 9.5), the predominant carbon component is therefore the bicarbonate anion (HCO_3^-). For further information on the equilibrium of carbon species in the water, the reader is referred to Appelo and Postma (1996) and Custodio and Llamas (1976).

Apart from the concentration of these ions, pH and electrical conductivity or total dissolved solids (*TDS*) need to be measured. The concentration of silica (SiO_2) can also be very interesting when dealing with waters circulating over volcanic or granitic rocks, as well as with thermal waters. In these cases, high silica concentration can be present. In the same way, the redox conditions may be responsible for high concentrations of Fe^{++}. Obviously, in cases of polluted waters, other components such as NH_4^+, total nitrogen, phenol, organic chlorinated compounds, pesticides, etc. can also be used for the characterisation of the reservoir water.

The values of conductivity, pH, temperature and alkalinity, which corresponds in most cases to the concentration of the ions HCO_3^- and $CO_3^=$, should be determined in the field due to their nonconservative character. For the analysis of cations, water samples should be acidulated by adding some drops of nitric acid (1:1) up to a pH value close to 2. The reason for that is to avoid the risk of precipitation of Ca^{++} and Mg^{++}, as well as of heavy metals. The precipitation of these components may happen during the storage period, for example, if an increment of the pH is produced due to the loss of dissolved CO_2. The total concentration of the anions HCO_3^- and $CO_3^=$ is determined in the field through the well known volumetric method based on the use of calibrated solutions of HCl or H_2SO_4 and methyl orange as indicator. This reagent changes its colour when the pH reaches a value of 4.3. This analysis

Fig. 5.17. Performing field chemical analyses of water at Piaui State (Brazil).

provides the so-called total alkalinity, given by the following expression:

$$Alkalinity = (m)HCO_3^- + (2m)CO_3^=$$

where *m* represents the molar concentration expressed in mol per litre of water. As mentioned before, if the pH is lower than 8.3, the alkalinity corresponds mainly to the HCO_3^- concentration.

The use of portable chemical analysis laboratories is recommended when the results are needed in the field to make decisions about the immediate actions of the ongoing investigation (see Figs. 5.17, 5.18 and 5.19).

5.4.3 Evaluation of the quality of the chemical analyses

Before proceeding with the interpretation of the results of chemical analysis, it is recommended to carry out an evaluation of their quality. To do this, the ionic concentrations, usually expressed in mg/l, must be converted into milliequivalents per litre (meq/l). This is done by dividing the mg/l values by the respective chemical equivalent, which is the quotient of the atomic or molecular weight divided by the valence. For the major components of water, 1 meq/l is equal to the following concentrations expressed in mg/l:

Cl^-	$SO_4^=$	NO_3^-	HCO_3^-	$CO_3^=$	Ca^{++}	Mg^{++}	Na^+	K^+
35.45	48.03	62.0	61.03	30.0	20.04	12.15	22.99	39.1

Fig. 5.18. Performing field chemical analyses of water at the Eastern Caprivi Region (Namibia).

Fig. 5.19. Performing field chemical analyses of water with a portable laboratory. The following components are determined: electrical conductivity, pH, Cl^-, $SO_4^=$, NO_3^-, HCO_3^-, $CO_3^=$, Ca^{++}, Mg^{++} and SiO_2.

The first quality condition is that the sum of the concentrations of all cations must be equal to the sum for the anions, with both being expressed in meq/l. This similarity of values is implied by the required electroneutrality of the water sample, that is, the sum of positive and negative charges associated with the dissolved ions must be the same. The overall quality of the analysis is sometimes expressed as a deviation from the electroneutrality (EN), expressed as a percentage, which is obtained by the following equation:

$$EN\,(\%) = \frac{\sum cations + \sum anions}{\sum cations - \sum anions} 100 \qquad (5.7)$$

where the sums of cations and anions are expressed in meq/l.

Consider that the terms of equation (5.7) are electrical charges, and that cations have positive charge while anions are negatively charged. Consequently, the plus and minus signs of this equation have to be inverted for calculation. Differences in the electroneutrality up to 2% are unavoidable. Differences between 2% and 5% are acceptable for most natural waters with typical *TDS* below 2 g/l. Differences higher than 5% are not acceptable, and the analytical procedures must be checked.

A second criteria to assess the quality of the chemical analysis is the comparison of the terms $\sum cations$ or $\sum anions$ with the EC. If the results are of reasonable quality, the following approximate relationship should be fulfilled:

$$\sum cations\ (\mathrm{meq/l}) = \sum anions\ (\mathrm{meq/l}) = 100(\mathrm{EC})\ (\mu\mathrm{S/cm}) \qquad (5.8)$$

This relationship is interesting, for example, when only anions or only cations have been analysed.

5.4.4 Common concentrations and the origin of the ionic components

The most frequent concentrations and the most probable origin of the major ionic components dissolved in natural waters are the following:

Chloride ion
Most frequent concentration: 2 to 80 mg/l.
Origin: atmospheric sources, evaporitic rocks (halite), sewage.

Sulphate ion
Most frequent concentration: 1 to 1000 mg/l.
Origin: gypsum, sulphide oxidation, sewage and polluted waters, and to a lesser extent, atmospheric sources.

Bicarbonate ion
Most frequent concentration: 20 to 400 mg/l.
Origin: dissolved CO_2 in soils; carbonate and silicate minerals in rocks and aquifers during weathering; dissolution of carbonates, such as limestone, dolomite,

aragonite, magnesite, etc.; oxidation of organic matter and volcanic exhalations in certain areas.

Nitrate ion
Most frequent concentration: 0.1 to 15 mg/l.
Origin: oxidation of organic matter, sewage, mineral fertilizers.[3]

Calcium ion
Most frequent concentration: 5 to 200 mg/l.
Origin: dissolution or exchange of carbonate rocks and gypsum, weathering product of feldspars, pyroxenes and amphiboles, and to some extent, atmospheric sources.

Magnesium ion
Most frequent concentration: 1 to 50 mg/l.
Origin: Some carbonates and silicates, such as dolomite, serpentine, pyroxene, amphibole, olivine and mica.

Sodium ion
Most frequent concentration: 2 to 50 mg/l.
Origin: atmospheric sources, dissolution of evaporitic rocks (halite), weathering of Na-feldspar, zeolite, etc., sewage and exchange of clay minerals with groundwater.

Potassium ion
Most frequent concentration: 0.5 to 10 mg/l.
Origin: feldspar and mica, sewage and polluted waters.

5.4.5 Solubility of minerals

All dissolution equilibria of ions present in water are described by the well-known mass action law. Let us suppose the following general reversible reaction:

$$aA + bB \Leftrightarrow cC + dD$$

The mass action law states that once the equilibrium between the right and the left side terms has been established, the following equation is fulfilled:

$$K = \frac{[C]^c[D]^d}{[A]^a[B]^b} \tag{5.9}$$

[3] In the case of nitrates derived from mineral fertilizers, or in the case of waters highly polluted with organic materials, concentrations higher than 1 g/l can be observed.

where K is the equilibrium constant and the parameters in brackets represent the ionic activities or "effective concentrations"[4] of the products taking part in the reaction expressed in mol per litre (1 mol is equal to the molecular weight expressed in grams).

The dissolution equations of mineral components are usually written as dissociation reactions. For example, the dissolution of gypsum is written in the following way:

$$CaSO_4 \cdot 2H_2O \Leftrightarrow SO_4^{=} + Ca^{++} + 2H_2O$$

Taking into account that the activity of a solid (gypsum, in this case) is equal to 1, the solubility constant, also called the solubility product, can be written in the following way:

$$K_{gypsum} = \left[SO_4^{2-}\right]\left[Ca^{2+}\right] = 10^{-4.60} \quad \text{at } 25°\text{C} \tag{5.10}$$

In this case, the molar concentration (mol per litre of solution) is the same for the ions Ca^{++} and $SO_4^{=}$. Therefore, the concentration of these ions in a saturated solution is given by $(10^{-4.6})^{1/2} = 10^{-2.3} = 5.0 \times 10^{-3}$ mol per litre. Taking into account that the molecular weight of gypsum ($CaSO_4 \cdot 2H_2O$) is equal to 172 g, its solubility in water at 25°C is $172 \times 5 \times 10^{-3} = 0.86$ g/l = 860 mg/l. This solubility is valid only for distilled water. The presence in the water of other ions increases the solubility of gypsum. For example, this solubility increases up to approximately 3 g/l in the case of a NaCl solution with a concentration of 4 g/l.

The presence in natural waters of Ca^{++} derived, for example, from the dissolution of calcite, shifts to the left the equilibrium given by equation (5.8), and therefore, the solubility of gypsum decreases. This phenomenon is known as the "common ion effect" and very frequently is responsible for the precipitation of chemical compounds dissolved in water. For example, if water saturated in calcite interacts with a geological unit bearing gypsum, $CaCO_3$ will precipitate.

The approximate solubility of some natural chemical compounds in distilled water at 25°C expressed in mol per litre is given below:

$KAlSi_3O_8$:	3×10^{-7}
$NaAlSi_3O_8$:	6×10^{-7}
$CaCO_3$:	6×10^{-5}
SiO_2 (opaline):	2×10^{-3}
$CaSO_4 \cdot 2H_2O$:	5×10^{-3}

[4] The activity or effective concentration of a determined ion depends on the presence of other ions in the solution. These produce an electrostatic shielding of the considered ion and consequently, a decrease in its reactivity. Due to this fact, the solubility of a chemical substance in water is higher when other strange ions are present. For the calculation of the activity, the reader is referred to Appelo and Postma (1996) and Custodio and Llamas (1976).

$Na_2SO_4 \cdot 10H_2O$:	2×10^{-1}
$MgSO_4 \cdot 7H_2O$:	3×10^{-0}
$Na_2CO_3 \cdot 10H_2O$:	3×10^{-0}
KCl:	4×10^{-0}
NaCl:	5×10^{-0}
$MgCl_2 \cdot 6H_2O$:	5×10^{-0}

As can be seen, the solubility in water of calcium carbonate (calcite) is very low, on the order of 6 mg/l, which corresponds to 2.4 mg/l of Ca^{++}. However, the concentrations commonly found in groundwater circulating over carbonatic rocks are much higher, reaching values up to 300 mg/l. The explanation for these high concentrations is the dissolution of carbonate rocks by the action of the carbonic acid derived from soil CO_2 ($CO_2 + H_2O \rightarrow H_2CO_3$). Calcium bicarbonate is produced according to the reaction:

$$CO_2 + CaCO_3 + H_2O \Leftrightarrow 2HCO_3^- + Ca^{++}$$

The solubility of calcium bicarbonate is much higher than that of calcium carbonate. The presence of CO_2, and therefore of H_2CO_3, in water generates an equilibrium between the species CO_2, HCO_3^- and $CO_3^=$ that is governed by pH changes, as illustrated in Fig. 5.16.

In a general way, it has been observed that water containing as major components calcium and magnesium bicarbonate shows, when saturated in these two components, EC values between 300 and 600 μS/cm (see equation 5.4), depending on the pH, temperature and the presence of other ions. When the major component is calcium sulphate (gypsum), the saturated solution will show conductivity values between 1500 and 3500 μS/cm, depending also on the presence and concentrations of other ions. EC values higher than 3500 μS/cm generally imply the presence of sodium chloride or, less frequently, sodium sulphate as additional or major components. Sodium sulphate is not present as a mineral component of common rock formations, but it is generated by ion exchange processes involving dissolved Ca^{++} and Na^+ from clayey rock formations.

From a practical point of view, there is a very interesting comparison of the solubility product given by equation (5.10) with the product of the effective concentration or activity values of the same ions in the analysed water samples. Let us designate the last product as *experimental ionic product (EIP)*. The saturation degree of the analysed water sample in the respective component is reflected by the ratio $R_S = (EIP)/K$. If $R_S = 1$, the solution is saturated, that is, an equilibrium has been achieved with the solid phase. If $R_S < 1$, the solution is undersaturated, and if $R_S > 1$, the solution is oversaturated. The value of R_S can show large variations. Therefore, the saturation degree of the solution is better expressed by the so-called *saturation index (SI)*, which is given by the following expression:

$$SI = log[(EIP)/K] \quad (5.11)$$

If $IS = 0$, the solution is in equilibrium with the solid phase. If $IS < 0$, the solution is undersaturated, and if $IS > 0$, the solution is oversaturated.

The value of the solubility product K depends in a very complex way on the temperature, pH and the concentration of other ions. For this reason, the saturation indices are usually obtained by means of computer programs that take into account these parameters. Examples of these programs are WATEQF, PHREEQE, PHREEQC, PHRQPITZ, etc. The last program has been specifically created for the calculation of saturation indices of highly saline water samples, such as brackish water. As mentioned above, the values of temperature and pH need to be measured in the field due to the importance of these parameters for the calculation of saturation indexes.

5.4.6 Applications

The determination of the chemical composition of water can provide very valuable information in leakage studies. As mentioned above, the evolution of the chemical composition of the reservoir water after being infiltrated into the ground provides information on the nature of the rock materials through which groundwater is flowing. The measurement of the chemical composition of borehole and spring water allows in many cases the identification of the existing preferential pathways.

In fact, the possibilities offered by this technique are difficult to foresee and can be evaluated only once the results of chemical analyses are available. For this reason, the performance of chemical analyses in a systematic way is recommended in all leakage studies. Sometimes, the more important hint to understanding the subsurface flow patterns problem arises from totally unexpected results. For example, in the case of the reservoir La Herradura, Cuba, the concentration of K^+ allowed the identification of the springs fed by the reservoir in contrast with other springs that were fed by the adjacent aquifer (Plata et al., 1999a). All these springs were located at the foot of the dam. A potassium concentration equal to 4 mg/l was found in the reservoir water, whereas the content of this ion in groundwater samples representing the discharge of the surrounding karstic aquifer was below the detection limit. It was found that only two small springs located at a small drainage channel and the small water losses of the deep sluice showed potassium concentrations similar to the value observed in the reservoir. These were the only emerging waters fed by the reservoir.

When dealing with reservoirs or lakes located in karstic regions, the impounded water in the reservoir is frequently saturated in calcite and other carbonate minerals (dolomite, aragonite, magnetite, etc.). On the other hand, karstic springs of such regions not related to the reservoir are generally oversaturated with respect to calcite and other carbonate minerals, as a result of the high CO_2 amounts dissolved by groundwater from the soil during its infiltration. When this oversaturated water emerges, the excess of CO_2 escapes to the atmosphere and, as a result, the precipitation of $CaCO_3$ occurs. Once this degassing process is completed, water remains with a value of the saturation index *SI* close to zero (see equation 5.11).

Obviously, reservoir water undergoes the same degassing process and will also show a saturation index close to equilibrium. Therefore, when this water is infiltrated in the ground because of the existing leaks, it cannot dissolve additional amounts of calcite. Dissolution will be possible only if a new CO_2 source is available, which normally is not the case at the bottom of the reservoir. Therefore, when the reservoir water emerges at the downstream springs, it will maintain the same saturation index close to the equilibrium value. This will not be the case for karstic springs that are fed by a karstic aquifer. These springs will still show oversaturation ($SI > 0$). This different behaviour helps in distinguishing between springs linked to the reservoir and those linked to the aquifer. Using this method, the authors could demonstrate that groundwaters emerging at the La Noguera springs, near the lake area of Ruidera, Ciudad Real, Spain, were fed by one of these lakes and not by the local karstic limestone aquifer of Jurassic age. The same method was used to prove that some springs located at the foot of the Cautillo dam, Cuba, were not related to the reservoir (Plata et al., 1999b).

Groundwater can be saturated in calcite, not only in the case of flow through carbonate rock formations, but also when calcite is present in small fractions in other sedimentary and detritic rocks, either as part of the rock matrix or as cement. When both the reservoir water and the spring water are saturated in calcite, the concentration of HCO_3^- might not provide clear evidence on the hydraulic connection between these two water bodies, in spite of the statement previously made. In such cases, the concentration of Cl^- and $SO_4^=$ ions can provide additional evidence, especially if the water is not saturated in gypsum. This was the case found at the Tous reservoir in Spain. A karstic spring, called Mourterall spring, with a flow rate higher than 250 l/s is located at about 800 m downstream of the dam. The groundwater flow feeding this spring was intercepted by some boreholes, which were drilled in the right abutment of the dam. The depth of this flow was located about 35 m below the bottom of the reservoir. The concentration of Cl^- and $SO_4^=$ ions proved that an important fraction of the emerging water was not fed by the reservoir. The following conductivity and concentration values were obtained:

Reservoir

Conductivity (25°C): 1078 μS/cm, Cl^-: 67 mg/l; $SO_4^=$: 293 mg/l

Mourterall spring

Conductivity (25°C): 907 μS/cm, Cl^-: 78 mg/l; $SO_4^=$: 113 mg/l

The ratio of $Cl^-/SO_4^=$ with values expressed in meq/l was 3.23 for the reservoir and 1.06 for the springs. The relatively low concentrations of both ions and the short distance between the reservoir and the spring exclude the possibility of their nonconservative behaviour. The high $SO_4^=$ concentration values were due to the dissolution of gypsum present in the Triassic and Upper Tertiary evaporitic rocks.

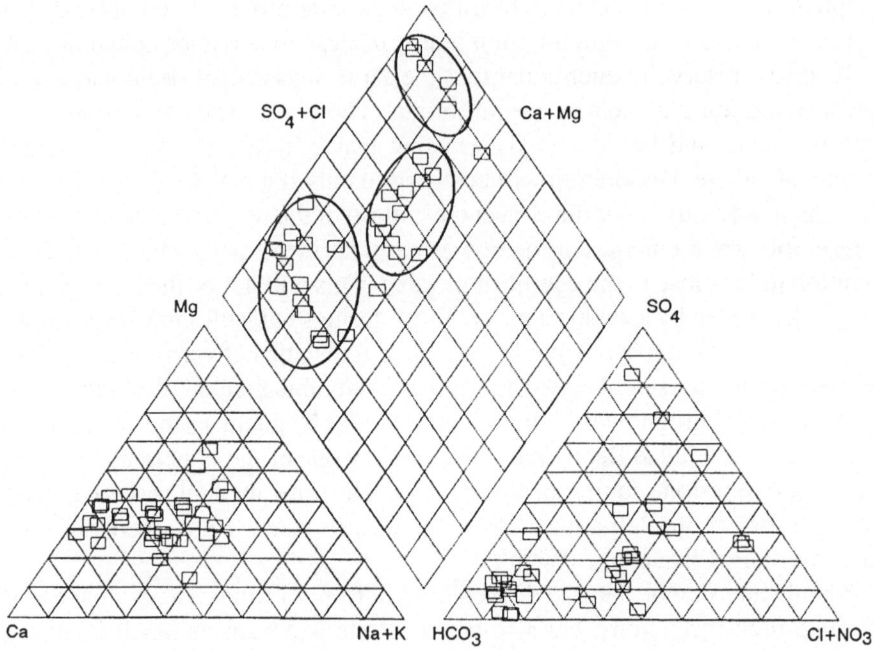

Fig. 5.20. Piper diagram obtained for the water samples collected at the Rules dam site (Granada, Spain).

Tritium contents measured in samples of both types of water demonstrated that the contribution of the reservoir to the springs could represent a maximum fraction of 10%.

The results of the chemical analyses are frequently illustrated in the form of different types of diagrams, such as Piper and Schöller diagrams, also called vertical column diagrams. Examples of such diagrams are shown, respectively, in Figs. 5.20 and 5.21, which correspond to water samples taken from the dam site of the Rules reservoir on the River Guadalfeo, Spain. The relevant samples were collected within the framework of a study carried out during the construction of the dam, designed to define the chemical and isotopic baselines of the dam site before filling the reservoir.

The Piper diagram illustrates the relative concentration values of the dissolved ions (major components) expressed in meq/l. The plotted values correspond to the percent contribution of each ion or pair of ions (cases of $Na^+ + K^+$; $Cl^- + NO_3^-$; $HCO_3^- + CO_3^=$) to the chemical composition of water. On the contrary, in the case of the Schöller diagram, the absolute values of the ionic concentrations are plotted.

The case illustrated in Figs. 5.20 and 5.21 shows the large variability observed in the chemical composition of several types of groundwater bodies existing at the dam site. In the Piper diagram, the points corresponding to groundwater circulating

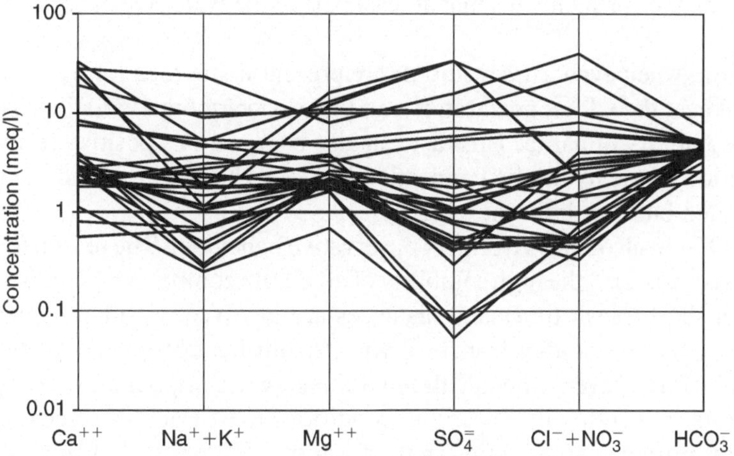

Fig. 5.21. Schöller diagram obtained for the water samples of the Rules dam site (Granada, Spain).

through the carbonate aquifer are as expected, close to the left corner of the diamond-shaped section. This is the position corresponding to water for which $CaCO_3$ and/or $MgCO_3$ are the major components. On the contrary, the points corresponding to the water circulating through the alluvial aquifer are located in the upper vertex of the same diagram, because of its high contents of $SO_4^=$ and/or Cl^- ions. The points located inside the intermediate ellipse correspond to mixtures of the above-mentioned types of water.

It is very important to point out that among all major ions usually dissolved in water, chloride is the most conservative ion. This behaviour indicates that it is a good tracer of water, that is, the behaviour of water and of the chloride ion is very similar, and it is not affected by water–rock interactions in most environments. In practical terms, this property means that this ion travels long distances through the terrain with no or very small variations of the Cl^- concentration. The main reason for this conservative character is that most minerals containing chlorine show a high solubility in water. The only exceptions to this behaviour are mercury, lead and thallium chloride, but these minerals are not present in natural waters in significant concentrations. Also, exchange reactions of Cl^- with other anions of the rock are not significant.

The conservative character of the Cl^- ion permits its use as a natural and, sometimes, as an artificial tracer. The conclusions on the relation between reservoir water and borehole or spring waters are more reliable when they are based on the Cl^- concentration. In other words, if the Cl^- concentration found at the spring water is the same as the one of the reservoir, the probability of a hydraulic connection between them is higher. A lower Cl^- concentration measured in the sample representing the spring means that at least part of the water is not coming from the reservoir.

However, when evaporitic materials are present in the rock formations involved in the leakage, the Cl^- concentration can undergo very high variations because of the high solubility of halite. This fact can also be taken as a positive feature for the investigation of the hydraulic connection between the reservoir and the emerging groundwater, but in a different way.

The NO_3^- ion also shows a rather conservative behaviour. The reason is similar to the one given for Cl^-: the high solubility of all nitrate compounds. This is especially true when short transit times and distances are involved, as often happens in the case of most leakage studies from reservoirs (denitrification processes are expected to be absent). However, when dealing with reservoirs impounding sewage and/or wastewaters, where highly anaerobic conditions exist (lack of oxygen), a major fraction of nitrogen can be found in the form of NH_4^+ or NO_2^-. If this is the case, the measurement of the total dissolved nitrogen can provide better results because a significant fraction of these nitrogen compounds can be oxidised to NO_3^- during the migration of the water through the involved rock formation.

The ion $SO_4^=$ can be more or less conservative when the concentration values are lower than about 200 mg/l. At higher concentrations, precipitation of this ion may occur if the concentration of Ca^{++} increases, for example, due to the dissolution of calcite (common ion effect).

Finally, it should be pointed out that cations in most cases are not good tracers of water because they can undergo exchange reactions with mineral phases present in the geological medium. This is especially true for earth-alkaline ions. Alkaline ions can show a rather conservative behaviour, especially for short distances and short transit times.

REFERENCES

Appelo, C.A.J. and Postma, D., 1996. *Geochemistry, groundwater and pollution*. A.A. Balkema, Rotterdam/Brookfield.

Cartwright, K. and McComas, M.R., 1968. Geophysical surveys in the vicinity of sanitary landfills in Northeastern Illinois. *Ground Water*, 6(5), 912–918.

Custodio, E. and Llamas, M.R., 1976. *Hidrología subterránea*. Editorial Omega S.A., Barcelona (in Spanish).

Fischer, H.B., List, E.J., Koh, R.C.Y., Imberger, J. and Brooks, N.H., 1979. *Mixing in inland and coastal waters*. Academic Press, New York.

Oeschger, H. and Gugelmannn, A., 1974. Das geophysikalische Verhalten der Umweltisotope als Basis für Modellrechnung in der Isotopenhydrologie. *Oesterr. Wasserwirtsch.*, Sonderausdruck, 26(3/4) (in German).

Plata, A., Araguas, L. and Garcia Agudo, E., 1998. *Workshop on leakage from dams*. IAEA and CNSTN, Vienna and Tunis.

Plata, A., Rodríguez, B., González, R., Romero, Z. and Llanes, A., 1999a. Estudio de fugas en la presa La Herradura, Cuba. In *Estudios de Fugas en Embalses*. IAEA, Vienna (in Spanish).
Plata, A., Kile, S., Linares, N. and Rodríguez, B., 1999b. Estudio de filtraciones en el embalse de Cautillo, Cuba. In *Estudios de Fugas en Embalses*. IAEA, Vienna (in Spanish).
Pérez Zabaleta, E., Montero, M., Aleixandre, J. and Iragüen, E., 1990. *Estudio con trazadores radiactivos del vaso del embalse de Beninar, Almería*. Internal report CEDEX, Madrid (in Spanish).
Wedepohl, K.H. (ed.), 1969–1976. *Handbook of geochemistry*. Springer Verlag, Berlin.

CHAPTER 6

Natural tracers of water: (B) Stable isotopes, tritium and carbon isotopes

6.1 BASIC CONCEPTS

In this chapter, the basic principles and the possibilities offered by the stable isotopes of water, deuterium (^{2}H or D), oxygen-18 (^{18}O), and tritium (^{3}H or T), as well as the carbon isotopes carbon-13 (^{13}C) and carbon-14 (^{14}C), are described in connection with the investigation of leaks from dams and reservoirs. Because these tools are not familiar to engineers and technical staff dealing with dam safety issues and the study of leakage problems, the theoretical background for their application is presented in the following sections.

An atom consists of a nucleus surrounded by electrons. Most of the mass of the atom is concentrated in the nucleus, which is formed by two kinds of particles: neutrons and protons. Protons are positively charged, and neutrons carry no electric charge. The number of protons is characteristic for each chemical element and is called the atomic number (Z). This number is equal to the number of electrons, which are negatively charged, so that the atom as a whole is neutral. Because repulsive electrical forces exist between protons, the presence of neutrons is required to stabilise the nucleus. In the case of light elements, the number of protons and neutrons is equal. The sum of neutrons and protons is called the mass number (A).

The term nuclide refers to each combination of neutrons and protons that can compose each element. The atomic nuclei containing different numbers of neutrons and protons are called isotopes. All isotopes of a given element present the same chemical behaviour because the chemical properties are controlled by the external configuration of the electrons, which is the same for all isotopes of a given element. In nature, each element is constituted by a mixture of different isotopes. However, the tiny differences of mass of the atoms or molecules containing other isotopes besides the most abundant one are responsible for a different physical behaviour, in which the structure and characteristics of the atomic nucleus are relevant. Isotopes are usually represented in the form ^{A}X, where X represents the symbol of the element, which implicitly incorporates the atomic number. Also the name of the element followed by the mass number is an accepted form of notation, for example, oxygen-18.

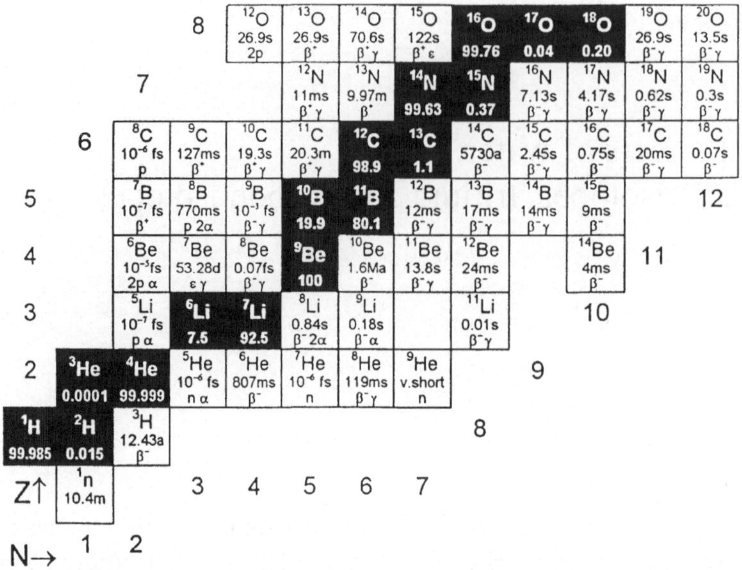

Fig. 6.1. Chart of the light element isotopes showing percent abundances of the stable isotopes (shaded black) and half-lives of radioisotopes (s = second, m = minute, d = day, a = year) with their principal and secondary decay modes (after Clark and Fritz, 1997).

Figure 6.1 shows the lowest end of the nuclide chart, where the known isotopes of the light elements are presented. In this chart, the number of neutrons (x-axis) existing in the nucleus of each element is plotted against the number of protons (y-axis). For a given element, the excess of neutrons with respect to the most abundant stable isotope (shown along a horizontal line) is represented by isotopes of increasing mass, e.g., ^{2}H and ^{3}H in the case of hydrogen (^{1}H); ^{17}O and ^{18}O in the case of oxygen (^{16}O), etc. As already mentioned, the excess or shortage of neutrons in most elements leads to differences in the physical behaviour of molecules or atoms containing these heavy isotopes.

Isotopes are classified in two categories: stable and radioactive. Stable isotopes are those that are commonly found in natural elements. Some elements contain only one stable isotope. However, when several stable isotopes are present in the natural element, their relative proportions remain constant, although small differences can originate when the element is involved in chemical reactions or phase changes. It is well known that the mass of some atoms affects the velocity of some chemical reactions produced by natural processes, resulting in changes in the ratio between the abundances of the heavy (e.g., ^{2}H, ^{13}C, ^{18}O, ^{15}N, etc.) and the light, that is, the most abundant isotope (e.g., ^{2}H/^{1}H, ^{13}C/^{12}C, ^{18}O/^{16}O, etc). These processes that result in changes of the isotope ratios are responsible for the so-called isotope fractionation.

On the other hand, radioactive isotopes are those that undergo spontaneous radioactive decay. That occurs mainly in the isotopes showing a large shortage or excess of neutrons compared with the stable isotopes of the same element. The radioactive decay is always associated with the spontaneous emission of energy in the form of α or β particles and gamma radiation. When such particles are emitted, the original isotope is transformed into a new one, which can be stable or radioactive. The type and energy of the emitted particles and gamma rays is characteristic of each radionuclide.

The number of radioactive atoms present in a sample or mass of a given radioisotope decreases with time as a consequence of radioactive decay. The extreme regularity of the decay process permits the use of these isotopes as "chronometers" for most natural processes. The number of atoms decaying per unit of time is called the activity of the radionuclide. The rate of decay is expressed by the *radioactivity decay law*, which is given by the following equation: $A = (dN/dt) = -\lambda N$, where λ is the decay constant, which is characteristic for each radionuclide, and N the number of radioactive atoms existing in the sample. The activity is expressed in becquerels, a unit named after the great scientist Henry Becquerel. One becquerel is equal to one decay event per second. Another common unit of radioactivity is the Curie, which is equal to 3.7×10^{10} becquerels.

The integration of the previous differential equation leads to:

$$N_t = N_0\, e^{-\lambda t} \quad \text{and therefore } A_t = A_0\, e^{-\lambda t} \tag{6.1}$$

where N_t and A_t are, respectively, the remaining number of atoms and activity after time t, and N_0 and A_0 are the initial values at $t = 0$.

The activity A_t of a radioactive sample is reduced to half of the initial activity A_0 ($A_0/A_t = 2$) after a period given by the expression $ln\, 2 = \lambda t_{1/2}$, which is derived from equation (6.1). The period of time defined by $t_{1/2}$ is called *half-life*, which is equivalent to $t_{1/2} = 0.6931/\lambda$. In the case of tritium, this value is equal to 4500 ± 8 days (12.32 years) (Lucas and Unterweger, 2000) and therefore, $\lambda = 0.05626$ years^{-1}.

It is worth mentioning at this stage that the three heavy isotopes of hydrogen and oxygen mentioned, deuterium, oxygen-18 and tritium, constitute an integral part of the water molecule, in the form of HDO, $H_2{}^{18}O$ and THO. This feature constitutes a substantial difference when compared with the dissolved tracers mentioned in previous chapters. Water molecules containing any of the three mentioned heavy isotopes behave, from the chemical point of view, as any other molecule formed exclusively with the most abundant isotopes (${}^{1}H_2{}^{16}O$). Therefore, their behaviour is almost ideal when compared with other tracers whose properties might be or certainly are different when compared to those of water molecules. Water molecules containing these heavy isotopes are perfect tracers to study the fluxes and interactions of different water bodies in the water cycle.

6.2 STABLE ISOTOPES OF WATER

6.2.1 Molecular species of water

Natural hydrogen, as shown in Fig. 6.1, is exclusively formed by two stable isotopes, ^{1}H (99.985%) and ^{2}H or deuterium (D) (0.01492%). Similarly, oxygen is formed by three isotopes ^{16}O (99.759%), ^{17}O (0.0374%) and ^{18}O (0.2039%). Because ^{1}H and ^{16}O are the most abundant species, the usual form to express the composition of water is $^{1}H_2{}^{16}O$. Out of nine other isotopically different water molecules that can be formed with the indicated isotopes, only three occur in nature in easily measurable concentrations, $^{1}H_2{}^{16}O$, $^{1}HD^{16}O$, $^{1}H_2{}^{18}O$. Taking the ocean as the major reservoir of water, and therefore, representing the "typical" water molecule, the respective abundances of the three most abundant molecules are:

$^{1}H_2{}^{16}O$	997.680 parts per million
$^{1}HD^{16}O$	320 parts per million
$^{1}H_2{}^{18}O$	2.000 parts per million

The common form to express the abundance of a given isotope is by using *isotopic ratios* (R), defined as follows:

$$R = \frac{\textit{Number of molecules}\ \mathrm{HDO}\ \textit{or}\ \mathrm{H_2{}^{18}O}}{\textit{Number of molecules}\ \mathrm{H_2O}} \qquad (6.2)$$

As previously indicated, isotope ratios are naturally modified by several processes along the water cycle, resulting in a characteristic labelling of all components of the water cycle with specific isotopic ratios. As mentioned, this natural labelling is the characteristic used as specific fingerprinting of water molecules, allowing the study of the origin and dynamics of different water bodies as explained in the following paragraphs.

6.2.2 δ-notation: isotope deviation

Isotope ratios R are not usually reported as absolute numbers. The isotopic abundances of an unknown sample A are reported as *deviations* of the relevant isotopic ratio, relative to the same ratio in a defined standard S of the same substance, according to the equation:

$$\delta_S(A) = 1000\frac{R_A - R_S}{R_S} \quad \text{(in per mil)} \qquad (6.3)$$

The main reasons for the adoption of such a notation are the following: (a) mass spectrometers are prepared to precisely measure small variations in isotopic ratios between two samples but are not suitable for obtaining absolute isotope ratios; (b) adoption of absolute ratios will imply the use of numbers with many decimal places; (c) absolute ratios are less relevant than relative changes in isotopic ratios. Therefore, a delta value (δ) was adopted, being a dimensionless number, not a unit, that is expressed as per mil (‰) deviation with respect to an internationally adopted standard.

Delta values can be negative or positive numbers, depending on the relative difference of the isotopic ratios of the sample and the adopted standard. As will be shown later, in most cases, δ-values of meteoric waters are negative because the adopted standard for water is a seawater sample.

The standard used as a primary reference to express most isotopic ratios in oxygen and hydrogen (not only for water) corresponds to the so-called V-SMOW "Vienna-Standard Mean Oceanic Water," which represents the mean isotopic composition of the ocean. The standard water was originally prepared by H. Craig (1961b) in the late 1950s, and called at that time SMOW, by mixing water samples from five different oceans, collected at different depths between 500 and 2000 m. The standard was distributed to all laboratories equipped with suitable instruments and unanimously adopted to express isotope ratios. The original reference material became exhausted after some years and was replaced by V-SMOW, prepared by the International Atomic Energy Agency (Vienna, Austria).

Specialised laboratories interested in defining the absolute ratios for most isotopes found that the isotopic ratios $^{2}H/^{1}H$ and $^{18}O/^{16}O$ of V-SMOW are equal to:

$$^{2}H/^{1}H = 155.95 \pm 0.08 \cdot 10^{-6}$$
$$^{18}O/^{16}O = 2005.20 \pm 0.45 \cdot 10^{-6}$$

6.2.3 Isotopic fractionation

As previously indicated, the isotopic ratios of heavy isotopes in different phases of the same substance or in different compounds containing the same element are distinct. These different isotopic ratios are produced by several processes, leading to the so-called *isotopic fractionation*. Isotope fractionation is the final consequence of the mentioned differences in mass between atoms or molecules containing heavy isotopes and those formed with the most abundant light isotopes. The differences in mass of the different molecules mean that:

1) Molecules containing heavier isotopes have a lower mobility. Because the kinetic energy of a molecule is exclusively provided by temperature, molecules containing heavier isotopes will present lower velocity (because of the excess of mass), leading to lower capacity to participate in chemical reactions or phase transitions and to lower diffusion velocities of these molecules.
2) Molecules containing heavier isotopes will show higher binding energies, that is, heavier molecules will tend to remain in the compound or phase of lower energy. In the case of evaporation, molecules containing heavier isotopes will tend to remain in the liquid phase, leading to lower concentration of heavy isotopes in the vapour phase.

As a consequence of the above, the vapour pressure of the different water molecules is controlled by the isotopes forming each molecule. For example, the vapour pressure (p) of the isotope species present in the water molecule is related in the following way:

$$p(^1H_2{}^{16}O) > p(^1H_2{}^{18}O) > p(^1HD^{16}O) \tag{6.4}$$

In the water cycle, this phenomenon controls the evaporation and condensation velocities of each isotopic species forming water or vapour and finally is responsible for the isotopic fractionation between the vapour generated over the ocean, the evaporating liquid water, and the precipitation over the continents. Molecules having larger vapour pressure values tend to evaporate faster and to condense slower than those having heavier isotopes and therefore, lower vapour pressures.

The magnitude of the isotopic fractionation, that is, the difference in the isotopic ratios between two compounds or phases, is expressed by means of the fractionation factor α, which is defined in the case of a phase transition, such as evaporation or condensation, by:

$$\alpha = \frac{R_l}{R_s} \tag{6.5}$$

where R_l stands for the isotopic ratio in the liquid phase, and R_s for the ratio in the solid phase. The factor α is usually very small, and in practical cases, the deviation form, ε is used rather than the fractionation factor. This quantity is referred to as the *enrichment factor* or simply *fractionation*:

$$\varepsilon(‰) = 1000(\alpha - 1) \tag{6.6}$$

In the case of enrichment, $\alpha > 1$, and in the case of depletion, $\alpha < 1$.

The change in the isotopic ratios between the liquid and the vapour phases during evaporation is expressed in δ-notation as follows:

$$\delta_{vapour/liquid} = 1000\frac{R_{vapour} - R_{liquid}}{R_{liquid}} = 1000\left(\frac{1}{\alpha} - 1\right) \tag{6.7}$$

Because the isotope ratio in the vapour phase is lower than in the liquid phase, due to the higher vapour pressure of the lighter molecules, α is higher than unity, and as a consequence, δ becomes a negative number. Because the evaporating flux is isotopically depleted in heavy isotopes, the remaining liquid phase becomes progressively enriched because molecules containing heavy isotopes tend to remain in the liquid water.

6.2.4 Isotopic fractionation under equilibrium and kinetic conditions

When an element or compound is transferred from one chemical species or from one phase to another, the isotope ratios of the reacting and the reactant phases change. For example, water vapour contains less heavy isotopes than the liquid

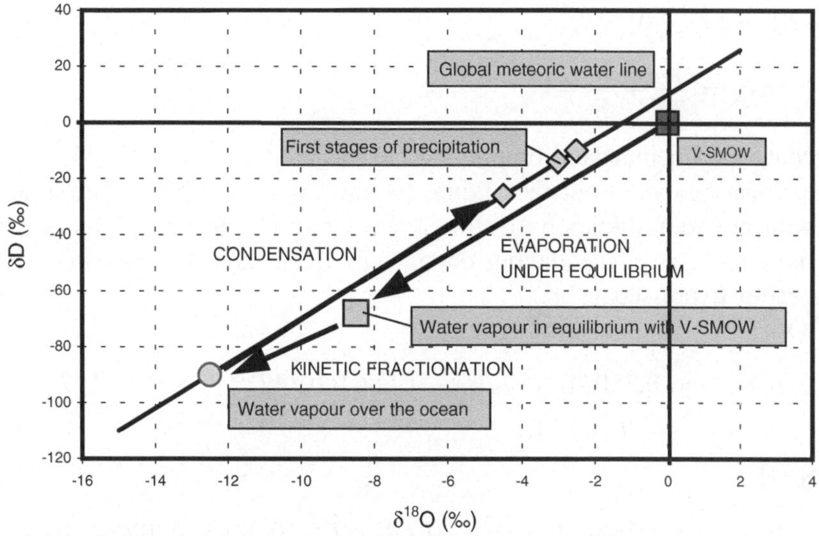

Fig. 6.2. Diagram δD–$\delta^{18}O$ illustrating different processes of isotope fractionation in the water cycle resulting in the Global Meteoric Water Line (see text for details).

water from which it has been formed. As a result, the newly formed phase, as in the case of water vapour newly formed from liquid water, is isotopically depleted with respect to the parent substance.

There are two types of isotope fractionation (see Fig. 6.2): equilibrium fractionation, referred as α_e, and kinetic fractionation, referred as α_k. The first type occurs when the chemical reaction or phase change is produced under thermodynamic equilibrium conditions. It means that the fractionation is controlled by the tendency of each isotopic species to be transferred from one phase to the other because the flux between the two phases is equal. The process takes place at a slow rate and velocity so that the condition of thermodynamic equilibrium is attained. This means that the proportions of the different molecular species are effectively controlled by the respective tendencies to be transferred from one phase to the other.

On the contrary, kinetic fractionation occurs when dealing with fast chemical reactions or phase changes, generally of an irreversible nature, such as precipitation of some phases or evaporation processes, in which the resultant phase is removed from the place where the reaction takes place. This is a common feature in many natural processes, such as evaporation over the ocean or precipitation of calcite in a water solution. In many instances, the magnitude of the kinetic effect depends on the diffusion coefficients of the involved molecular species.

Equilibrium fractionation factors are exclusively controlled by temperature and can be calculated and experimentally reproduced in the laboratory. In the case of liquid water and water vapour, the equilibrium fraction factor α_e is generally given by equations of the form:

$$ln\,\alpha_e = \frac{A}{T^2} + \frac{B}{T} + C \tag{6.8}$$

Recent experimental data compiled by Horita and Wesolowski (1994) to express the fractionation (α_{l-v}) between liquid (l) and vapour (v) phases are in accordance with previous studies by Bottinga and Craig (1969) and Majoube (1971). The commonly accepted equations obtained by Horita and Wesolowski (1994) are (temperature expressed as °K):
For $^{18}O/^{16}O$:

$$\begin{aligned} ln\,\alpha_{l-v} &= 0.35041(10^6/T^3) - 1.6664(10^3/T^2) - 6.7123(1/T) \\ &\quad - 7.685 \cdot 10^{-3} \end{aligned} \tag{6.9}$$

For $^2H/^1H$:

$$\begin{aligned} ln\,\alpha_{l-v} &= 1.1588(T^3/10^9) - 1.6201(T^2/10^6) - 0.79484(T/10^3) \\ &\quad - 116.04 \cdot 10^{-3}T + 2.9992(10^6/T^3) \end{aligned} \tag{6.10}$$

The experimental values obtained for the equilibrium fractionation factors for oxygen-18 and deuterium are shown in Fig. 6.3.

In many natural processes, such as evaporation from open surfaces, the conditions for thermodynamic equilibrium are not satisfied, for instance, during evaporation from the open water surface into an unsaturated atmosphere, with relative humidity h below 100%. In this case, additional differences in the transfer of

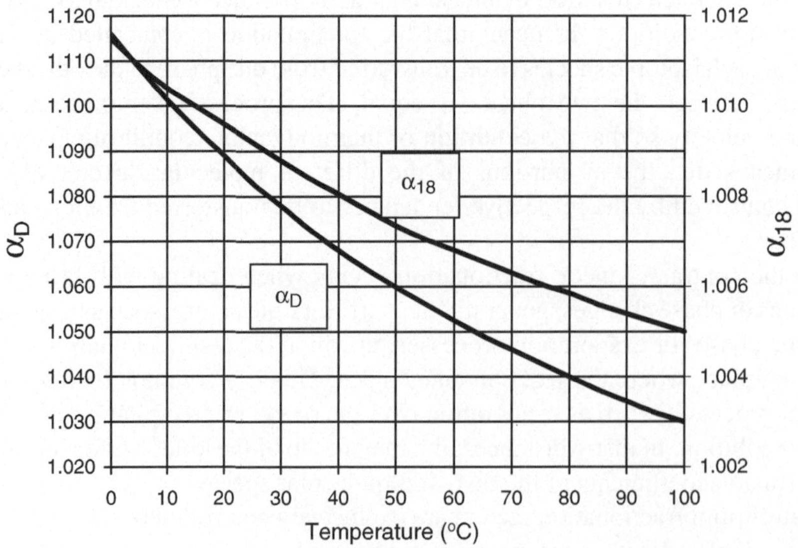

Fig. 6.3. Variation with temperature of the equilibrium fractionation factors α_D and α_{18}.

molecules containing light and heavy isotopes lead to additional fractionation due to the higher diffusivity of lighter molecules. Under these conditions, molecules containing lighter isotopes preferentially escape from the water surface when compared to molecules containing heavier isotopes, and the overall fractionation α_t, including these kinetic effects, is larger than expected only under equilibrium conditions.

In nature, evaporation of water takes place under relative humidity lower than 100% and with removal of the newly formed vapour by the dominating wind regime. These factors control the magnitude of the total isotope fractionation of the vapour formed over the ocean, which subsequently is transported to higher latitudes.

6.2.5 The Rayleigh condensation model

The Rayleigh model was originally developed to predict the changes in the composition of liquid mixtures during distillation. This model has often been used to describe the isotope separation taking place between two phases, such as water vapour and newly formed rain, when one phase is transformed into the other and removed from the cloud system. In reality, the process is assumed to occur under pseudo-adiabatic conditions because a small amount of energy is removed from the cloud by precipitation.

Let us consider a parcel of vapour M with a given initial isotopic ratio R. The removal of a fraction of water vapour dM as rain with the associated total fractionation α can be expressed by the following mass balance equation:

$$\frac{dR}{R} = (\alpha - 1) \cdot \frac{dM}{M} \tag{6.11}$$

The integration of the above differential equation, adopting $R = R_o$ and $M = M_o$, leads to the general form of the Rayleigh equation:

$$\frac{R}{R_o} = \left(\frac{M}{M_o}\right)^{\alpha-1} \tag{6.12}$$

where the fraction M/M_o represents the remaining mass of vapour in the system.

For example, if 10% of the original water vapour parcel is removed by condensation at a temperature of 5°C, ($M/M_o = 0.9$) and, using the isotopic fractionation for oxygen-18 ($\alpha = 1.01069$), calculated according to equation (6.9), the value $R/R_o = 0.998874$ is obtained. This means that the remaining vapour becomes isotopically depleted by 1.13‰.

The Rayleigh model can be used with reasonable accuracy to predict the isotope composition of single rain events, as well as the expected composition of the water vapour generated from evaporation, by using $1/\alpha$ instead of α. However, the fractionation factors to be included have to consider the mentioned kinetic conditions generally occurring under natural conditions. The evaluation of the magnitude of the kinetic isotope fractionation during evaporation has been accomplished by three different approaches:

1) Study of the molecular diffusion characteristics of the relevant isotope species in air (Merlivat, 1978), arriving at mean values for each fractionation factor; $\varepsilon_k = 25.1‰$ for deuterium and $\varepsilon_k = 28.5‰$ for ^{18}O.
2) Wind tunnel experiments (Vogt, 1976) relating the kinetic fractionation to h_d, the relative humidity of the atmosphere on the top of the viscous layer, normalized to the temperature of water. The experimental values usually adopted obey the following equations:

$$\varepsilon_k(‰) = 12.4(1 - h_d) \text{ for deuterium, and} \tag{6.13}$$

$$\varepsilon_k(‰) = 14.3(1 - h_d) \text{ for oxygen-18} \tag{6.14}$$

3) Empirical models to be applied for different conditions in the marine environment (Merlivat and Jouzel, 1979). The general formula to be applied has the form:

$$\varepsilon_k = (k/k - 1)(1 - h_{10}) \tag{6.15}$$

where h_{10} is the relative humidity of the atmosphere at 10 m height.

Under smooth conditions at the surface of the ocean, that is, wind speed lower than 7 m/s, $k_{18} = 6.2‰$, and $k_D = 5.5‰$. For rough wind conditions (wind speed above 7 m/s), $k_{18} = 3.5‰$ and $k_D = 3.1‰$.

6.2.6 Relation between δD and δ^{18}O in precipitation

Under equilibrium conditions, the isotope depletion of the evaporating water vapour follows a parallel behavior for both isotopes, resulting in more depleted values at lower temperatures with a constant ratio close to 8, which basically is defined by the following expression (see Figs. 6.2 and 6.4):

$$S = \frac{d(\delta D)}{dT} \Big/ \frac{d\delta^{18}O}{dT} = \frac{\alpha_D - 1}{\alpha_{18} - 1} \tag{6.16}$$

where S is the slope of the straight line obtained when the deviation δD is plotted versus the deviation δ^{18}O (Dansgaard, 1964).

The condensation of water vapour generally occurs under equilibrium conditions because the atmosphere is saturated in water vapour, and therefore, relative humidity is in almost all cases very close to 100%. Under these conditions, the expected slope relating the contents of deuterium and oxygen-18 (a value close to 8) is maintained in the newly formed raindrops.

However, as previously mentioned, the evaporation process does not occur under equilibrium conditions due to the lower relative humidity and different wind conditions prevailing in the ocean. Due to these changing factors, the evaporation rates and the relevant isotope fractionation factors affecting the three water molecules of interest (H_2O, DHO, $H_2{}^{18}O$) are different from this theoretical situation. In fact,

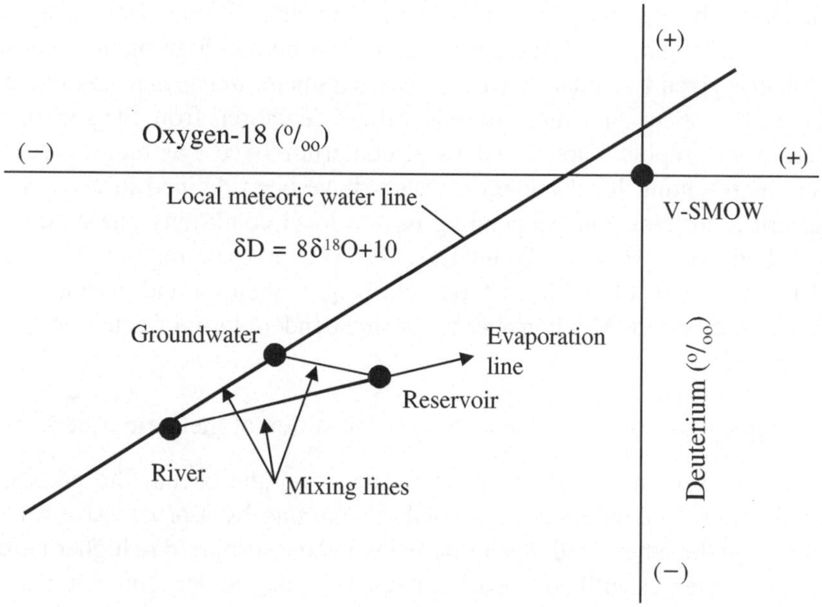

Fig. 6.4. Theoretical δD–δ^{18}O relationship illustrating the possible isotope contrasts between the different types of water interacting in the vicinity of a reservoir. The diagram is commonly used for the identification of the origin of emerging waters downstream of the dam site.

the kinetic fractionation is relatively more pronounced for oxygen-18, resulting in additional isotopic depletion of the water vapour (see Fig. 6.2). The water vapour generated over the ocean, as well as all subsequent rain events, will depart from the theoretical line of slope 8 ($\delta\text{D} = 8\delta^{18}\text{O}$) if the whole process would happen under equilibrium conditions. The kinetic fractionation is responsible for the departure of the meteoric waters from the mentioned line. Under present climatic conditions in the intertropical areas of the marine environment (25°C and 85% relative humidity and slow wind velocity), where most of the evaporation at global scale takes place, the kinetic fractionation accounts for about 6‰ for oxygen-18. However, the kinetic effect for deuterium is comparatively lower (~5‰), resulting in the displacement of the water vapour above the previous line (see Fig. 6.2). Subsequent rains formed by this vapour align along a line of the form:

$$\delta\text{D} = 8 \cdot \delta^{18}\text{O} + 10 \tag{6.17}$$

which is known as the Global Meteoric Water Line (GMWL), after Craig (1961a). This line indicates the magnitude of the kinetic fractionation, which results in the displacement of the GMWL of +10‰ in the intercept with respect to the situation in which only equilibrium fractionation would intervene. This is valid for most areas were vapour is formed under the influence of the open ocean. However, in some closed basins, water vapour is generated with a different isotope signature, resulting in an alignment of meteoric waters in a line with an intercept different from the one given by equation (6.17). The parameter *deuterium excess*, defined as $\Delta D = \delta D - 8 \cdot \delta^{18}O$, is used to reflect these deviations from the general trend in selected geographical areas. Values of deuterium excess as high as +22‰ in samples representing local meteoric waters have been defined in areas such as the eastern Mediterranean. Depending on the local conditions, the definition of the so-called Local Meteoric Water Line (LMWL) may be required for a meaningful interpretation of stable isotope data in precipitation and groundwater. In the case of the eastern Mediterranean, the independent term of equation (6.17) is therefore +22‰.

6.2.7 Temperature control of the isotope composition of meteoric waters

Once a water vapour parcel has been formed over the ocean, the process that eventually leads to condensation is adiabatic cooling by a progressive ascent of humid air. On the other hand, the air parcel may be transported to higher latitudes, where it can interact with colder air masses, bringing colder temperatures to the vapour and inducing condensation. In these two ways, temperature is the key parameter controlling the degree of rainout and therefore, the magnitude of the isotope fractionation reflected in the isotope composition of the raindrops and of the remaining vapour phase.

Due to the critical role of temperature in controlling the evolution of a given air mass, the isotope composition of each rain event will be controlled by the mechanism leading to condensation. Therefore, in temperate climates, the seasonal temperature variations play a critical role in generating a marked seasonal isotope signal in precipitation. The same key parameter is also hidden in other so-called "isotope effects," described in the following paragraphs, such as the *altitude effect* or the *continentality effect* (Rozanski et al., 1993).

Temperature is also responsible for depleted isotope values found in deep groundwater samples collected in confined aquifers. These aquifers can contain water recharged under colder climatic conditions, and therefore, the rains that recharged those systems showed lower isotope ratios than expected for such climatic conditions. Stable isotope data, supported by carbon-14 dating of these old waters, are used to identify the so-called paleo-waters, generally infiltrated during past glacial periods.

6.2.8 Other factors controlling the isotope composition of meteoric waters

The isotopic composition of precipitation at a given place is affected by a number of factors, which are summarised in the following sections. It can be stated that the isotopic composition of a particular water sample depends on its "history," from the moment a water vapour parcel is formed over the ocean to the place where condensation occurs. However, for a specific site, this "history" is statistically repeated year by year, so that the monthly and annual distributions of the average isotopic composition show more or less constant characteristic δ-values. This is especially true for groundwater because, in this case, the aquifer is averaging the δ-values of the individual events of precipitation, corresponding to very long periods of time.

6.2.8.1 *Conditions prevailing during the formation of clouds and humid air masses*

The relative humidity prevailing during the formation of water vapour in the marine environment has a marked effect of the isotope signal of the initial air masses. Under common climatic circumstances of temperature and relative humidity (~85%) over the ocean, deuterium excess is close to +10‰ (Merlivat and Jouzel, 1979). However, in areas where the deficit in relative humidity is larger, such as in closed basins, the newly generated vapour shows deuterium excess values significantly higher. Well-known examples of such features are found in areas such as those surrounding the Mediterranean Sea and the China Sea. As mentioned above, in these environments, values of ΔD as high as +22‰ define the Local Meteoric Water Line.

6.2.8.2 *Mixing of atmospheric vapour of oceanic origin with water vapour of local origin*

This process is of relevance in large continental basins, where a large fraction of precipitation derives from the recycling of water vapour. The isotopic depletion of rain over some areas of the basin is lower than the predicted values from the net influx of water vapour from the oceanic sources. This effect has been described in large tropical forests, such as the Amazon basin (Salati et al., 1979). Precipitation generated in these areas shows δ values more positive than modelled assuming a rainout of the net influx of oceanic water vapour.

6.2.8.3 *Latitude effect*

Usually, the average isotopic composition of precipitation corresponds to more negative δ-values as latitude increases. This is a consequence of the increasing distance to the source area of the ocean where most of the water vapour is generated, which corresponds to the subtropical trade-wind belts. The condensation of water vapour along the transit to high latitude areas gradually produces more and more depleted δ-values in precipitation due to the preferential condensation of the heavy water molecules.

6.2.8.4 *Altitude effect*

When a given air mass is forced to rise along a mountain range, the progressive decrease in temperature results in enhanced condensation, forming the so-called orographic rains. The induced rainout is translated into a progressive depletion in the stable isotope composition of precipitation with altitude, which of course is transferred to groundwater. This phenomenon is known as the *altitude effect* and represents the main isotope feature used in the study of the origin of water systems by the methods of isotope hydrology. The isotope altitude gradients vary from region to region, but the typical values are −0.25‰ per 100 meters elevation for $\delta^{18}O$ and −2.4‰ per 100 meters for δD. Taking into account the typical analytical uncertainty for the respective isotope analysis (±0.1‰ for $\delta^{18}O$ and ±1 for δD), the isotope signature may, theoretically at least, differentiate recharge areas with a difference in altitude in the order of 100 meters.

6.2.8.5 *Partial evaporation of falling raindrops*

Especially in arid and semiarid regions, the early stages of precipitation are produced under unsaturated conditions in the atmosphere, leading to partial evaporation of raindrops and the consequent isotope fractionation of the rain. Obviously, this effect is more pronounced at lower values of relative humidity. This effect also explains why in temperate climates summer rains of low intensity tend to show generally enriched isotope contents with clear evidence of evaporation when the values are plotted in a δD–$\delta^{18}O$ diagram. In these cases, rains tend to appear below the LMWL. In the case of more intense rains, the atmosphere eventually reaches saturation and the expected values of the two isotopes in the δD–$\delta^{18}O$ diagram (in or close to the LMWL) are observed.

6.2.8.6 *Evaporation from surface waters*

Evaporation of surface waters such as rivers and lakes is a major process leading to significant changes in the isotope composition of such waters. Evaporation over land takes place usually under a relative dry atmosphere, that is, low relative humidity, leading to the progressive isotope enrichment of the remaining waters in the lake or the reservoir. As previously mentioned, evaporation of surface waters leads to the preferential enrichment in $\delta^{18}O$ with respect to δD. As a result, the isotope evolution of a parcel of water left to evaporate to dryness shows a straight line in the δD–$\delta^{18}O$ diagram but with a slope lower than 8, which is the typical value for nonevaporated meteoric waters. The typical slope for evaporation depends on the mean value of relative humidity at which the evaporation process takes place. Under common conditions, the evaporation slopes range between 3.5 and 5.5. All samples exposed to evaporation show a deuterium excess, calculated as $\Delta D = \delta D - 8 \cdot \delta^{18}O$, lower than the typical value for the nonevaporated meteoric waters; this value can even become negative in some lakes or ponds submitted to significant evaporation. This effect is illustrated in Fig. 6.5, which corresponds to groundwater samples collected in the Caprivi region, Namibia. Most groundwater samples in the study area are aligned along an evaporation line that follows the equation $\delta D = 5.5\delta^{18}O - 11$.

Figure 6.6 illustrates the progressive enrichment in $\delta^{18}O$ due to evaporation under controlled conditions (25°C and different values of relative humidity) for a water sample with an original $\delta^{18}O$ content of −7.5‰. The enrichment is expressed

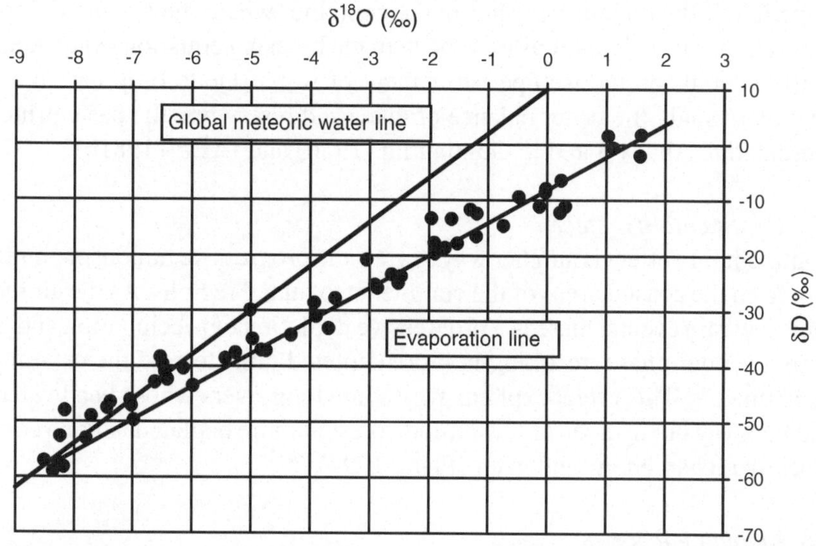

Fig. 6.5. Diagram relating the δD and $\delta^{18}O$ values for the groundwater of the Caprivi region in Namibia. Most of the points are situated along an evaporation line with a slope of 5.5.

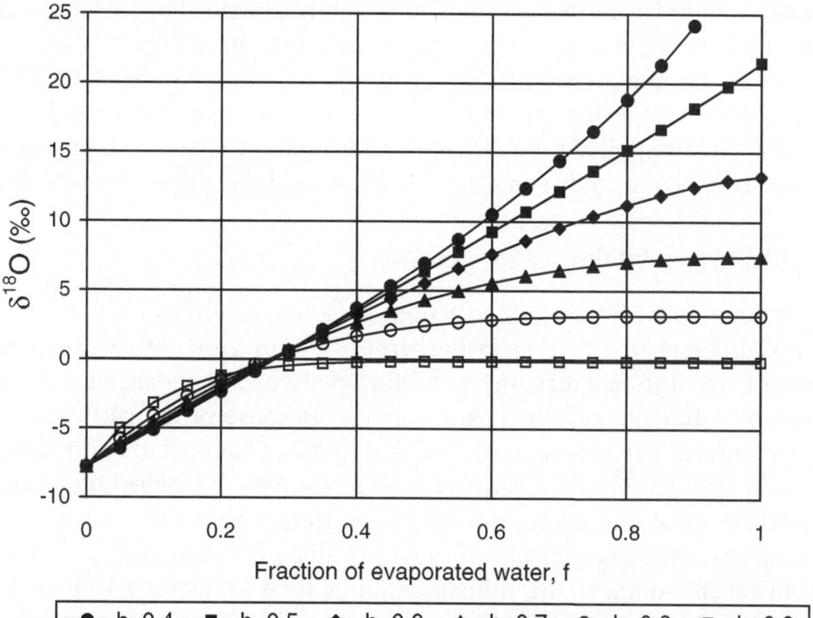

Fig. 6.6. Isotopic enrichment of oxygen-18 calculated by evaporation in natural conditions of a water parcel with an original $\delta^{18}O$ value equal to -7.5‰ for different values of the relative humidity.

as a function of the remaining water in the container where the sample is exposed to evaporation. The calculations have been made by using equation (8) of Annex 3. For further details on the isotope enrichment of water due to evaporation and its application to study the water balance of lakes and reservoirs, the basic principles are presented in Annex 3 and in Gonfiantini (1986) and IAEA (1981b).

6.2.8.7 *Continentality effect*

The continentality effect is another way to view the progressive rainout of air masses moving from the coastal areas of the continents to inland regions. As the air masses produce new rain events, there is a progressive depletion of precipitation and of the remaining vapour phase, resulting in more depleted rains toward the central parts of the continents. This effect explains why some long rivers whose headwaters are located far away but in areas of low altitude present more depleted isotope contents than locally recharged groundwater (Plata, 1994).

6.2.8.8 *Amount effect*

Finally, another expression of the magnitude of the rainout experienced by air masses is defined for tropical regions, where the annual variations in mean air temperature are minimal and do not explain the observed seasonality in the stable isotope composition of precipitation. Heavy rains are systematically more depleted than rains produced in the dry season. This is simply explained by the larger fraction of vapour condensed during events of heavy rains. In tropical regions, the existence of one or several rainy seasons is the result of large-scale events of hemispheric scale, such as the displacement of the Intertropical Convergence Zone (ITCZ). The same isotope tools applied for hydrogeological studies or the investigation of dam leakage are applied to study some aspects of atmospheric circulation and climate.

6.2.9 Offered possibilities

Stable isotopes of water are primarily used in isotope hydrology to identify water bodies of different origins. Under the term origin of groundwater, hydrogeologists refer to the different recharge areas, flow paths and/or geochemical evolution followed by different water bodies of interest. Because of the particular fingerprint incorporated in the water molecule during precipitation and infiltration, each water body with a defined recharge area and flow path is labelled by an isotopic characteristic value that allows differentiation from other water bodies. Because of the conservative character of this labelling along any flow path or almost any geochemical environment, the ultimate origin of groundwater can be traced.

Obviously, the basic condition for ensuring the usefulness of the technique in the study of dam leakage problems relies on the existence of isotopic contrast between the different water types involved in the study area. In most cases, the involved water bodies are the following: (a) reservoir water and the feeding rivers and minor tributaries, (b) groundwater present in the vicinity of the dam derived from local recharge sampled in boreholes and wells, and (c) emerging springs downstream of the dam. The main question to be answered in the early stages of the study is related to the origin of the water emerging downstream of the dam. The origin may represent: (a) exclusively local groundwater, (b) a real leak connected to the reservoir, or (c) a mixture of both components. In some cases, the situation may be more complex due to the presence of runoff derived from recent rain events, mixed in different proportions with the emerging groundwater or subsurface flows.

Because the isotope composition of the river is, in almost all cases, different from the local isotopic composition of precipitation, the required contrast in the isotope composition exists among the involved waters. Figure 6.4 illustrates a theoretical δD–δ^{18}O relationship showing the possible isotope contrasts between the different types of water interacting in the vicinity of a dam site. This diagram is commonly used for the study of the origin of emerging waters downstream of the dam site. This requires the isotope characterisation of the involved types of waters, i.e., the end-members of possible mixing diagrams. Chemical and isotope data are used in this way to prove the degree of hydraulic interconnection between water bodies and to calculate the mixing proportions, as has been shown previously.

In a more detailed form, the factors resulting in such isotopic contrast between the water types that may be present at a dam site can be related to the processes described in the following paragraphs.

River water impounded in the reservoir represents a mixture of direct runoff and the eventual discharge of aquifers located at higher elevations in areas situated inland with respect to the reservoir. Usually the mountainous areas are characterised by enhanced precipitation with respect to low-altitude areas. In such cases, several of the isotope effects previously mentioned, such as *altitude*, *continentality* and *amount*, operate, generating different isotopic signatures, mainly because they contribute to a higher degree of rainout. The depleted isotope contents in the incoming river waters are common in reservoirs located at low elevation but close to mountainous areas (see Fig. 6.4). This isotopic contrast is marked in many reservoirs described in this book, such as Colbún (Chile), Sabaneta (Dominican Republic), El Tejo and Beninar (Spain), Aulouz (Morocco), etc. The investigations carried out in some of these reservoirs are described in Chapter 12, where all other relevant results obtained by using different techniques are also presented.

Once river waters are incorporated into the reservoir, the intensity of the evaporation undergone along the river increases in the reservoir due to the impoundment over longer periods and stagnation of flowing waters. The fraction of water losses due to evaporation may be relatively small if the turnover time (ratio between the inflow and the total volume stored in the reservoir) is small. Moreover, the relative evaporation will be influenced by the ratio between the area exposed to evaporation and the total volume of the impounded waters. In any case, evaporation affecting reservoir waters results in isotope enrichment of the remaining water volume. The enrichment is proportional to the fraction of water losses, as shown by the Rayleigh model. This process produces a displacement in the δD–δ^{18}O diagram of the points corresponding to reservoir water along an evaporation line, which, as previously mentioned, lies below the LMWL and has slope lower than 8. All points located on this line are characterised by deuterium excess values lower than +10‰.

In the La Sierpe reservoir (Panama), the origin of some leaks emerging in the roof of the power plant building was unknown. Both the reservoir and the local groundwaters were the potential sources. Reservoir water is conducted to the power plant station through an underground pressure pipeline several km in length. It was suspected that a leak from such a conduit could be the reason for the observed water discharge. On the other hand, a small creek fed by precipitation in the nearby area runs close to the power plant station and could feed the discharge point. Stable isotope data in the three samples provided the following results:

$$\text{Leak in the power plant: } \delta D = -72.5‰, \delta^{18}O = -10.65‰$$

$$\text{La Sierpe reservoir: } \delta D = -42.1‰, \delta^{18}O = -6.76‰$$

$$\text{Chiriquicito creek: } \delta D = -63.2‰, \delta^{18}O = -9.19‰$$

These results clearly proved that the leaks were not related to the reservoir, but to the nearby creek. In this case, the isotopic contrast was enormous, taking into account that the analytical uncertainty is around ±0.1 for $\delta^{18}O$ and ±1‰ for δD. The reservoir waters maintained the indicated values, as verified after conducting a second sampling campaign. On the contrary, the isotope composition of the creek showed large variations, depending on the precipitation events generating the runoff in its catchment area. The high contrast in the isotopic composition of water was explained, in this case, both by the high elevation from which the creek water was coming and by the isotopic enrichment derived from the evaporation of the reservoir water.

In reservoirs and lakes with low and intermediate average depth and long turnover times of the impounded water, evaporation plays an important role in defining the isotope signal of the reservoir waters. The isotope composition of the stored water can show a significant enrichment compared to the river waters feeding the reservoir but with a δD–$\delta^{18}O$ characterised by deuterium excess values much lower than +10‰, which is typical for most river waters and local groundwater. The situation is schematically depicted in Fig. 6.4. This isotopic contrast between the existing water types provides strong arguments for the identification of the possible interconnection between the involved water types. Successful examples of this application are the studies conducted in the waste landfill site in Valladolid and the Aznalcóllar dam, both in Spain; Lake San Marcos in Ecuador, and Lake Managua in Nicaragua. Stable isotopes were used to prove the absence of interconnection in the cases of Valladolid and Managua, while the connection was identified for the other two cases (Aznalcóllar and San Marcos).

The Koka reservoir in Ethiopia is an extreme case of isotopic enrichment of the stored water due to evaporation. The mean water thickness with full reservoir amounts to only some 8 meters, and the effective evaporation amounts to a value close to 3 m/year. This is due to the high average temperature (around 30°C) and the high elevation of the lake (1600 m.a.s.l.). Under these circumstances, oxygen-18 enrichment values of several per mil can be expected. Also, the electrical conductivity and the concentration of representative chemical components of the reservoir water will undergo an important increase throughout the year after the rainy season, which takes place from June to September. The study of the yearly evolution of both parameters will allow the calculation of isotopic and chemical water balances, from which precise information on the evaporation rate, the eventual discharge of groundwater (as suspected) and the eventual water losses can be derived. Such a study is presently underway.

In the case of rivers regulated by a series of reservoirs, the effects of evaporation in each reservoir accumulate along the direction of flow. The effect of such sequence of evaporating stages is the progressive enrichment of river waters facilitating the distinction between water related to the reservoir from that derived from local recharge. In small- and medium-size reservoirs located in rivers with large flow rates, the turnover time of water could be lower than one month. In such cases, the isotope signal of the reservoir could present isotope changes due to variations in the isotope signal of precipitation, for example, by the seasonality in the isotope values of precipitation later transmitted to the reservoir. Under such conditions, it is advantageous to monitor whether the isotope changes produced in the reservoir are reproduced in the emerging springs downstream. Obviously, the study for such an investigation requires prolonged monitoring of the reservoir waters and the springs to obtain information on the origin of water and its transit time.

Reservoirs with a water column above 50 m may develop an isotope stratification similar to the thermal and saline stratification described in Chapter 2. This is the consequence of the lack of effective mixing of the water entering at the tail of the reservoir and the impounded waters. In many instances, recent waters are displaced in the body of the reservoir according to the stability and properties of the water column, as described in Chapter 2. If such isotope separation exists, the δD and δ^{18}O values of the leaks may indicate the depth at which the infiltration takes place. This situation was observed in the Cautillo reservoir, Cuba. The isotope composition of the leaks was compatible with the infiltration of water from the epilimnion but not from the hypolimnion. Therefore, this information was used to exclude the deepest 14 m of the reservoir as the suspected area of infiltration.

As a conclusion, it can be stated that the stable isotope composition of water provides in almost all studies dealing with the nature of leaks from dams valuable information on the hydraulic connections in the vicinity of the dam site. However, in some cases, it is difficult to judge in advance the usefulness of the technique because the existing isotope contrast might be insufficient to conclude on the investigated hydrological aspects. Nevertheless, a preliminary reconnaissance campaign might be useful, and it is highly recommended, especially taking into account the properties of these natural tracers of water.

The application of techniques based on the measurement of stable isotopes of water during preconstruction studies deserves special mention. The combined use of these tools with tritium, the chemical composition of water and other methods will surely provide valuable information on the origin and dynamic processes of the groundwater bodies existing in the dam site. Some examples of the obtainable information are described in Chapter 11.

6.3 TRITIUM

6.3.1 Generalities

As mentioned in Section 6.1, tritium is the hydrogen isotope of mass 3, and the only radioactive isotope of this element. Tritium is converted by radioactive decay into ^{3}He through the emission of a β-negative particle, with a maximum energy of 18.6 KeV and a half-life of 4500 ± 8 days (12.32 years; Lucas and Unterweger, 2000).

The tritium concentration is expressed in Isotope Hydrology in *tritium units* (T.U.). One T.U. is equivalent to one atom of tritium in 10^{18} atom of ^{1}H. This low activity expressed in other common radioactivity units corresponds to:

$$1\,\text{T.U.} = 3.2 \cdot 10^{-3} \quad \mu\text{Ci/m}^3 = 118.4\,\text{Bq/m}^3$$

Tritium is present in water bodies, both from natural and artificial sources. Natural sources of tritium are the nuclear reactions occurring in the upper atmosphere by interaction of atmospheric gases, mainly nitrogen, with free neutrons. These are also produced by nuclear reactions of cosmic rays with the atmospheric gases. Tritium produced in this way is rapidly oxidized and forms water vapour, entering the water cycle. This natural source of tritium produces in precipitation concentrations between 2 and 20 T.U., depending on the latitude and distance to the coast. Natural production of tritium has been evaluated on the order of 0.25 atoms $\cdot$ cm$^{-1} \cdot$ s^{-1}. The steady state global inventory of natural tritium in the atmosphere has been estimated to be about 1.3 TBq (tera-becquerels), equivalent to 3.6 kg of the radionuclide.

Tritium derived from man-made activities is mainly the result of the thermonuclear atmospheric explosions produced between 1952 and 1963 (nuclear weapon tests). Other sources of tritium, although of minor entity are: (a) emission as byproduct of tritiated compounds produced by nuclear power plants and plants that reprocess irradiated nuclear fuels, and (b) emissions made by industrial companies producing luminescent compounds activated with tritium, which are used for various purposes but mainly for watch production (Rozanski et al., 1991).

The amount of tritium artificially injected into the atmosphere as the result of the thermonuclear bombs was on the order of 7.4×10^5 TBq/Mt or 2.1 kg/Mt (Mt stands for megaton). Therefore, the total amount of tritium injected into the atmosphere from October 1952 to April 1963 was above 600 kg. Because of the low density of the hydrogen gas, most of the tritium produced was directly injected into the stratosphere, especially in the Northern Hemisphere, where most of the explosions were made. Bomb tritium was distributed worldwide, although substantially higher concentrations were recorded in the precipitation of the Northern Hemisphere during the years following the maximum number of explosions. The concentrations observed in the Southern Hemisphere were significantly lower.

The International Atomic Energy Agency, in cooperation with the World Meteorological Organization launched a Global Network for Isotopes in Precipitation (GNIP) in 1960 with the aim of monitoring the temporal and spatial variation of the concentration of this isotope in precipitation. This network, whose primary interest was tritium, shifted the interest to stable isotopes in the 1980s due to the decreasing trend of tritium contents at the global scale. The time series values of tritium contents, as well as the stable isotope deviations of $\delta^{18}O$ and δD and relevant meteorological data, in monthly composite samples are available for more than 500 stations at http://isohis.iaea.org, as well as in published reports by the IAEA (1969, 1970, 1971, 1973, 1975, 1979, 1981a, 1983, 1986, 1990 and 1992).

6.3.2 Measurement of natural tritium

Tritium is generally measured by the liquid scintillation technique. Due to the low concentrations of tritium generally found in precipitation and groundwater, the measurement of this radionuclide requires the implementation of an electrolytic enrichment procedure. A volume of water of several hundred millilitres is submitted to electrolysis with a voltage close to 3 volts using sodium peroxide as electrolyte. Under these conditions, the light molecules of water (tritium-free molecules) are preferentially decomposed. The electrolytic enrichment process continues until the remaining water volume reaches a value close to 12 ml. About 80% of the tritium contained in the original sample is recovered in this final sample. Therefore, if the initial volume was 400 ml, the concentration increase amounts to a factor (enrichment factor) of $0.8 \cdot 400/12 \approx 27$. The final sample is measured, after distillation to separate the electrolyte, with the liquid scintillation detector. The recent developments in liquid scintillation counting and the use of larger volumes of samples for electrolytic enrichment have lowered the detection limit to ± 0.1 T.U. for routine measurements. This limit is sufficient for most purposes, especially for water samples of the Northern Hemisphere, where the tritium concentration in precipitation is higher.

Other even more precise analytical methods are available to measure very low tritium contents. These methods are based on the measurement of the concentration of the daughter isotope ^3He by mass spectrometry after degassing the sample and measuring the amount of helium generated as a result of the tritium decay during a long period of time. Tritium contents as low as 0.001 T.U. can be measured with this technique.

6.3.3 Tritium distribution models

Tritium contents in precipitation over the Northern Hemisphere have been simulated by spatial and temporal extrapolation by Doney et al. (1992). A much simpler approach was followed by the senior author of this book, who produced a composite input function from 1952 to 1999 using the GNIP database. This input function, valid for most places located in the midlatitudes, is shown in Table 6.1. Columns 3 and 4 in Table 6.1 report, as an example, the reconstruction of the complete input function for the Madrid station, based on the available data (1970–1999) for this particular station.

The data presented in Table 6.1 show clearly that tritium concentration, even considering radioactive decay, during the last decades has always been higher than the values measured at present in precipitation. In Fig. 6.7, the Northern Hemisphere model (column 2 of Table 6.1) and the experimental values obtained for the Madrid station (column 3) are plotted, together with data obtained at the Gibraltar station and several stations in Portugal. Figure 6.8 shows the values of column 5 (referred to year 2000) corrected by radioactive decay. It can be seen that for the last 20 years, these corrected values have been approximately constant around a value of 6.5 T.U. for Madrid. A similar situation will be found in most GNIP stations, although the average tritium concentration value will be different.

Therefore, tritium values found in groundwater samples above this value can be explained only as the result of rainwater infiltrated during the 1960s and 1970s, when significantly higher tritium values were observed in precipitation. Following this line of interpretation, tritium-free groundwater samples imply the presence of water recharged before 1952. Finally, a tritium concentration between 0 and the characteristic present value in precipitation is interpreted as a mixture of two types of groundwater: a recent component containing tritium and an older component recharged before 1952.

Evidently, the shape of the input function for tritium was controlled by the timing and magnitude of the detonation of thermonuclear devices. The number and magnitude of the bomb tests increased from 1952 to 1963, when the three recognised nuclear countries at that time signed the Limited Test Ban Treaty. Afterward, only France and China conducted atmospheric tests. The maximum concentration of tritium was observed in the Northern Hemisphere in 1963, reaching, at some high-latitude stations, values as high as 10,000 T.U. The maximum concentration in the Southern Hemisphere was observed one year later due to the slow motion characterising the interhemispheric transport in the atmosphere. Obviously, the maximum tritium concentrations found in the Southern Hemisphere were significantly lower, by almost one order of magnitude.

Figure 6.9 shows the evolution during the last 50 years of the tritium content in the precipitation of selected GNIP stations. The plotted values correspond to average annual nonweighted tritium concentration values. As expected, the shape of the distribution curves is similar because the thermonuclear tests responsible for

Table 6.1. Mean annual tritium concentration in precipitation for the Madrid station and normalised values for the Northern Hemisphere, using the GNIP database for more than 40 stations. Column 5 shows the estimated tritium input function for the period 1953–1999 for Madrid. The last column contains the tritium values of columns 3 and 4 corrected by decay referred to the year 2000.

Year	Northern Hemisphere	Tritium contents in Madrid (T.U.)		
		Experimental	Estimated	Corrected by decay
1953	8.4		12.8	0.9
1954	91.5		139.5	10.7
1955	13.2		20.1	1.6
1956	58.5		89.5	7.7
1957	37.6		57.2	5.2
1958	186.8		284.7	27.4
1959	146.6		223.4	22.7
1960	50.7		77.3	8.3
1961	53.6		81.7	9.3
1962	276.6		421.5	50.6
1963	1000.0		1524.0	193.6
1964	538.1		820.1	110.2
1965	273.0		416.0	59.1
1966	183.2		279.0	41.9
1967	106.3		161.9	25.7
1968	81.1		123.5	20.7
1969	76.8		117.0	20.8
1970	65.4	104.6		19.6
1971	78.5	153.2		30.4
1972	41.8	51.4		10.8
1973	37.0	64.8		14.4
1974	41.6	67.1		15.7
1975	35.4	60.4		15.0
1976	27.8	31.2		8.2
1977	24.8	39.5		11.0
1978	27.5	44.7		13.1
1979	16.5	21.9		6.8
1980	14.5	33.5		11.0
1981	15.7	23.9		8.3
1982	10.9	14.5		5.3
1983	9.5	11.6		4.5
1984	7.5	18.6		7.6
1985	7.3	11.5		5.0
1986	7.2	14.4		6.6
1987	6.7	10.4		5.0
1988	6.3	12.3		6.3
1989	7.4	10.8		5.8
1990	5.3	8.0		4.6
1991	5.4	9.0		5.4
1992		7.5		4.8
1993		9.1		6.2
1994		9.9		7.1
1995		6.9		5.2
1996		5.9		4.7
1997		5.8		4.9
1998		8.2		7.3
1999		7.5		6.9

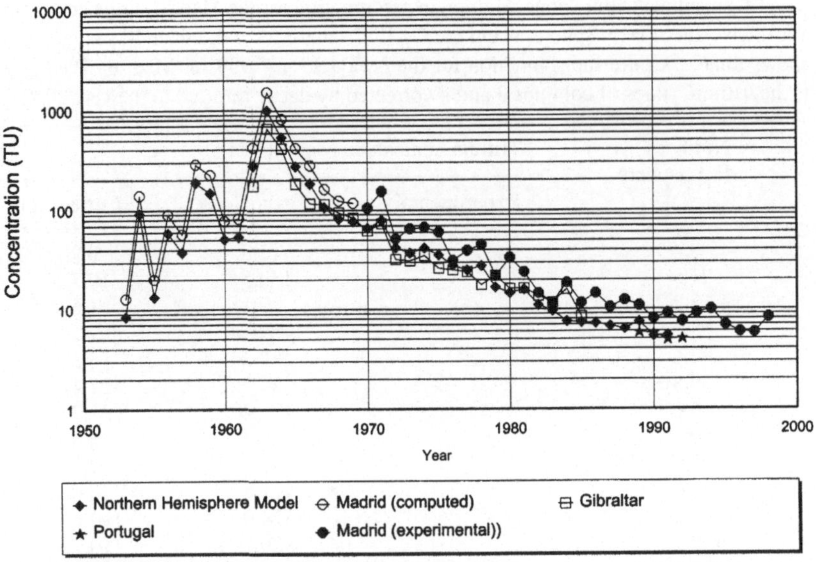

Fig. 6.7. Tritium concentration values measured in the precipitation of the Madrid station after 1970 (annual average values) in comparison with the model developed by the senior author for the Northern Hemisphere. Data obtained for the Gibraltar station and for some stations located in Portugal are also plotted.

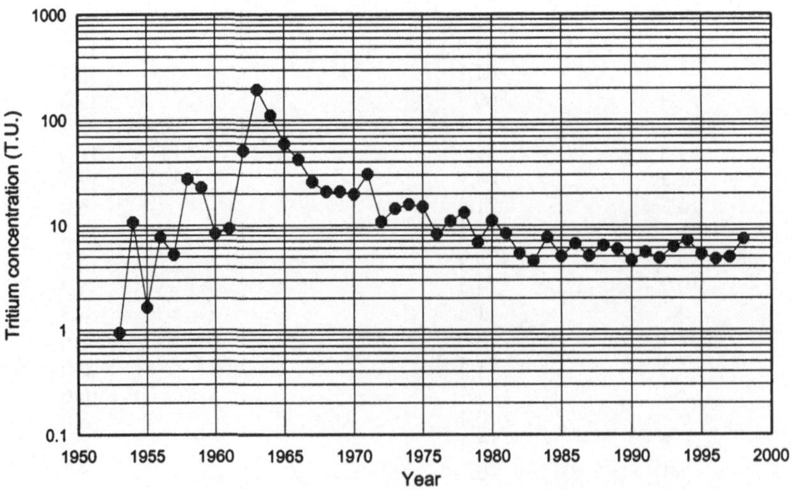

Fig. 6.8. Radioactive decay-corrected values of the experimental and calculated tritium concentrations for the precipitation of the Madrid station (last column of Table 6.1).

the tritium contamination were the same. The lower concentration in the Southern Hemisphere is present even in the precipitation of recent years. This is partially due to the higher ratio in this hemisphere between the surface of the ocean and continental land.

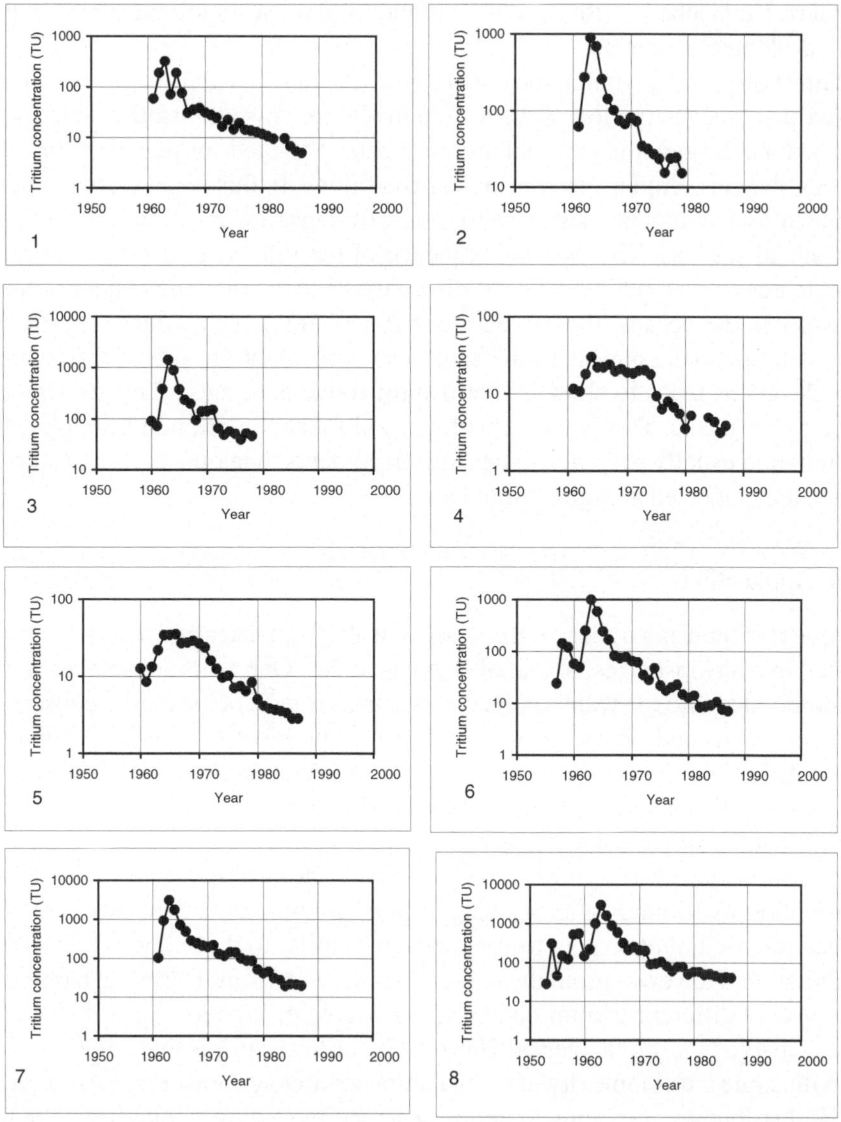

Fig. 6.9. Evolution of the tritium contents in precipitation for selected GNIP stations. The tritium contents, expressed as tritium units (T.U.), represent average nonweighted annual concentrations. (1) Hong Kong (China); (2) Tokyo (Japan); (3) Khartoum (Sudan); (4) Malan (South Africa); (5) Kaitoke (New Zealand); (6) Valentia (Ireland); (7) Vienna (Austria); (8) Ottawa (Canada).

Ongoing monitoring programs of tritium in precipitation indicate that the rain-out of thermonuclear tritium is almost complete and that current values are considered to be indicative of the natural tritium production. Despite the low tritium values found in most areas, especially in the intertropical zones and the

Southern Hemisphere, tritium will continue to play an important role as a tracer for hydrological systems.

It must be pointed out that the data provided by the GNIP programme allow the approximate reconstruction of the tritium model for any other station in the world. This is done through the extrapolation of values obtained for neighbouring GNIP stations showing similar meteorological conditions. In this respect, it is important to take into account that coastal stations usually show lower tritium concentrations than inland stations. This is a consequence of the influence of locally generated cloud masses with practically no tritium content (due to the long residence time of the water in the oceans, these are almost tritium-free). This effect is particularly important on small islands. In the same way, precipitation of stations located at high elevations tends to show higher tritium contents because they are less influenced by the ocean. This is the case of the Addis Ababa station in Ethiopia, where precipitation usually presents higher tritium contents than other nearby stations or even stations located at higher latitudes.

6.3.4 Applications

The use of tritium as a tracer for the study of leaks from lakes and reservoirs may be similar, in many instances, to that of stable isotopes. One of its advantages is that it is incorporated into the water molecule, and therefore, it behaves as a conservative tracer in the ground. Because of this property, the identification of groundwater not related to the reservoir can be accomplished with additional confidence. It is very unlikely that a parcel of water not related to the reservoir will present similar chemical and isotopic characteristics.

In most instances, reservoir waters are the result of direct runoff with a minor contribution by groundwater discharged by shallow aquifers. In this case, these surface waters should present tritium contents similar to those observed for recent precipitation. However, groundwater, even if derived from nearby recharge areas, may present different tritium contents, for example, higher values if the waters were recharged during the period 1960–1970, or lower if it is even older.

To illustrate the complexity of some natural aquifer systems, Fig. 6.10 shows the spatial distribution of tritium contents measured in shallow groundwater samples collected at a toxic waste landfill site near Valladolid, Spain in 1999. The area occupied by the facility is about 50,000 m^2. The leachate of the landfill showed about 30 T.U., reflecting the influence of industrial sources of tritium, most probably tritium from luminescent products. The rest of the samples were collected in boreholes and showed variations of natural tritium. Values between 5 and 7 T.U. were interpreted as recent waters, that is indicating rapid turnover time of the water circulating in some areas of the landfill. However, values as high as 12.7 T.U. were found in a spring discharging a perched calcareous aquifer. This tritium content is higher than the average concentration of the last 18 years' precipitation (see last column of Table 6.1). Therefore, such values can only be explained by the presence

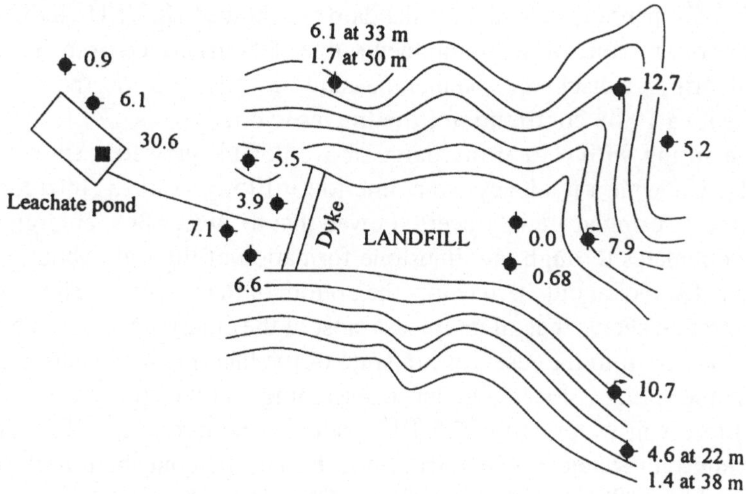

Fig. 6.10. Distribution of tritium concentration values obtained for several groundwater types in the vicinity of a landfill for industrial wastes near Valladolid, Spain.

of water recharged during the 1960s and 1970s. Tritium-free waters were found in some samples collected from boreholes that had been drilled in low-permeability strata (marl and clay deposits). Intermediate values between 0 and 6 T.U. were also found. These are interpreted as mixtures of old and recent waters.

The situation found in the previous case is not unusual, reflecting the complexity of the flow patterns in some areas of reduced dimensions. The experience gained by conducting many hydrological studies indicates that most assumptions about flow conditions have to be proven by direct methods, even if common sense indicates that the aquifer system is very simple. The example of the Rules dam described in Chapter 11 illustrates the previous statements.

In other cases, the investigation of the origin of groundwater samples may benefit from other sources of human-generated tritium. This was the case of the study of leaks in the vicinity of the Tous dam in eastern Spain, which is located in limestone formations of Cretaceous and Tertiary ages. The main source of tritium was, in this case, the discharge of waste waters from a nuclear power plant located upstream. Due to these discharges, the waters of the river Júcar are characterised by anomalously higher tritium contents than local precipitation. The mean tritium concentration in the reservoir varies between 25 and 40 T.U. Such high tritium contents cannot be found at present in any groundwater system exclusively fed by precipitation in Spain. The following conclusions were drawn about the origin of the discharges emerging at the dam toe and at other points close to the dam:

1) Most of the karstic springs showed tritium contents ranging from 2.8 to 8 T.U., clearly indicating that only a minor fraction, in the most favourable situation, can be related to the reservoir waters.

2) A spring discharging several l/s at the dam toe showed 18.7 T.U., indicating the presence of a mixture of two components, one of them derived from the reservoir.
3) A small drip in a discharge channel measured 25 T.U. and was the only sample found with a 100% contribution from the reservoir.
4) A large spring with a total discharge above 350 l/s showed a tritium content of 5.2 T.U., with a tendency over time toward lower values. Interconnection experiments using artificial tracers showed that the water flow emerging in this spring circulates through the limestone formation of the right abutment of the dam at a depth of about 50 m below the bottom of the reservoir. The low tritium concentration shows that all or at least most of the emerging water in the spring is not coming from the reservoir, in spite of the fact that the groundwater flow feeding the spring is located under it. Assuming that the spring water is a mixture of reservoir water, with 27.5 T.U. (value found in one specific campaign), with old groundwater in which tritium is absent, the contribution of the reservoir would be $100(5.2/27.5) = 18.8\%$. Obviously, the reservoir contribution would be lower if the groundwater component contains a tritium concentration different from zero T.U.

6.4 CARBON-13

6.4.1 Fundamentals of the technique

In vegetated areas, the total inorganic dissolved carbon (TDIC) present in groundwater mainly derives from the incorporation of CO_2 stored in the upper soil horizons to the infiltrating waters. The partial pressure of this gas in most soils is substantially higher than the corresponding pressure in the atmosphere due to organic matter decomposition and root respiration. As a consequence, the contribution of atmospheric CO_2 to the TDIC is relevant only in arid environments where lack of vegetation reduces the contribution of soil CO_2.

Dissolved CO_2 in the infiltrating waters induces the dissolution of soil carbonates because the addition of CO_2 displaces the pH toward more acidic values and enhances the capability of dissolving carbonates by the infiltrating waters according to the following equation:

$$CO_2 + CO_3Ca + H_2O \Rightarrow 2HCO_3^- + Ca^{++} \qquad (6.18)$$

Regarding carbon isotopes, Fig. 6.1 indicates that carbon has two stable isotopes, ^{12}C and ^{13}C, and the radioactive ^{14}C. Carbon-13 represents about 1.107% of the total carbon. Within the carbon cycle, the fractionation processes described in previous sections for the stable isotopes of water control the relative concentration of carbon-13 in different substances. Due to the differences in mass between the two stable isotopes, plants preferentially assimilate the light isotope of carbon during the photosynthetic function. As a result, the ratio $^{13}C/^{12}C$ in plants and soil CO_2

(biogenic) is significantly lower than the same ratio in atmospheric CO_2. Therefore, the CO_2 incorporated by the infiltrating waters is isotopically depleted with respect to the atmospheric CO_2.

The isotope abundance of ^{13}C is also expressed using the δ-notation in a similar way to the abundance of deuterium and oxygen-18. The reference standard to express the relative deviations is a marine fossil carbonate known as Vienna-Pee Dee Belemnite (V-PDB). Therefore, carbonates of marine origin show $\delta^{13}C$ values close to 0‰ because they share the same origin as the internationally accepted reference material. Atmospheric CO_2 ranges between −6.5‰ and −8‰, and biogenic CO_2 is significantly depleted due to the previously mentioned preferential uptake of ^{12}C. Most plants show ^{13}C values between −22‰ and −27‰.

This enormous isotopic contrast between the different sources of alkalinity to groundwater can be used to find out the origin of groundwater discharging close to a dam site. However, the complex geochemistry of carbon may complicate in some instances the interpretation of the measured isotope deviations.

When infiltrating, groundwater dissolves soil carbonates by the reaction shown in equation (6.18). Half of the carbonate, theoretically, derives from gaseous soil CO_2 and half from the solid carbonates. This reaction, if dominant, will lead to $\delta^{13}C$ values in recent groundwater close to −12‰ to −14‰. However, other processes, such as isotope exchange between gaseous CO_2, solid carbonates and dissolved CO_2, disturb the simple model depicted. In any case, $\delta^{13}C$ in TDIC of groundwater samples representing aquifer systems with short residence times are expected to present relatively depleted isotope values.

On the other hand, carbon species dissolved in surface waters tend to equilibrate with atmospheric CO_2 due to isotope exchange processes. Obviously, this process is more intense in small and turbulent rivers, where the available surface for isotope exchange allows the completion of the process in a short period of time. As a result of this process, most surface waters, including reservoir and lake waters, show values close to −7.5‰. In most instances, if the suspected springs are formed by a leak from the reservoir, the $\delta^{13}C$ isotope signal found in the TDIC should reflect, under normal circumstances, a typical value expected for surface waters equilibrated with atmospheric CO_2. Of course, this $\delta^{13}C$ value can be directly measured in the carbonate species dissolved in the reservoir water. On the contrary, if the investigated spring is discharging locally recharged groundwater, the corresponding $\delta^{13}C$ value should be more depleted, generally between −15‰ and −12‰.

However, this general geochemical evolution may not be true due to the previously mentioned complex geochemistry of carbon. Groundwater circulating with long turnover times over carbonate rocks (limestone, dolomite or even sedimentary rocks with calcareous cements) and therefore long periods of contact between the dissolved carbon and the solid carbonate may reflect the consequences of marked isotope exchange. Depending on the intensity of this exchange, the $\delta^{13}C$ signal progressively becomes more enriched if the carbonates are of marine origin. Under extreme conditions, $\delta^{13}C$ of TDIC in deep groundwater samples may be as enriched as 0‰. This process may produce similar $\delta^{13}C$ values for groundwaters with different origin and geochemical evolution.

A $\delta^{13}C$ deviation close to −7.5% may erroneously indicate that a spring is discharging reservoir water. However, the opposite, that is, a value on the order of −13‰ in the spring water, can not be attributed to the reservoir.

6.4.2 Case studies

The described methodology was used to identify the origin of emerging springs downstream of the dam in several studies conducted by the authors in Costa Rica, Cuba and Spain. The results obtained by using this technique are presented below.

El Arenal reservoir, Costa Rica The Arenal reservoir is located in Costa Rica, about 150 km south of the capital city San José, in a mountainous zone with high volcanic activity, where the mean annual precipitation ranges between 3,000 and 4,000 mm. The total capacity of the reservoir is 3.5×10^6 m^3. Both at the dam toe and on the abutments several springs were identified that could be related to leaks from the main dam, causing considerable concern about the safety issues derived from the possible failure of the dam. Just about 4 km downstream from the dam, the flow rate measured in the river was around 4 m^3/s.

Due to the hydrogeological and climatic characteristics of the site, the chemical and stable isotopes of water failed to provide a conclusive answer about the origin of groundwater emerging in the relevant springs. In some instances, chemistry proved that some points were not related to the reservoir, but these conclusions were not supported by oxygen-18 and deuterium measured in water samples. The conclusive answer was obtained by measuring the $\delta^{13}C$ in the TDIC (Cordero and Plata, 1999).

The measured values are presented in Table 6.2. Surface water from the reservoir showed values close to −8.66‰. A lower value (−9.66‰) was found at the bottom of the reservoir. Both values are more depleted than the expected values for water equilibrated with atmospheric CO_2. This difference could be explained either by the discharge of groundwater close to the sampling points or by the intense biological activity in the lake.

The creeks in the vicinity of the reservoir from which samples were collected are fed by springs, most of them recharged at higher elevations than the reservoir itself. The values of $\delta^{13}C$ in these samples presented large variability, depending on the distance between the sampling point and the emerging spring. Long distances usually corresponded to almost equilibrated values with atmospheric CO_2, while short distances were reflected by intermediate δ-values. These samples correspond to the points marked with an asterisk (*).

All springs and sampling points along drainage galleries, drainage systems in the tunnels, and drainage wells showed significantly depleted $\delta^{13}C$ values when compared to the typical values of the reservoir and rivers, indicating that most of these waters were fed by groundwater infiltrated in the nearby mountainous areas. It was

Table 6.2. Analytical results of $\delta^{13}C$ measured in the TDIC in water samples collected in the vicinity of the El Arenal reservoir (Costa Rica).

Sample code	$\delta^{13}C$ (‰)	Sample code	$\delta^{13}C$ (‰)
Reservoir (5 m depth)	−9.00/−8.32	Leak 14 × A	−14.95
Reservoir (bottom)	−9.66	Drainage 32A, Gal RM	−19.83
Creek Sucia*	−11.35	Drainage 36B, Gal RM	−18.93
Creek Pueblo Nuevo*	−8.89	Drainage 42, Gal RM	−20.06
Creek Volcán 1RM*	−7.41	Drainage 44T, Gal RM	−19.32
Creek Hania (spring)*	−12.72	Drainage 48T, Gal MD	−19.85
River Aguas Gatas (RM)*	−9.52	Spring Gal LM	−18.02
River Chiquito (RM)*	−8.89	Weir V-16, Gal LM	−15.78
Creek Rugama (LM)*	−9.76	Drainage A-1, Gal LM	−16.42
River Caño Negro (RM)*	−9.43	Drainage A-5, Gal LM	−18.62
River Arenal upstream*	−8.52	Drainage 9A, Gal LM	−19.28
River Arenal downstream*	−8.51	Drainage 55, Gal LM	−18.39
Spring Flaqué 1*	−11.96	Manometer 4 tunnel	−22.68
Spring Flaqué 3	−15.59	Manometer 6 tunnel	−17.36
Drainage well 3	−17.56	Manometer 9 tunnel	−15.06
Drainage well 4	−18.00	Manometer 17 tunnel	−16.04
Drainage well 8	−16.29	Manometer 18 tunnel	−14.30

RM = right margin; LM = left margin; Gal = gallery.

surprising to verify that most relief wells, built to release pressure due to eventual leaks from the reservoir, were effectively discharging groundwater not related to the reservoir. The main conclusion derived from applying this methodology was that basically all critical points were not related to the reservoir. This was later supported by the negative results of several interconnection experiments conducted in the vicinity of the dam, where 8 l of rhodamine WT were injected. The tracer was not detected in the springs located downstream of the dam nor in the main channel.

La Herradura reservoir, Cuba Carbon-13 in TDIC was determined in 19 water samples collected during a field campaign in March 1997 at La Herradura reservoir, Cuba. The measured values are presented in Table 6.3. In this case, the $\delta^{13}C$ values obtained from the samples collected in the reservoir were −6.53‰ and −6.87‰, that is, values slightly more enriched that that indicated for the mean present δ-value for atmospheric CO_2 (~ -7.5‰). This enrichment can be explained, in this case, by the dissolution and exchange of TDIC with the carbonate rocks outcropping along the riverbed and in the reservoir. The $\delta^{13}C$ value for these marine carbonates is close to 0‰.

The results presented in Table 6.3 clearly show that all samples, except borehole 22 at 6 m depth, are significantly more depleted in $\delta^{13}C$ than the isotope deviation observed in reservoir waters. These results proved that these waters were not coming from the reservoir. This conclusion was also supported by the chemical

Table 6.3. Analytical results of $\delta^{13}C$ measured in the TDIC in water samples collected in the vicinity of the La Herradura reservoir (Cuba).

Sample code	$\delta^{13}C$ (‰)	Sample code	$\delta^{13}C$ (‰)
Reservoir	−6.53/−6.87	Borehole 14, top	−14.43
Drainage well 11	−13.74	Borehole 22, 6 m depth	−3.70
Drainage well 26	−15.28	Borehole 22, 34 m depth	−13.61
Drainage well 27	−15.34	Borehole 25	−15.25
Drainage well 28	−15.64	Well Caseta	−15.29
Spring RM	−14.19	Well PS 12-2, 20 m	−15.36
Spring LM	−13.23	Weir, LM	−11.29
Drainage prism	−13.48	Weir, Central area	−13.68
Borehole 12 (5 m depth)	−14.41	Weir, RM	−14.02

Table 6.4. Analytical results of $\delta^{13}C$ measured in the TDIC in water samples collected in the vicinity of the Arquillo de San Blas reservoir (NE Spain).

Sample code	$\delta^{13}C$ (‰)
Reservoir, surface	−6.87
Reservoir, 33 m depth	−7.38
San Blas Artesian well	−9.20
Ambel spring	−9.02
Leak, LM	−10.78
La Señorita spring	−9.02
Azud de carburos spring	−9.05
River in Azud Las Cadenas	−8.66
El Balsón spring	−9.27
San Blas spring, RM	−10.18

RM = right margin; LM = left margin.

composition of the water. On the contrary, the isotopic composition of reservoir water showed similar δ-values as the local groundwater and, for this reason, did not provide any definitive conclusion on the origin of these waters.

Arquillo de San Blas reservoir, Spain The Arquillo de San Blas reservoir is located in the river Guadalaviar, northeast Spain. In a stretch located between 4 and 6 km downstream of the dam, there is pronounced increase in the flow rate of the river (about 900 l/s) due to the discharge, mainly in the riverbed, of several springs. Some of these springs were known to local people before the construction of the dam, but there were some hints that part of the discharges were related to leaks from the reservoir.

The results of the isotope analyses carried out on samples collected in the vicinity of the dam are presented in Table 6.4. The $\delta^{13}C$ value of the surface water collected in the reservoir was −6.87‰, and a bit more depleted value was obtained for the sample collected at depth (−7.38%). Both values are in good agreement with the equilibration process with atmospheric CO_2. As in the previous cases, $\delta^{13}C$ of all other samples collected in the vicinity of the dam site was isotopically more negative than the previous values representing the reservoir waters, indicating that the contribution of reservoir water to these groundwaters is minor or negligible. Among these samples, the lowest difference from the reservoir was obtained in the river Azud Las Cadenas, where a mixture of different groundwater discharges (with about 700 l/s) joins the river water derived from the reservoir. This conclusion was fully supported by an interconnection experiment performed with artificial tritium injected in the reservoir. The details about this experiment and other studies conducted at this reservoir are presented in Chapter 12.

6.5 CARBON-14

Carbon-14 is a classical tool for "dating" old groundwater. The method is based on the presence in the TDIC present in water of small fractions of cosmogenic carbon-14 produced in the upper atmosphere by nuclear reactions of the type ^{14}N (n,p) ^{14}C (neutron–proton reaction). Most of the carbon species dissolved in water are originally incorporated during the dissolution of biogenic CO_2 (modern carbon), with a $\delta^{13}C$ value close to −25‰. This CO_2 reacts with the carbonate minerals of the surrounding rock formations to produce HCO_3^-. Taking into account that this rock carbonate typically contains no carbon-14 and shows a $\delta^{13}C$ value close to 0‰, the bicarbonate ions originally formed will show intermediate values both for carbon-14 and $\delta^{13}C$. The $\delta^{13}C$ value can be used to estimate the initial ^{14}C content of the infiltrating water. Once this infiltrating water becomes isolated from the source of carbon-14, the ^{14}C content decreases according to the half-life of this isotope (5,730 years). Therefore, the actual ^{14}C concentration found in a water sample collected from a given aquifer will provide information on the "age" of the water, that is, the time elapsed since infiltration. However, some chemical reactions occurring in the aquifer can alter the ^{14}C content of groundwater, interfering with the dating procedure. For further information on this technique, the reader is referred to the specialised literature on this topic (e.g., Kalin, 2000; Mook, 1980; Pearson and Hanshaw, 1970).

It is obvious that water impounded in reservoirs is recent and mainly derives from recent precipitation. Therefore, the ^{14}C does not play any role in the identification of such water. However, in specific cases, old groundwater can also be found at the dam site due to the discharge of a local aquifer. For example, this can occur when old water was discharging along the river before the dam construction. The hydraulic barrier formed by the reservoir water can divert the groundwater flow toward the dam site, and the old groundwater is forced to emerge downstream of the dam or to flow through the boreholes drilled in the area. This was the case found by the authors in the reservoirs Giribaile and Bellús, Spain. In both cases, thermal water springs existed at points close to the dam. The same thermal water was identified in some boreholes drilled inside the dam galleries. Chemical composition, water temperature and the null content of tritium helped in the identification of this type of water. The performed carbon-14 measurements demonstrated, in both cases, that the emerging waters are several thousands of years old.

REFERENCES

Bottinga, Y. and Craig, H., 1969. Oxygen isotope fractionation between CO_2 and water and the isotopic composition of marine atmosphere. *Earth Planet. Sci. Lett.*, 5, 285–295.

Clark, I.D. and Fritz, P., 1997. *Environmental isotopes in hydrogeology*. Lewis Publishers, Boca Raton, 328 pages.

Cordero, C. and Plata, A., 1999. Estudio de fugas de agua en la presa Arenales, Costa Rica: Aplicaciones de técnicas de trazadores. In *Estudio de fugas en embalses*, IAEA, Vienna (in Spanish).

Craig, H., 1961a. Isotopic variations in meteoric waters. *Science*, 133, 1702–1703.

Craig, H., 1961b. Standard for reporting concentrations of deuterium and oxygen-18 in natural waters. *Science*, 133, 1833–1834.

Dansgaard, W., 1964. Stable isotopes in precipitation. *Tellus*, 16, 436–468.

Doney, S.C., Glover, D.M. and Jenkins, W.J., 1992. A model function of the global bomb tritium distribution in precipitation, 1960–1986. *J. Geophys. Res.*, 97(C4), 5481–5492.

Gonfiantini, R., 1986. Environmental isotopes in lake studies. In *Handbook of Environmental Isotope Geochemistry, Vol. 2, The Terrestrial Environment*, P. Fritz and J.C. Fontes (Eds.), Elsevier Science Publisher, New York, 113–168.

Horita, J. and Wesolowski, D.J., 1994. Liquid–vapour fractionation of oxygen and hydrogen isotopes of water from the freezing to the critical temperature. *Geochim. Cosmochim. Acta*, 58, 3425–3437.

International Atomic Energy Agency, 1969. Environmental isotope data no. 1: World survey in isotope concentration in precipitation (1953–1963). *Technical Reports Sereis, No. 96*, IAEA, Vienna.

International Atomic Energy Agency, 1970. Environmental isotope data no. 2: World survey in isotope concentration in precipitation (1964–1965). *Technical Reports Series, No. 117*, IAEA, Vienna.

International Atomic Energy Agency, 1971. Environmental isotope data no. 3: World survey in isotope concentration in precipitation (1966–1967). *Technical Reports Series, No. 129*, IAEA, Vienna.

International Atomic Energy Agency, 1973. Environmental isotope data no. 4: World survey in isotope concentration in precipitation (1968–1969). *Technical Reports Series, No. 147*, IAEA, Vienna.

International Atomic Energy Agency, 1975. Environmental isotope data no. 5: World survey in isotope concentration in precipitation (1970–1971). *Technical Reports Series, No. 165*, IAEA, Vienna.

International Atomic Energy Agency, 1979. Environmental isotope data no. 6: World survey in isotope concentration in precipitation (1972–1975). *Technical Reports Series, No. 192*, IAEA, Vienna.

International Atomic Energy Agency, 1981a. Statistical treatment of environmental isotope data in precipitation. *Technical Reports Series, No. 206*, IAEA, Vienna.
International Atomic Energy Agency, 1981b. Stable isotope hydrology: deuterium and oxygen-18 in the water cycle. *Technical Reports Series, No. 210*, IAEA, Vienna.
International Atomic Energy Agency, 1983. Environmental isotope data no. 7: World survey in isotope concentration in precipitation (1976–1979). *Technical Reports Series, No. 226*, IAEA, Vienna.
International Atomic Energy Agency, 1986. Environmental isotope data no. 8: World survey in isotope concentration in precipitation (1980–1983). *Technical Reports Series, No. 264*, IAEA, Vienna.
International Atomic Energy Agency, 1990. Environmental isotope data no. 9: World survey in isotope concentration in precipitation (1984–1987). *Technical Reports Series, No. 311*, IAEA, Vienna.
International Atomic Energy Agency, 1992. Statistical treatment of environmental isotope data in precipitation. *Technical Reports Series, No. 331*, IAEA, Vienna.
Kalin, R.M., 2000. Radiocarbon dating of groundwater systems. In Cook, P. and Herczeg, A.L. (Eds.), *Environmental Tracers in Subsurface Hydrology*, Kluwer Academic Publishers, Boston, 111–145.
Lucas, L.L. and Unterweger, M.P., 2000. Comprehensive review and critical evaluation of the half-life of tritium. *J. Res. Natl. Inst. Stand. Technol.*, 105, 541–549.
Majoube, M., 1971. Fractionement en oxygène-18 et en deuterium entre l'eau et sa vapeur. *J. Chim. Phys.*, 197, 1423–1436.
Merlivat, L., 1978. Molecular diffusivities of $H_2{}^{16}O$, $HD^{16}O$, and $H_2{}^{18}O$ in gases. *J. Chem. Phys.*, 69, 2864–2871.
Merlivat, L. and Jouzel, J., 1979. Global climatic interpretation of the deuterium–oxygen 18 relationship for precipitation. *J. Geophys. Res.*, 84(C8), 5029–5033.
Mook, W.G., 1980. Carbon-14 in hydrogeological studies. In *Handbook of Environmental Isotope Geochemistry, Vol. 1, The Terrestrial Environment*, P. Fritz and J.C. Fontes (Eds.), Elsevier Science Publishers, New York, 49–73.
Pearson Jr., F.J. and Hanshaw, B.B., 1970. Sources of dissolved carbonate species in groundwater and their effect on carbon-14 dating. In *Isotope Hydrology*, IAEA, Vienna, 271–286.
Plata, A., 1994. Composición isotópica de las precipitaciones y aguas subterráneas de la Península Ibérica. *Monografía CEDEX,* Madrid, 139 pp.
Rozanski, K., Gonfiantini, R. and Araguás Araguás, L., 1991. Tritium in the global atmosphere: distribution patterns and recent trends. *J. Phys.*, 17, S523–S539.
Rozanski, K., Araguás Araguás, L. and Gonfiantini, R., 1993. Isotopic patterns in modern global precipitation. In *Climate change in continental isotopic records*, Vol. 78, P.K. Swart, K.C. Lohmann, J. McKenzie and S. Savin (Eds.), American Geophysical Union, Washington, 1–36.
Salati, E., Dall'olio, A., Matsui, E. and Gat, J.R., 1979. Recycling of water in the Amazon basin: an isotopic study. *Water Resour. Res.*, 15, 1250–1258.
Vogt, H.J., 1976. *Isotopentrennung bei der Verdunstung von Wasser.* Thesis. Institute of Environmental Physics, University of Heidelberg, 58 pp. (in German).

CHAPTER 7

Artificial tracers

7.1 INTRODUCTION

Artificial tracers are absolutely necessary tools for the investigation of leakage from reservoirs. This type of study usually requires a detailed knowledge of the groundwater flow patterns in the concrete zone affected by the leaks. This information can only be obtained through the application of direct techniques such as those based on the use of artificial tracers. Practical experience shows that in almost all cases, the study of the nature and precise flow path of the leaks cannot be determined without applying techniques based on artificial tracers.

In most cases, the complexity of the hydrological framework, the lack of information on how the techniques based on artificial tracers should be applied, and the difficult interpretation of the results obtained have contributed to limit the application of these techniques. In other cases, interconnection experiments are conducted with insufficient or inadequate definition of the hydrogeological settings and/or the properties of the tracer to be used, leading to negative results and calling into question the validity of the artificial tracer methods.

In most cases, detailed planning in advance of the concrete experiment to be performed is necessary, which requires detailed knowledge of the geological and hydrogeological characteristics of the involved area, as well as the behaviour and detection limit of the tracer to be applied. In many instances, artificial tracers are applied as the proof of the suspected interconnection between two points, as indicated or suggested by techniques based on variations in the concentrations of natural tracers, as discussed in previous chapters.

The main objectives of the experiments based on tracer injection to a water body for the investigation of dam leakage can be summarised in the following list:

1) Performance of interconnection experiments between the infiltration area in the reservoir and the existing springs, or between boreholes and springs.
2) Localisation and measurement of groundwater flows through boreholes drilled in the affected zone.

3) Execution of tracer tests inside the reservoir to locate preferential seepage or infiltration areas.

Hundreds of substances have been used as artificial tracers to be incorporated or injected into diverse water bodies. The most important tracers used in hydrology can be classified in the following groups:

1) Suspended solids (particulate tracers).
2) Ionic tracers.
3) Gaseous fluoro-compounds.
4) Fluorescent tracers.
5) Radioactive tracers.

Obviously, tracers for water labelling have to fulfil several conditions, such as the following:

1) Same hydraulic behaviour as water, that is, limited or null sorption by the solid medium as water circulates in channels or aquifers, no exchange with certain minerals of the rock formations, not forming salts that may precipitate along the flow path, etc.
2) Not being naturally present in water, at least in concentrations that may interfere with their measurement.
3) High solubility in water.
4) Easily measurable at low concentrations, so that large water volumes can be labelled using a small amount of tracer.
5) For some experiments, the possibility to be measured in situ.
6) Nontoxic, inexpensive and easily available.

In reality, a tracer that fulfils all these conditions does not exist. Therefore, the most appropriate tracer for each particular tracing experiment has to be selected. In the present chapter, the characteristics and the possibilities offered by the most interesting tracers available for water tracing are discussed from the point of view of their application for dam leakage studies.

7.2 SUSPENDED SOLIDS

For obvious reasons, tracers consisting of suspended solids can be applied only for the investigation of groundwater flows through well-developed fractures, solution channels and/or cavities in consolidated rocks, such as karstified limestone or gypsum formations. In porous or microfissured media, these tracers will be filtered by the terrain and, therefore, will not travel long distances from the injection point.

Different products have been suggested and/or used as suspended solid tracers. The most important are club moss (lycopodium) spores, bacteria, fluorescent microspheres and, to a lesser extent, starch. The latter can be used in the form of small particles and is qualitatively detected in water by dyeing these particles with an iodine solution. An intense blue colour will be developed in the presence of such particles.

The use of bacteria as a water tracing agent was tested more than a century ago (Pfhul, 1897). A comprehensive study of the possibilities offered by this type of tracer has been made recently by a working group commissioned for the investigation of a German aquifer (DVGW, 1983; UBA, 1985). Bacteria for water tracing should, at least theoretically, be easy to grow, nonpathogenic, not present in water, easy to determine, absent of its own motion, lacking in sorption capacity, long-lived, etc. *Serratia marcescens* (Käss et al., 1983), *Escherichia coli* (Pschyrembel, 1964), *Bacillus stearothermopholus* (Satar et al., 1972) *Pseudonomas violácea* (Lukas, 1959), *Bacillus acidocaldarius* (Darland and Brock, 1971) and *Saccharomyces cerevisiae* (Wood and Ehrlich, 1978), among others, have been suggested and/or used as tracers.

Schröter (1987) has suggested the possibility of using fluorescent microspheres as a tracer of water fluxes. Such microspheres are used in physiological studies to simulate the propagation of pathogenic bacteria. They are produced by the firm Polysciences Inc., Northampton, UK and sold under the name "Fluoresbrite^{TM} Plain Microspheres" (see Käss, 1998). Electrically neutral microspheres are made of polystyrene with a diameter between 0.05 and 90 μm and a density of 1.05 g/cm^3. Pyronin GY (green-yellow) and PC-red (polychromatic-red) are used as fluorescent dyes. The detection can be directly made in collected water samples using a fluorometer or, even better, by using an appropriate filtering device. Important advantages of the microspheres with respect to bacteria are that they do not need to be cultivated for the injection and their detection is very simple and fast.

Spores from the club moss (*Lycopodium*) have been used since 1956 as a suspended qualitative tracer for the investigation of water pathways in fractured or karstified rocks (Brandecker et al., 1965; Gardner and Gray, 1976; Maurin and Zötl, 1964; Richter, 1944; Zojer and Zötl, 1975; Zötl, 1957). The diameter of these spores is about 30 nm, and their shape corresponds to an arched tetrahedron. Its density is about 1.1 g/cm^3. The main advantage of this tracer is the possibility of being continuously accumulated during a period of time as long as desired in a sampling device fitted with an appropriate filtering system. The detection of the spores is carried out by means of microscope counting (Dechant and Hacker, 1988). An extensive description of the use of spores for hydrological purposes has been made by Käss (1998).

This tracer can be of special interest for interconnection experiments when the expected transit time of water is very long and high dilution is expected. The tracer can be injected in a sinkhole existing in the reservoir or in a borehole tapping the fractured rocks. Obviously, the identification of the appropriate injection point can create some unexpected difficulties. If grab samples are collected at the sampling point (usually a spring) for a sufficiently long period of time, the time of the first appearance and the moment of maximal concentration should be determined.

Due to the suspended character of the tracer, the measured transit times are usually shorter than the transit time derived from experiments using conservative tracers. Logically, the breakthrough curves show very long tails because of the retention of spores in small cracks.

One interesting option is the use of coloured spores for multitracing experiments. *Lycopodium* spores can be dyed with different colours using both nonfluorescent and fluorescent dyes (Dechant, 1967; Dechant and Hacker, 1988). The last option is much easier and faster to determine in water samples (Käss, 1982) using a fluorescence microscope. Acridine orange, Pyronin GY and Tinopal CBS-X have been used as dying agents, providing, respectively, a yellow, red and blue-white fluorescence (Käss and Reichert, 1988).

The main disadvantage of club moss spores is that they are not easily available. Probably, the tracer group most experienced in the use of this kind of tracer is the Institute Joanneum of Geothermic and Hydrogeology, Graz, Austria. Many hydrological problems of karst hydrology have been tackled by this group since 1956.

7.3 IONIC TRACERS

7.3.1 Generalities

Strong electrolytic ions are used as ionic chemical tracers. This category includes those elements and compounds that are present in water in ionic form, resulting in their separation into charged species possessing a positive (cations) or negative (anions) charge. The ionic charge affects its migration through the aquifer by numerous mechanisms resulting from water–rock interaction processes. Many of these tracers show unwanted properties due to their fixation in the terrain by ionic exchange reactions and sometimes, by precipitation, as it is the case for most cations, with the exception of some alkaline ions. Everts et al. (1989) have studied the mobility of numerous tracers under laboratory and field conditions.

The best ionic tracers are, undoubtedly the halogen ions chloride (Cl^-), bromide (Br^-) and iodide (I^-). As indicated in the previous chapter, chloride and the other halogens behave in a rather conservative manner in the terrain due to the fact that almost all their chemical compounds are highly soluble in water. The less soluble compounds formed with certain cations such as Pb^{++}, Pb^{+++}, Hg^{++} and Tl^{++} do not create any problem due to the extremely low concentrations of these cations generally found in natural waters. Fixation processes of halogen ions are also of negligible relevance, as is their capability to be absorbed or incorporated by biotic systems. Therefore, these tracers can travel long distances in aquifer systems with very small losses and negligible interaction with the geological media. Other ions rarely used for special purposes are Li^+ and NO_3^-.

As a general rule, anions show better behaviour for tracing water that cations. Anion exchange on clayey materials increases with decreasing pH and with the presence of organic matter (Mattson, 1929). Most cations easily interact with clay minerals through the process of cation exchange.

Ionic tracers can be detected by using standard analytical methods applied in hydrochemistry or by measuring the increase in the electrical conductivity (EC) of the flowing water. In the latter case, rather high concentrations of salt solutions must be injected in the system being studied to produce significant changes in the EC at the sampling points. These concentrated solutions will be denser than the original groundwater and will tend to sink to the bottom of the aquifer and therefore, follow a more distinct flow path, on the average, than the parcel of water intended for labelling.

7.3.2 Chloride ion

Chloride ion is used for water tracing in the form of NaCl. In most cases, common (table) salt is used as a tracer. Due to the above-mentioned reasons, its behaviour is almost ideal, even in the case of granular media (porous aquifers). The solubility of common salt in water is about 350 g/l for a water temperature around 20°C. The tracer can be injected in almost any water body as a saturated water solution or as a solid material, depending on the experimental requirements. If the tracer concentration is high enough, it can be measured/estimated in place by means of a conductivity meter. Standard chemical analytical methods allow the determination of concentrations as low as 0.05 mg/l. The most commonly used analytical method is the volumetric determination with $AgNO_3$, using the chromate ion as indicator. The portable water analysis system commonly used by the authors uses mercuric nitrate as titration solution and diphenylcarbazone as reagent.

The main limitation for the use of chloride as tracer derives from the fact that this ion is always present in natural waters, sometimes, even at very high concentrations. In general, this tracer can be used only when its concentration in natural waters is less than 50 or 100 mg/l and the water volume to be labelled is relatively small. For practical reasons, only water volumes lower than 1,000 m^3 can be labelled using a reasonable amount of solid product. When the tracer has to be determined in place by EC measurements, the natural EC of water is always a limiting factor.

The authors are routinely using common salt as a tracer for different types of applications. A few examples are described below.

Tracer tests in boreholes The authors have conducted for many years as a routine technique flow tests in boreholes based on the labelling of the whole water column with a saturated solution of common salt. The different variants of the method are described in detail in Chapter 8.

Determination of flows in shallow groundwater The measurement of flows in shallow groundwater at the vicinity of a reservoir can easily be performed by means of a simple technique based on the dilution of the injected tracer. Small holes are dug in the area of interest where it is suspected that water is escaping from the reservoir. The water column contained in these holes is labelled with the solution of common salt (or solid) in such a way that the EC value reaches several times the natural EC. After the tracer injection, its concentration or simply the EC in the labelled zone is controlled during a period of time long enough to determine the slope of the dilution curve representing the variation of concentration as a function of time. Let us call Q the flow rate that circulates through the cross section of the hole, where a water volume V exists. The tracer concentration C will vary with time according the following equation:

$$C_t = C_0\, e^{-(Q/V)t} \tag{7.1}$$

The value of the flow rate Q is determined from the experimental setting. Obviously, the exponential tracer dilution rate is fulfilled only if the tracer concentration in the volume V is kept homogeneous at any time during the experiment. Usually, this condition is fulfilled for low values of the ratio Q/V (slope of the exponential curve).

The senior author has successfully applied this technique at the Maguaca reservoir, Dominican Republic. The area of the upper layer where water was escaping from the reservoir at its maximum level was clearly identified.

Identification of groundwater discharge in small, close lakes A similar dilution technique was successfully used by the senior author for the identification of groundwater discharge in a small lake near the town of Ruidera, Central Spain. The small lake was located on a highly karstified formation composed of limestone and dolomite. No surface flow existed during the duration of the experiment. The EC of the small lake increased from about 450 μS/cm to about 1,550 μS/cm. During the following months, the variation with time of the EC and the concentration of Cl^- were periodically measured in the lake water to determine the eventual discharge of groundwater at the bottom of the lake. The EC of the local groundwater ranged from 500 to 600 μS/cm and the Cl^- ion concentration from 10 to 15 mg/l. No tracer dilution was observed. On the contrary, the EC of the lake water increased slowly with time due to evaporation. The experiment showed that no significant groundwater discharge was taking place at the bottom of the lake.

Measurement of small flow rates Common salt can be used as a tracer for the measurement of small flow rates (up to 1 m^3/s), for example, in springs located downstream of the dam. The method is described in detail in Annex 2. The technique is very convenient when other alternative methods are not available or cannot be applied, for example, due to the high turbulence and/or to the irregular cross section of the flow. The tracer method can provide more precise results than the method based on the use of a current meter.

7.3.3 Iodide ion

Rodda (1977) and Wagner (1977) have described the use of the iodide ion as a tracer for water systems. The tracer is injected in the form of NaI or KI. The geochemical behaviour of this ion is similar to the that indicated for the chloride ion, although it is sorbed to a greater extent than either Br^- or Cl^-. An important advantage of this tracer is derived from its low concentration in natural waters, usually less than 0.01 mg/litre. A concentration as low as 0.05 mg/l (in the form of I^-) can be determined by atomic absorption spectrometry. This ion acts in the determination as a catalyst controlling the redox system Ce^{++}–Ce^{+++}. Using this analytical method, water volumes up to 10^5 m^3 can be labelled with this tracer. However, due to the relatively high price of the tracer, most researchers prefer the use of bromide as an alternative tracer.

7.3.4 Bromide ion

Bromide offers the best possibilities to be considered an excellent tracer for groundwater studies (Davis et al., 1985; Jabro et al., 1994; Porro, 1993; Seaman et al., 1995). The concentration of Br^- in natural waters is roughly 1/300 of that of the Cl^- ion commonly present in groundwater, that is, usually less than 0.1 mg/l. The tracer is used in the form of NaBr or NH_4Br. The solubility of NaBr is 850 g/l at 10°C. Advantages of this tracer are its low tendency to be absorbed in the soil material and its conservative character regarding microbiological uptake.

The analytical determination of Br^- ion is rather cumbersome. It can be determined using a specific ion electrode with a lower limit of detection of about 0.05 mg/l. This sensitivity is too low if a large volume of water needs to be labelled.

Concentrations in the parts per billion (ppb) range can be determined by gas chromatography (Ando and Sayato, 1983). The bromide ion is treated with citric acid and potassium permanganate to form pentabromoacetone. This is measured by gas chromatography with no interference by chloride.

Very high detection sensibility can be achieved using the isotope dilution method described by Heumann and Seewald (1985). However, this method requires the use of a bromide solution as a dilution product, enriched in ^{79}Br or ^{81}Br, and an appropriate mass spectrometer.

The bromide ion has been used as a tracer for the investigation of many hydrological problems. Bronswijk et al. (1995) have investigated the behaviour of this tracer in heavy clay soils. Leap (1992) has studied the relative retardation of tritium and bromide in dolomite. Molz et al. (1985) have used this tracer in single-well experiments based on the injection of the tracer and its subsequent recovery by pumping. In an experiment carried out in the United States, about 3,000 kg of NH_4Br were injected in a single well.

7.4 GASEOUS FLUORO-COMPOUNDS

During recent years, gaseous fluoro-compounds (fluorocarbons and especially sulfurhexafluoride) have deserved the increasing attention of many investigators as tracers for surface and groundwater systems. These compounds appear to offer a number of advantages over most conventional tracers for many groundwater applications (Thompson et al., 1974; Wilson and Mackay, 1993, 1996). They are extremely nonreactive and do not break down chemically under normal groundwater conditions. They are easily detectable in water in very low concentrations with an optimum working range of between 1 and 100 parts per trillion ($1\ \text{ppt} = 1 \times 10^{-12}$ g/ml), although much lower concentrations are measurable in the case of certain tracers under optimal analytical conditions. The capability of measuring tracer concentration in this range enables an enormous volume of water to be labelled easily with a small quantity of tracer. For example, 1 g of tracer (about 0.5 ml) labels 100,000 m^3 of water at the maximum working concentration of 100 ppt. Another important consideration is that many of the tracers have very low toxicity.

The detection of fluoro-compounds is carried out using a gas chromatograph system equipped with an electron capture detector. The extreme detection sensitivity is due to the high electron-capturing cross section shown by most halocarbons. They are detected in picogram quantities. The tracer has to be extracted from the water sample and then measured. The extraction procedure can be carried out as follows: (1) Pure nitrogen is added by means of a syringe to a closed bottle or vial containing the water sample in order to remove interfering air from the headspace. (2) After shaking for 2 or 3 minutes, the gas is collected with the same syringe and injected into the chromatograph for analysis.

Sulfurhexafluoride is the fluoro-compound most commonly used as a tracer for hydrological applications. It is a nontoxic gas that sublimates at −64°C. Its solubility in water at 25°C is only 32 mg/l. The detection limit in water using the gas chromatography method is on the order of 10^{-16} mol/l (Watson et al., 1987). However, for the evaluation of its possibilities as a tracer of water, the natural presence in all surface waters and air, as well as the worldwide increasing concentrations of this compound, has to be taken into account as a limiting factor.

Fluorocarbons are injected at concentrations of a few tenths of a part per billion. The ability to use and detect tracers at very low concentrations has important practical and economic implications, enabling large volumes of water to be labelled very inexpensively, using a small quantity of tracer.

Fluoro-compounds have to be used as dissolved gases and, therefore, are subject to loss by volatilisation from open water samples. Most interesting fluorocarbons for water tracing are CCl_2F_2 (freon-12), CCl_3F (freon-11), $CClBrF_2$, CBr_2F_2, $CBrF_2$–$CBrF_2$ (freon-113), fluorobenzoate, etc. The solubility in water at 1 atm and 25°C of the first two compounds is 0.028% and 0.11%, respectively. Ilgenfritz et al. (1988) have suggested the use of fluorobenzene, and Bowman and Gibbons (1992) have suggested the use of difluorobenzoates as effective nonreactive tracers in soil and groundwater tracing experiments.

One important advantage of fluorocarbons is their potential in experiments requiring the combined use of several tracers. The different fluorocarbons can be separately measured without significant interference using an electron capture gas chromatograph detector.

A limiting factor for the use of fluoro-compounds stems from the fact that they are organic compounds and as such, they become dissolved in or strongly sorbed by many other organic materials. Therefore, they will generally be unsuitable for tracing water in the subsurface when coal, peat, oil shale, and other natural organic substances are present in rocks or soils.

7.5 FLUORESCENT TRACERS

7.5.1 Generalities

Fluorescent substances are surely the tracers most commonly used in hydrology. These tracers have been used for many decades as qualitative tools for the investigation of the hydraulic connection between different water bodies. Dole (1906) was the first scientist to recommend the use of dyes for the study of groundwater systems. However, extensive use of fluorescent dyes as tracers for hydrological purposes did not begin until the 1960s. In many cases, their detection was exclusively made by visual inspection. In a latter stage, lamps of ultraviolet rays were used to facilitate the qualitative detection of the tracers. However, these tracers have become important tools for water tracing with the development of filter and spectral fluorometers. These measuring systems permit, in most cases, the quantitative determination at very low tracer concentrations.

A high number of fluorescent products are known. However, only a few of them are commonly used for water tracing. These are uranine, rhodamine B and rhodamine WT (water tracing). Uranine is the sodium salt of fluorescein. It is used instead of fluorescein because of the higher solubility in water. Uranine and rhodamine B are used in solid form, while rhodamine WT is supplied as a liquid solution with a concentration of the active ingredient of approximately 20%. As its name indicates, this tracer has been specifically developed for water tracing.

Fluorescent tracers obey their name because they emit a characteristic fluorescence when they are excited by a light beam of a certain wavelength. Both the wavelength of the excited light beam and the emitted fluorescence are characteristic for each fluorescent product.

7.5.2 Measuring techniques

7.5.2.1 *Filter fluorometers*

In the case of filter fluorometers, the excitation of the fluorescence is made by the light of a lamp (usually, halogen lamps) appropriate for the fluorescent product to be measured (see Fig. 7.1). The light emitted by this lamp is filtered by one or two coloured filters, which provide a light beam of a wavelength range appropriate also for the fluorescent product to be measured. These filters are called excitation filters. The fluorescence emitted by the water sample containing the tracer is also filtered by the so-called emission filter to modify the wavelength of the emitted ray in a convenient way so it can be measured by the photomultiplier tube used as a detector. For the measurement of the tracer concentration, the fluorescent light intensity emitted by the sample is compared with the intensity of a reference light beam created by the same lamp after crossing the so-called reference filter. Both light beams are directed alternately to the photomultiplier tube and the difference between the electrical signals generated by these light beams provides the tracer concentration in the water sample. Obviously, the measuring system needs to be calibrated using standard solutions. The filter and, sometimes also the lamps, have to be changed when the tracer to be measured is changed.

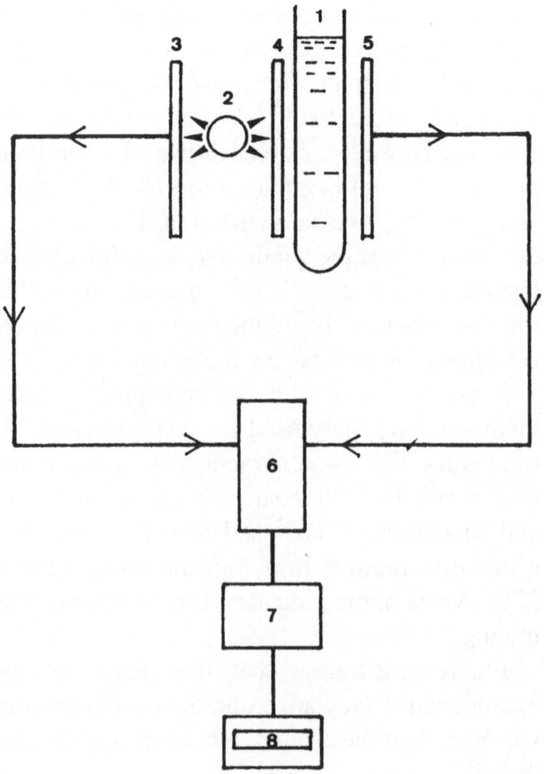

Fig. 7.1. Simplified view of a filter fluorometer. 1: Water sample; 2: lamp; 3: reference filter; 4: excitation filter; 5: emission filter; 6: photomultiplier tube; 7: signal processor; 8: digital display.

Filter fluorometers are very convenient for field work because of their versatility, simplicity of use, and portable character. These fluorometers can be connected to a car battery, facilitating field measurements. The measurements can be performed on discrete water samples or in a continuous way. In the first case, the water sample is placed inside a glass test tube, which is introduced into the measuring cell of the instrument. A sealing cap stops the external light from entering the measuring chamber. The water volume needed for this measurement can vary between 5 and 40 ml, depending on the fluorometer type and on the size of the required test tube. The duration of the measurement is short, usually between a few seconds and tens of seconds.

For the continuous measurement of the tracer concentration, flow-through cells have been developed, in which a small flow rate of water is pumped through the measuring cell using a peristaltic pump. The tracer concentration can be recorded on paper or digitally stored in a portable computer for further processing. This technique is used when the dynamic measurement of the tracer concentration is needed. This might be the case, for example, when determining in a continuous way the tracer distribution inside a reservoir or when recording the variation with time of the tracer concentration in a given control station without collecting water samples. The modern fluorometers are fitted with electronic systems allowing the automatic correction of the temperature effect in the fluorescence emitted by the water flowing through the measuring cell.

The continuous measuring technique presents some problems, however. The most important is related to the possibility that the tracer concentration in water reaches values higher than the upper limit of the measuring range within which the fluorometer response maintains its linearity. A further problem of the technique is the difficulty in calibrating the measuring system when water contains interfering pollutants or a high concentration of suspended sediments.

Fluorometers have to be calibrated using appropriate standards in order to get quantitative values of the tracer concentration. Usually, the available measuring systems provide a linear response within a range between 0 and 100 ppb (or mg/m^3) for rhodamine B and uranine, and between 0 and 500 ppb for rhodamine WT (20% of active ingredient). Within this range, the fluorometer can be calibrated using a single and unique standard solution. Between 100 and about 1000 ppb (5000 ppb for rhodamine WT), the fluorometer response is not linear, but it can be used after its calibration using a set of appropriate standards. However, due to practical reasons, the measurement of such high tracer concentrations is better carried out by diluting the water sample up to a concentration within the linear range of measurement. Concentrations higher than 1000 ppb cannot be measured. Even with very high concentrations, the response of the fluorometer decreases when the concentration increases due to the darkening (opacity) of the water sample. The light transmission decreases at increasing concentrations. The previously stated values are valid for water samples with a thickness of about 20 mm. If measuring cells with a smaller thickness are used, the linear range is somewhat higher because the darkening effect is smaller.

The sensitivity of filter fluorometers can be as low as 0.02 ppb for rhodamine B and uranine and five times lower for rhodamine WT when the fluorescent product is dissolved in distilled water. However, in the case of natural waters, this sensitivity can be much lower due to the background fluorescence of water. Further information on this point will be provided below.

7.5.2.2 *Fluorometers for in situ measurement*

Recently, the firm Turner Designs, Inc. has developed a fluorometer for in situ, direct measurement of rhodamine B and WT. The sensor is introduced into the water body at the required depth, and the optical signal produced by the tracer is transmitted to the measuring unit located at the surface via a fibre optic cable. This kind of fluorometer can be of the highest interest for the investigation of dynamic processes of water inside reservoirs as well for the determination of water flows inside boreholes using the techniques described in Chapter 8. The main advantage of fluorescent tracers compared with saline or ionic tracers for this type of study lies in the fact that no significant change in the water density is produced. Due to their recent development, the authors do not have experience with the performance of this type of fluorometer. The outer diameter of the probe is 2.5 inches, therefore allowing its introduction inside most boreholes and wells to perform the flow tests described in Chapter 8.

7.5.2.3 *Spectral fluorometers*

Obviously, filter fluorometers work with light beams whose wavelength occupies a relatively wide range of the spectrum, depending on the quality of the filter used. On the contrary, spectral fluorometers use a monochromatic light beam for the excitation of the fluorescence of the sample to be measured. The wavelength of this light beam can be regulated to the most appropriate value, depending on the fluorescent tracer to be determined. In the same way, the spectral distribution of the emitted fluorescence, which is a characteristic of each fluorescent product, can be recorded. These instruments permit, in many cases, the quantitative measurement of admixtures of two different fluorescent tracers. This feature can be of interest, for example, when multitracing experiments have to be performed (Behrens, 1982). Spectral fluorometers of double monochromatic light beam allow the separate measurement of both tracers.

7.5.3 Types of fluorescent tracers

7.5.3.1 *Generalities*

A number of fluorescent products have been used as tracers for hydrological purposes. The most important dyes are showed in Table 7.1. Most of them are products derived from xantene. Their behaviour is to some extent related to their molecular structure. All dyes contain in their molecules radicals that provide, to some extent, a polar character. This is the reason why these dyes present a certain tendency to be

Table 7.1. Main characteristics of the fluorescent tracers most commonly used for hydrological purposes. The indicated detection limits are referred to the measurement of a solution of the dye in distilled water using a spectral fluorometer. For the measurement of natural water solutions using a filter fluorometer, the stated detection limits can be between 50 and 1000 times lower.

Product	Formula	$\lambda_{excitation}$ (nm)	$\lambda_{emission}$ (nm)	Detection limit (ppb)	Soil fixation	Photo-decomposition
Uranine	$C_{20}H_{10}O_5Na_2$	492	513	0.002	Very low	High
Eosine	$C_{20}H_6Br_4O_5Na_2$	515	535	0.001	Low	Very high
Piranine	$C_{16}H_7O_{10}S_3Na_3$	460[1]	512	0.008	Very low	High
		407[2]	512			
Sulforhodamine G	$C_{25}H_{30}O_7N_2S_2Na_2$	535	555	0.005	Medium	Low
Sulforhodamine B	$C_{27}H_{30}O_7N_2S_2Na_2$	560	584	0.007	Medium	Low
Rhodamine B	$C_{28}H_{31}O_3N_2Cl$	555	582	0.006	Strong	Very low
Rhodamine WT	$C_{29}H_{30}O_5N_2Na$	558	583	0.006	Medium	Very low

[1] For pH ≥ 10; [2] for pH ≤ 4.5.

fixed in the terrain. Cohesive forces between fluorescent products and soil materials are responsible for this ability to be fixed. The relative tendency of the most relevant fluorescent products to be fixed is indicated in Table 7.1. Due to the ionic character of their molecules, their fixation in the soil materials strongly depends on the pH of water.

On the other hand, this kind of tracer presents a limited stability in many hydrological systems. Low pH values can produce the decomposition of the tracer in a total or partial way. In the same way, solar light produces their decomposition at different rates that are characteristic of each dye. This point is discussed below.

As mentioned above, the most frequently used fluorescent tracers for hydrological purposes are uranine, rhodamine B and rhodamine WT. The toxicity of water solutions containing these products has been analysed by different authors (e.g., Smart, 1982, 1984). Some manufacturers even issue toxicity certificates for different fluorescent products. In general, it is accepted that for permanent consumption of drinking water, a concentration up to 1 ppb (mg/m^3) is tolerable. The toxicity of uranine is lower than that of other dyes, such as rhodamine.

7.5.3.2 *Uranine*

As mentioned above, uranine is the sodium salt of fluorescein. It is used instead of fluorescein due to its higher solubility in water: 500 g/l versus 20 g/l at a temperature of 20°C. The concentrated uranine solutions show an intense red colour, while the diluted solutions have a green colour. The human eye can detect a concentration as low as 20 ppb (Anguita, 1969). In low alkalinity waters with a low pH, this dye can be somehow altered by the dissolved CO_2. Sunlight, some types of bacteria and ferric oxide produce its partial decomposition (Feuerstein and Selleck, 1963; Molinari, 1969).

The pH of the water sample has a drastic influence on its detection. At pH values lower than 7, the tracer partially decomposes. The decomposition is, to a great extent, reversible but not quantitatively. Usually, the tracer is measured with values of the pH higher than 10. In the case of natural waters, this high pH value can produce the precipitation of calcium and magnesium hydroxide. When it happens, the turbidity of the water samples seriously interferes with the measurement of the tracer. To obtain a pH value of the water sample higher than 10 while avoiding this turbidity problem, two procedures have been proposed. The first one consists of keeping the water samples for a few hours inside a closed box where a concentrated ammonium solution is also located. The ammonium vapours emitted increase the pH of the samples to the appropriated value. The second solution consists of adding to the water samples a small amount of a sodium salt of EDTA. The pH changes to the desired pH value, and the earth-alkaline ions remain dissolved in the form of EDTA complexes.

The detection sensibility of uranine decreases slightly with increasing temperature, as will be shown later. Due to this fact, it is convenient that all water samples that have to be compared each other should be measured at the same temperature.

The possibilities offered by this tracer for surface water studies are seriously limited by its photodecomposition or photolysis. This is a consequence of the action of ultraviolet radiation on the dye molecules. Further information on this point is presented below. According to the results obtained by some authors, the photodecomposition of uranine can be very rapid, reaching the so-called half-time decomposition $t_{1/2}$, in several hours. The senior author has found in an experiment carried out in Namibia $t_{1/2}$ values of only 25 minutes. The high intensity of the solar radiation, combined with a high temperature (about 36°C), was responsible for the observed extreme photodecomposition rate. On the contrary, this tracer showed a rather conservative behaviour when used by the authors in a tracing experiment carried out in the River Magdalena, Colombia. Ten kilograms of uranine were injected together with 6 Ci of bromine-82 and 10 Ci of tritium. Uranine showed an identical behaviour as the other two tracers during a three-day period. It provided the same transit time of water in the river as well as the same value of the flow rate (about 1600 m^3/s). The reason for this favourable behaviour was the high concentration of suspended sediments carried by the river, together with the fact that cloudy weather was prevailing during the experiment.

As will be indicated later, uranine can also be decomposed by strong oxidising agents, such as ozone or free chlorine. Due to the previous considerations, it can be stated that uranine is more appropriate for the investigation of groundwater flows, particularly in fractured and/or karstified rocks. Its relatively low adsorption on clayey materials allows also its use for short-distance experiments in granular or porous media.

The dissolution in water of uranine is always a cumbersome process due to the formation of foam and the presence, in the commercially available products, of small fractions of insoluble materials. The formation of foam renders difficult the precise measurement of the solution volume, what can be important in experiments in which a fixed, well-known amount of tracer has to be injected. The problem is usually solved by keeping the prepared solution at rest for some minutes or hours or, even better, slowly pouring the solution into a second glass container.

7.5.3.3 *Rhodamine B and WT*

Rhodamine can be detected at similar concentrations as uranine (Marichal and Benoît, 1961). The detection limit for rhodamine WT is about five times higher because it contains only about 20% of active ingredient. Rhodamine is more stable than uranine at pH values above 5.5 units. At lower pH values, the fluorescence decreases strongly, becoming negligible at a pH of 1.5 units. Rhodamine shows a higher tendency than uranine to be adsorbed in clayey materials. The photolysis effect (decomposition by solar light) is low, especially for rhodamine WT. For these reasons, rhodamine B and WT are more appropriate than uranine for surface water studies.

Rhodamine B and WT are adsorbed by soil materials, especially by clayey materials (Ptak and Schmid, 1996; Sabatini and Austin, 1991). In spite of that, these dyes can be used for groundwater studies as qualitative tracers, especially when dealing with fractured or karstified rocks. An important disadvantage of rhodamine is its high temperature coefficient, which amounts to 2.6% per °C. The intensity of fluorescence decreases with increasing temperature. For this reason, temperature correction is necessary when discrete samples are measured. To keep the required correction low, it is recommended to equilibrate all samples to be measured, including standards, at the same temperature, for example, at room temperature. Even the heat emitted by the fluorometer lamp can produce a small shift in the readings if the sample is kept for a long period of time (for example, several minutes) inside the measuring chamber. Modern fluorometers are equipped with automatic temperature correction systems when they are used for continuous-flow measurements.

The solubility in water of rhodamine B is approximately 50 g/l. However, the solubility in an acetic acid solution of 30% concentration in volume increases up to 400 g/l (Gioda et al., 1989). In any case, the formation of foam always creates serious problems. Rhodamine B solutions should be kept in glass containers to avoid adsorption in the walls (as occurs in plastic containers).

It is important to point out that chlorinated tap water usually decomposes the product in a nonreversible way due to its oxidant character. Therefore, this kind of water never should be used to dissolve rhodamine B or to prepare standard solutions.

As indicated in Table 7.1, rhodamine WT shows some advantages with respect to rhodamine B. Just the fact of being supplied in a liquid form renders its use easier, especially when an exact amount of tracer has to be injected. This is usually the case when this tracer is used for flow rate measurements or for interconnection experiments. The main disadvantage of rhodamine WT is its high price in comparison to rhodamine B. One has to take into account that 1 kg of rhodamine B is equivalent in its tracing capability to 5 litres of rhodamine WT.

Rhodamine can be extracted from an aqueous solution using the organic solvent isoamilic alcohol. This fact allows the detection of extremely low tracer concentrations (Guizerix et al., 1963). Rhodamine B is a carcinogenic product at high concentrations. Concentrations higher than 100 ppb are dangerous for human consumption.

7.5.4 Further information

7.5.4.1 *Photolysis effect*

All fluorescent tracers are more or less sensitive to light. The reduction of fluorescence F occurs according to an exponential law similar to radioactive decay, given by the following equation (Moser and Sagl, 1967):

$$F_t = F_0\, e^{-\alpha I t} \tag{7.2}$$

where α is the specific photolytic coefficient, I is the light intensity and t is the duration of the exposure.

According to Behrens and Teichmann (1982), the photolytic coefficient is equal to 550 for uranine, 7.7 for rhodamine B and 4.6 for rhodamine WT. This means that the sensitivity of uranine to light is 70 and 120 times higher for uranine than for rhodamine B and rhodamine WT, respectively. This is the reason why uranine is not appropriate, except in very particular cases, for experiments carried out in surface waters.

The half-life concentration for uranine (time needed for a concentration reduction of 50%) can range from less than 1 hour to several hours or tens of hours. Values around 10 hours have been obtained in Central Europe. On the other hand, Behrens and Demuth (1990) have indicated that the photolytic coefficient also increases with the temperature. This explains the above-mentioned results obtained by the senior author in Namibia. The photolytic coefficient for uranine is about 100 times higher than for rhodamine B or WT.

7.5.4.2 *Turbidity effect*

The turbidity of water samples affects the fluorescence produced by the dissolved dye. Usually, a higher reading is obtained when the turbidity increases due to the light scattering inside the water sample. This effect depends on the nature of the materials responsible for the turbidity. Organic compounds such as humic acids and colloidal ferric hydroxide are, in many cases, responsible for this turbidity. A false positive result can be obtained in the presence of these products.

If a spectral fluorometer is available, a positive fluorescence signal derived from turbidity can be recognised through the recording of a synchronscan spectrum. If the measurement is carried out using a filter fluorometer, this effect can be recognised through the filtration of the water samples using a membrane filter (Käss, 1964). All suspended particles over 25 μm can be eliminated. However, it should be considered that if membrane filters made of cellulose nitrate or acetate are used, some fraction of the dye may be adsorbed.

7.5.4.3 *Temperature effect*

It is well known that the fluorescence intensity of any water sample decreases at increasing temperature (Dunn and Vaupel, 1965; Wilson, 1968), according to the following law:

$$F_t = F_0\, e^{-\tau(t-t_0)} \qquad (7.3)$$

where τ is the coefficient temperature, t is the temperature during the measurement and t_0 is a reference temperature.

Moser and Sagl (1967) have obtained a temperature coefficient equal to 0.004 (0.4% per degree centigrade). For rhodamine B, values between 0.017 and 0.028 have been reported (Dunn and Vaupel, 1965; Moser and Sagl, 1967; Smart and Laidlaw, 1977; Wilson, 1968). For rhodamine WT, values close to 0.026 have been measured (Alexander et al., 1988; Smart and Laidlaw, 1977; Smart and Smith, 1976; Wilson, 1968).

7.5.4.4 *Quenching effect*

Fluorescence quenching occurs with all fluorescent materials (Foster, 1951) due to the absorption of the fluorescence spectra in some dissolved components, such as the ions Cd^{++}, Cu^{++}, Co^{++}, NO_3^-, etc. However, this effect is commonly of minor importance.

7.5.4.5 *Effect of oxidising products*

Strong oxidising products used for water sterilisation, such as chlorine and ozone, can rapidly destroy fluorescent dyes. This is the reason why treated tap water never should be used for the preparation of standard solutions. However, the addition of sodium thiosulfate can reverse this effect.

7.5.4.6 *Natural background fluorescence*

Natural background fluorescence is a limiting factor for the detection of fluorescent dyes in natural waters. The natural fluorescence spectrum covers a relatively wide range of wavelength, ranging from 440 to 528 nm (Käss, 1965; Smart and Laidlaw, 1977). The maximum fluorescence intensity is around a wavelength of 480 nm. Many products can be responsible for this natural background, for example, humic acid, detergents, hydrocarbons, algae, tuffaceous limestone (Käss, 1965), etc. Fluorescence intensities up to 2 mg/m^3 can be found.

Obviously, the effect of this natural background fluorescence can be corrected by using background samples from the same type of water involved in the experiment. It is also recommended to prepare the standard solutions using the same water.

7.5.5 Enrichment methods

7.5.5.1 *Using activated carbon*

Some authors have suggested the possibility of using activated carbon to enrich fluorescent tracers dissolved in the water. The technique is performed by placing inside the water body (reservoir, borehole or spring) a small sack containing some grams of activated carbon in grains. The sack is made of permeable nylon gauze that allows the water to flow through the activated carbon. This material has to be free from dyes that can be adsorbed by the activated carbon and interfere with the measurement of the tracer. The sack is kept submerged in the water body for a certain period of time, which will depend on the dynamic characteristics of the investigated water flow. For the investigation of the tracer breakthrough curve, the sacks are replaced from time to time. The fluorescent tracer eventually incorporated into the water flow is partially adsorbed in the activated carbon.

The tracer amount adsorbed in the activated carbon is determined after elution of the fluorescent dye with an appropriate solution. Two kinds of elution solvents have been used:

1) Reagent formed by a 1:1 mixture of pure ethyl alcohol 96% with a 15% potassium hydroxide aqueous solution (Bauer, 1967, 1972). The detection limit of uranine in the eluate is around 0.01 μg/l.

2) An aqueous solution of N-N-dimethyl-formamide prepared by mixing 80% of this product with 20% of distilled water (Perlega, 1976). Using this method, uranine fluorescence is quenched in the eluate because of the pH, while rhodamine B or WT can be measured directly. After adding to the eluate one drop of 25% ammonia, uranine fluoresces again. The detection limit is around 0.1 ppb for uranine and 0.05 ppb for rhodamine B.

The eluted tracer solution is measured by a fluorometer. If the amount of activated carbon used in all bags is the same and the elution is carried out using the same amount of eluting solution, the measurements can provide relative values of the tracer concentration in the water flow.

The described technique sounds highly attractive for interconnection experiments using fluorescent tracers, in which the tracer is injected at a point of the water flow and its arrival at a control point is investigated. In principle, the technique offers the following advantages:

1) The tracer uptake is made in a continuous way, that is to say, 24 hours a day. Therefore, there is no risk of the tracer passing through the measuring station without being detected.
2) In cases of unknown and very long transit times of water, the activated carbon bags can also be placed at the control points for very long periods of time, avoiding the need to visit the control site very frequently. Apart from that, the exposed bags can be stored for long periods of time before their measurement. However, they have to be kept in the dark to avoid the photodecomposition of the dye.
3) In principle, the tracer dissolved in water is fixed to the activated carbon in a cumulative way. It produces a concentration effect of the tracer that increases its labelling possibilities. In other words, using activated carbon, lower tracer concentrations in water can be measured than in the case of direct measurements.

In spite of these advantages, the experience of the authors in using activated carbon for the detection of fluorescent tracers is rather negative. Activated carbon usually contains adsorbed products that interfere with the measurement. These products have to be eliminated before the activated carbon is used for fluorescent tracer adsorption. This can be done by means of the procedures described by Benischke (1987) and Bauer (1972). The activated carbon with a grain size of 1.5 mm is repeatedly shaken in tap water to remove small particles of carbon dust. Then the carbon is heated at a temperature between 110°C and 150°C for several hours. After that, the nylon sacks are filled with about 20 g of the treated carbon. However, the activated carbon can also absorb impurities contained in the water flow in which the tracer has to be detected, leading to a false result. Obviously, if a spectral fluorometer is available, it is possible to identify whether the produced fluorescence is derived from the injected tracer.

In the opinion of the authors, further studies are necessary for the evaluation of the real possibilities offered by the fluorescent tracers detection technique. Experiments carried out under natural conditions using activated carbon, combined with the direct measurement of the tracer, would provide valuable information in this respect. The same argument can be applied for multitracing experiments using both fluorescent and conservative tracers, such as artificial tritium.

7.5.5.2 *Use of exchange resins*

It is important to point out that apart from activated carbon, some ion exchange resins have also been used for the fixation of fluorescent tracers. Pang et al. (1998) have described the combined use of rhodamine WT and *Bacillus subtilis* endospores as a bacterial tracer to estimate the transport parameters along preferential flow paths of sewage in an alluvial aquifer. Nylon mesh bags containing 0.4 g of XAD-7 resin were used to trap rhodamine WT. The tracer trapped at each bag was extracted using 5 ml of acetone and analysed for rhodamine WT using a Shimadzu RF-1501 spectrofluorometer with an excitation wavelength of 543 nm and an emission wavelength of 573 nm. The authors indicated a detection limit of 0.05 μg/l.

Laane et al. (1984) have described a method for enriching rhodamine B and rhodamine WT in natural waters using Sep-pack disposable cartridges containing 70 μm particles of Bondapak C18 (Waters Ass. Milford, U.S.A.). Prior to use, the Sep-pack cartridges are rinsed with 5 ml of a methanol–ethylacetate mixture (80:20) and then with 10 ml bidistilled water. Then, the filtered natural water sample is passed over the cartridge at a rate of 5 ml/min. The Sep-pack is washed with 10 ml bidistilled water and dried for about 10 minutes. The dye is then extracted using 5–10 ml of the mixture of methanol–ethylacetate. The extracted solution is then measured with a filter fluorometer fitted with appropriate excitation and emission filters or using a fluorescence spectrophotometer. An enrichment factor higher than 100 is possible according to the indications of the authors. Further studies under natural conditions are necessary for a precise evaluation of the possibilities offered by this detection technique.

7.5.5.3 *Liquid–liquid extraction method*

Uranine can be extracted from an aqueous solution with ether in acid medium. The diluted uranine solution (for example 1 l) is acidified with sulphuric or nitric acid. Later, a small volume of diethylether is added (for example, 50 ml) and shaken. Under these conditions, uranine passes to the organic phase. The decanted ether phase is then shaken with alkaline distilled water, and uranine passes again to the aqueous phase (Lukas, 1959).

In the same way, rhodamine can also be enriched using isoamylalcohol as extraction solvent (Rodier and Benoît, 1961). Two litres of clear water (after centrifugation, if needed) are poured into a separating funnel. Then 5 ml of 2 M hydrochloric acid and 75 ml of isoamylalcohol are added. After shaking, the solution is left to settle. The supernatant alcohol phase is collected, and 0.5 grams of anhydrous sodium sulphate is added to clarify the opalescent turbidity. The clear solution is measured at 551 nm. The authors have indicated a value of 1.0 μg/l as the quantitative detection limit.

7.5.5.4 *Enrichment through evaporation*

Different authors have reported that a solution of fluorescent dyes can be concentrated by evaporation without significant losses (Fabiol and Pouzancre, 1981; Richter, 1944; Shontton and Habeeb, 1955). This method for volume reduction offers the possibility of increasing the tracer concentration in water by a factor between 10 and 20, allowing the measurement of water samples with concentrations below the detection limit (Käss, 1976). The formation of a precipitate or colloidal turbidity may occur during the evaporation process. If this is the case, the evaporated sample has to be filtrated. This concentration procedure can be important at the first stage of the interconnection experiment, to facilitate the detection of the first arrival of the tracer.

7.6 ARTIFICIAL RADIOACTIVE TRACERS

7.6.1 Generalities

The use of radioactive tracers for the investigation of dam leakage can be necessary in some instances. Due to the difficulties involved with their use, which are mainly derived from the restrictions imposed by legal regulations in many countries, the use of these tracers is restricted to cases when the nonradioactive tracers described in previous sections cannot provide the appropriate response to the investigated problem. The most interesting radioactive tracers are bromine-82, iodine-131, gold-198 and tritium.

Advantages of radioactive tracers in comparison with nonradioactive tracers are the following:

1) The operational life span of the tracer is limited, due to its spontaneous decay. After a period of time equal to five or six times its half-life, its activity becomes negligible.
2) These tracers may be measured in place, except in the case of tritium. Many hydrological processes can be studied only with tracers that fulfil this condition.
3) These tracers have very high measuring sensitivity, which permits the labelling of huge water volumes using small quantities of tracer solutions.
4) In the case of artificial tritium, this radioisotope offers the only existing possibility to carry out an isotopic labelling of water, that is to say, with a tracer incorporated into the water molecule (Florkowski and Wojtas, 1979; Payne, 1970).

Disadvantages of these tracers are:

1) Official authorisation for the acquisition and use of the radioactive material is required in most countries.
2) In the case of short-lived radioisotopes, they have to be acquired and used at a specific time.
3) In some cases, heavy shielding lead containers have to be use for their transport, such as for bromine-82.
4) In some cases, the use of radioactive tracers may result in some kind of risk for the population and for the field team involved in their use.

In the next sections, the characteristics and most important uses of the four above-mentioned radioactive tracers of water are described.

7.6.2 Bromine-82

Bromine-82 is used in the form of ion Br^-, usually as NH_4Br. This radionuclide decays with a half-life of 36 hours by emitting beta particles and several types of gamma rays, with energies ranging from 0.55 to 0.78 MeV.[1] Due to these high values of the gamma ray energy, heavy lead shielding containers are needed for its transport (between 300 and 500 kg in weight), and special handling tools are required when the tracer activity is higher than about 100 mCi.[2] As mentioned in Section 7.3.1, the Br^- ion is a good tracer for hydrological applications.

Bromine-82 is usually measured using a scintillation detector with a crystal of NaI (Tl) submerged in the water body (see Fig. 7.2). The lower detection limit under these conditions is equal to approximately 0.04 $\mu Ci/m^3$. Therefore, 1 Ci of bromine-82 allows the labelling of a water volume of 25×10^6 m^3. Activities of various curies can be used.

Bromine-82 is usually used for short duration tracing experiments, when high water volumes are involved. For dam leakage studies, this is the ideal tracer for the investigation of deep flows inside the reservoir using the technique described in Chapter 10. Its short half-life is an important advantage in this case, because the tracer concentration attains negligible values after four or five days.

Due to its short half-life, this radioisotope can be used only in countries where a nuclear reactor for the irradiation of NH_4Br is available.

7.6.3 Iodine-131

Iodine-131 is use as a tracer in the form of ion I^- (NaI or KI solution). This radionuclide decays by the emission of beta particles with a half-life of 8.05 days.

[1] The mega-electron-volt (MeV) is equivalent to 10^6 electron-volt (eV). The electron-volt is defined as the energy acquired by an electron when it is accelerated with a potential of 1 volt under vacuum. 1 eV is equivalent to 1.6×10^{-12} ergs.

[2] One milli-curie (mCi) is equivalent to 10^{-3} curies. One curie is equivalent to 3.7×10^{10} becquerels or disintegration per second.

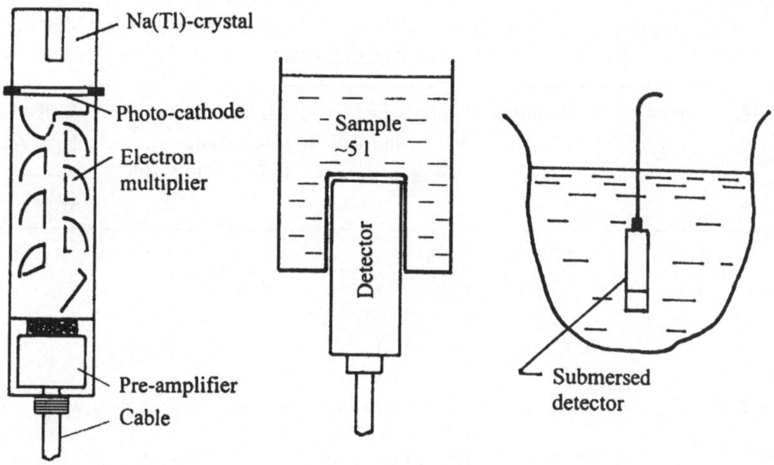

Fig. 7.2. Schematic view of the components of a scintillation detector and two measuring arrangements.

The characteristics of this radionuclide are shown in Table 7.2. It is a gamma ray emitter of medium energy that can be safely transported with lead containers weighing between 5 and 10 kg. This radioactive material is easily available in most of countries due to its wide use for medical purposes.

Iodine-131 is probably the radioactive tracer more frequently used for hydrological purposes. Its half-life makes it appropriate for experiments of short and medium duration (up to 1 month). Typical applications in the field of dam leakage are interconnection experiments, identification of deep flows at the bottom of the reservoir and flow tests in boreholes (see Chapter 8).

The lower detection limit for direct measurement using a scintillation detector with NaI (Tl) crystal submerged in the reservoir is about 0.2 mCi/m^3. Therefore, 1 Ci of this tracer can label a water volume of about 5×10^6 m^3. Several curies can be used with no or little difficulties.

The senior author has frequently used a high sensitivity measuring technique based on the coprecipitation of ^{131}I in the form of AgI. The technique is very convenient in the case of interconnection experiments when the dilution process of the tracer is unknown. Water samples with a volume between 10 and 25 l are periodically collected at the point where the tracer appearance is suspected (spring or borehole), according to a previously planned sampling program. The measuring procedure is illustrated in Fig. 7.3. It consists of the following steps:

1) About 100 mg of nonradioactive I^- is added to the water samples, followed by some drops of a 1:1 NO_3H solution.
2) After shaking, a solution containing the Ag^+ ion is added in excess to precipitate both the nonradioactive I^- ion and the $^{131}I^-$ (coprecipitation).

Table 7.2. Characteristics of the radionuclides more commonly used for water tracing. All radionuclides shown decay by emission of a beta particle.

Radionuclide	Chemical form	Half-life	Gamma-ray energy in MeV (Yield in %)	Air kerma-rate (Γ constant)[1] $\mu Gy \cdot m^2 \cdot h^{-1} \cdot GBq^{-1}$	Half-value layer (cm)[2]	
					Water	Lead
Tritium	HTO	12.32 y	None			
Iodine-131	I^-	8.05 d	0.08 (2%) 0.36 (82%) 0.28 (5%) 0.64 (9%)	66.8	6.5	0.25
Bromine-82	Br^-	35.7 h	0.55 (71%) 0.62 (43%) 0.70 (28%) 0.78 (83%) 1.04 (27%) 1.32 (27%) 1.47 (17%)	432.4	ca. 9	ca. 0.85
Gold-198	$AuClO_4H$	64.8 h	0.41 (99%) 0.68 (1%)	67.6	6.1	0.31

[1] The kerma is defined as the sum of the initial kinetic energies of all the charged ionising particles liberated by uncharged ionising particles (such as gamma rays) in a given material. The unit of kerma is the Gray (1 Gy = 1 joule $\cdot$ kg^{-1}). GBq means giga-becquerels (10^9 Bq).
[2] Thickness of water or lead that reduces the kerma rate to the half.

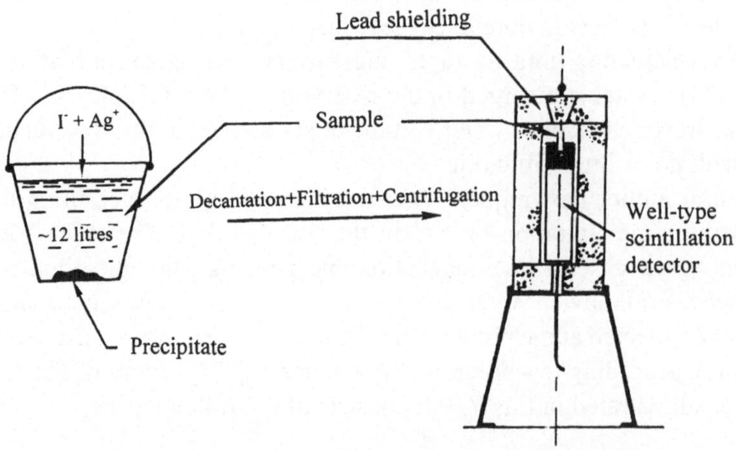

Fig. 7.3. Illustration of the analytical technique for the determination of very low ^{131}I concentrations in water solutions, based on its coprecipitation with AgI and the measurement of the precipitate using a well-type scintillation detector.

3) After a shaking again for two or three minutes, the sample is left to permit the formation of a precipitate. When the precipitate is completely decanted, the supernatant is taken away by means of a plastic tube (siphoning). The precipitate is filtered and put in a test tube of appropriate diameter using a spatula. For the pursued objective, it is not necessary to recover the total amount of the precipitate. Then the precipitate is centrifuged to place it in a small volume at the bottom of the test tube.
4) Finally, the precipitate is measured using a well-type scintillation detector. The test tube is placed in the hole of the NaI (Tl) crystal. Under these conditions, the counting efficiency is very high due to geometrical reasons and to the lack of the absorption effect of water.

For the measurement, the window of the single-channel analyser is adjusted to the most appropriate width for the measurement of the gamma rays, 0.36 MeV energy. This adjustment drastically reduces the counting background. This low background, together with the high counting efficiency, makes possible the detection of ^{131}I concentrations as low as 0.005 $\mu Ci/m^3$. This means that 1 Ci of the tracer can be used to label 200 Hm^3 of water.

If several samples are simultaneously processed, the required time for the analysis can be reduced to less than 5 minutes per sample. With the described procedure, the tracer recovery is not quantitative. However, this fact is not usually of great concern for the objectives pursued in the interconnection experiment. The precipitation technique is only used at the first stage of the experiment to minimise the risk of missing the tracer arrival at the observation point. Once the tracer has reached the sampling point, this measuring technique is replaced by the direct measurement technique, assuming the tracer concentration allows it.

7.6.4 Gold-198

Gold-198 is used in the form of chloro-auric acid ($HAuClO_4 \cdot 4H_2O$). This is obtained by dissolving gold irradiated in the nuclear reactor with aqua regia formed by three parts of HCl and one part of HNO_3. The characteristics of this radionuclide are shown in Table 7.2. Like iodine-131, it is a medium energy gamma emitter. Lead containers weighing between 5 and 10 kg can be used for safe transportation.

This radionuclide is strongly absorbed by solid materials of the earth's crust. Therefore, it is used exclusively as absorbable tracer to locate the strata or points through which the water infiltrates in the ground. This tracer has been used in several cases to identify the exact infiltration points through joints and fractures in concrete dams. Repair works can then be conducted by specialised teams in small areas without emptying the reservoir.

7.6.5 Tritium

Tritium for water tracing is used in the form of tritiated water (THO). It is the only radioactive tracer that exists for the isotopic labelling of high water volumes with the tracer incorporated into the water molecule. All other tracers are dissolved substances or transported tracers. For this reason, tritium offers the highest warranty for a good tracing of the water flow.

As mentioned in Section 6.3.1, tritium is a pure beta particle emitter. The maximum energy of the emitted beta particles reaches only 18 KeV. Due to this low energy and to the lack of gamma rays, tritium content in water cannot be measured in situ. Water samples have to be collected in the field and measured in the laboratory. For this reason, tritium is not appropriate for experiments that require in situ measurement, such as direct measurement of water flows inside the reservoir or boreholes.

Tritium is measured by the liquid scintillation counting technique. Usually, 10 ml of water are mixed with 12 ml of an appropriate scintillation cocktail, such as Picofluor LLT, Instagel, Ultima Gold LLT, etc. The detection limit for direct measurement of the water sample using a modern low-background scintillation spectrometer is about 20 tritium units (T.U.), that is, about 0.064 μCi/m^3. Therefore, 1 Ci of tritiated water can be used to label a water volume of about 16×10^6 m^3. Using the enrichment technique described in Section 6.3.1 (IAEA, 1981), tritium concentrations around 1 T.U. can be measured with a standard deviation of 0.12 T.U. In this case, 1 Ci of tritium would allow the labelling of 320×10^6 m^3 of water. However, the tracing possibilities of tritium are limited by the presence of natural tritium in most surface and shallow groundwaters. Tritium content in recent surface water will usually range from 1.5 to almost 15 T.U., depending on the geographical characteristics of the study area. This concentration has to be taken into account for the planning of interconnection experiments. Concentrations about 50% higher than the natural ones detected at the sampling point can be sufficient to identify the presence of the injected tracer, assuming that such values are found in many water samples or that increasing concentration values are observed with time.

On the other hand, the half-life of tritium (12.32 years) is much higher than the observation time required by most of the hydraulic processes where the use of this tracer may be necessary.

7.6.6 Comments on radiological risk

The use of radioactive tracers for hydrological purposes may involve, in some cases, a potential hazard and risks for the public health. The basic rules of radiological safety recommended by the IAEA in 1982 (IAEA, 1982) are based on the recommendations made by the International Commission of Radiological Protection (ICRP, 1977). These rules describe a dose limitation system that includes the following elements:

1) Justification: to avoid unnecessary exposure, no experiment should be authorised by the competent authority unless evidence exists that the benefits resulting from the application are sufficient to justify the eventual risks and detriments that might result from the exposure to ionising radiation.

2) Optimisation: the planning, design and subsequent execution of the experiment shall be done in such a way that exposures are as low as reasonably achievable, economic and social factors being taken into account.
3) Limitation of individual risk: this is carried out by controlling the radiation doses in a group of individuals most likely to receive the highest doses from the practice. No individual will be unnecessarily exposed above the prescribed limits as a result of controlled experiments. The prescribed limits are the sum of the resulting equivalent dose for external radiation during a one-year period and the equivalent dose for the whole life derived from the ingestion of radioactive nuclide during the same period of time.

The dose limitation system establishes a difference between radiation workers and members of the public (ICRP, 1992). Concerning the radiation workers, the 1990 recommendations of the ICRP stated the following:

"For occupational exposures, the effective dose is limited to 0.1 Sievert (Sv)[3] in a five year period (average annual dose of 0.02 Sv) with a limit of 0.05 Sv in any single year and includes annual limits on the equivalent dose to skin and hands of 0.5 Sv and to the lens of the eye of 0.15 Sv. The latter are important in relation to external radiation of the body. In the case of internal exposure, the annual limit of intake is based on a committed effective dose of 0.02 Sv. Thus, the annual limit of intake (ALI) for any radionuclide is obtained by dividing the annual average effective dose limit (0.02 Sv) by the committed effective dose, E(50), resulting from the intake of 1 Bq of that radionuclide."

For members of the public, the previous values have to be reduced by a factor of 10. These reduced values are to be applied to the critical group of the population. A comprehensive description of the dose limitation system can be found in the cited reference (IAEA, 1982).

The ALI values (including ingestion and inhalation) indicated by the ICRP (1992) for the four isotopes recommended by the authors as tracers for hydrological purposes are shown in Table 7.3. On the other hand, water consumption for the so-called standard man is around 2.2 l per day, according to the statement of the ICRP. This is equivalent to 803 l per year. This value allows the calculation of the maximum concentration permitted in the water for permanent human consumption. The calculated values are showed in the third row of Table 7.3.

For risk evaluation and eventually licensing of particular uses, the definition of the term *trivial concentration* of radionuclides in water can be very helpful. Trivial

[3] The Sievert is the unit of equivalent dose and corresponds to an absorbed energy of 1 joule per kilogram. This unit replaced the old unit called rem (röntgen equivalent man), which is equivalent to 1 mSv (milli-Sievert). The röntgen is defined as the gamma radiation dose that produces in 1 kg of dry air under normal pressure and temperature conditions a number of ion pairs such that the associated electrical charge amounts to 2.58×10^{-4} coulombs. It is equivalent to about 2×10^9 ion pairs per cm^3 of air. For water and soft tissue, the röntgen is equivalent to an absorbed dose of about 100 ergs per gram.

Table 7.3. ALI values and derived maximum concentrations for the isotopes recommended as tracers for hydrological purposes.

Radionuclide	Annual limit of intake (MBq/year)	Derived maximum concentration (MBq/m^3)	Derived trivial concentration (kBq/m^3)	Minimum detectable concentration (kBq/m^3)
Tritium	100	125	1496	18.5[1] 1.1[2]
Iodine-131	0.08	0.10	1.15	14.8[3] 0.37[4]
Bromine-82	4	5	60	7.4[3] 0.18[4]
Gold-198	1	1.24	14.9	12.9[3]

[1] Direct measurement using a liquid scintillation detector.
[2] Liquid scintillation counting after electrolytic enrichment.
[3] Direct measurement using a scintillation detector with a 2 × 2 inches NaI(Tl) crystal submersed in the water with saturation geometry.
[4] Measurement with a NaI(Tl) well-type scintillation detector after precipitation from 10 litres water using the technique described in Section 7.6.3.
MBq and kBq mean, respectively, mega-becquerels and kilo-becquerels.

concentrations can be derived from the concept of trivial exposure to ionising radiation that has been suggested by different authors (IAEA, 1988; Menhold, 1984) and defined as the level of dose that is small in comparison to the natural background radiation (about 2 mSv per year). An effective dose equivalent in the order of one to a few percent of natural background (i.e., 20 to 100 μSv per year) has been proposed. The mean value of this range is 60 μSv per year, which corresponds to 1.2% of the limit for annual effective dose equivalent for members of the public (5 mSv). According to the IAEA recommendation (IAEA, 1982), the same proportionality is valid for the ALI values. Therefore, about 1.2% of these values could be considered as trivial. The values obtained for the isotopes used as tracers for water are shown in the fourth row of Table 7.3.

The last column of Table 7.3 shows the minimum detectable concentration of the four radionuclides for specific counting conditions. It can be observed that in all cases the detection limits are lower than the derived maximum concentration and, in some cases, lower even than the trivial concentration. This means that the use of these radionuclides for water tracing can be carried out with negligible risk and hazard for the population if an appropriate dilution factor of the tracer is achieved.

Particularly favourable is the case of tritium. An appropriate concentration range of this radionuclide for water tracing experiments using the direct measurement technique is between 20 and 2000 T.U., which corresponds, respectively, to 2.36 and 236 kBq/m^3. This range is far below the trivial concentration indicated in Table 7.3 (1496 kBq/m^3).

The exhaustive description of the potential risks that may be derived from the use of radioactive tracers for hydrological purposes would be very cumbersome due to the different techniques and scenarios involved in their use. In the opinion of the authors, the potential risk and hazard is very low in most cases. This is mainly due to the fact that these radionuclides can be detected at very low concentrations, as mentioned before. The risks can be kept negligible through appropriate planning of the experiment to be carried out. This statement is based on the following reasons:

1) In most cases, the hydraulic behaviour of the tracer to be injected in the water body can be estimated from the approximate values of the involved hydraulic parameters. Even if this estimation cannot be done, it will always be possible to establish half-quantitative hypotheses of the behaviour of the tracer for the most unfavourable conditions prevailing during the experiment.
2) The experiments carried out with tracers have in general a transitory character. In other words, they always have a short duration and only very seldom are repeated in the same place. In most cases, the experiment finishes when the tracer cannot be detected due to its high dilution and/or its radioactive decay. In the case of experiments carried out in surface water flows, this happens after a few hours or, in less frequent cases, days. Therefore, in this case, the exposure of the population to radiation can take place only for a short period of time, therefore having a low contribution to the annual dose.
3) Most of the studies have a local character, that is, the experiments are carried out in small areas (several square kilometres at the most), so that an adequate control of the use of the labelled water body can be made.
4) The experiments are usually carried out in water bodies that are not used for human consumption. Therefore, any human consumption is sporadic and fortuitous. In some cases, the experiments are conducted on nonpotable or polluted waters, for which the risk derived from the dissolved pollutants is much higher than the one corresponding to the radioactive tracer.

REFERENCES

Alexander, E., Davis, M. and Dalgleish, J., 1988. Dye tracing through thick unsaturated zones. *Proc. 5th Int. Symp. Underground Water Tracing, Athens IGME*, 181–188.

Ando, M. and Sayato, Y., 1983. Microdetermination of bromide in water by gas chromatography. *Water Research*, 17(12), 1823–1827.

Anguita, F., 1969. La fluoresceina en el trazado de aguas subterráneas (I y II). *Cimbra, Revista de Ingeniería Técnica de Obras Públicas*, Madrid, nº 51–52, 23–26 and 5–10 (in Spanish).

Bauer, F., 1967. Erfahrungen beim Uraninnachweis mit Aktivhohle. *Steir. Beitr. Hydrogeol.*, 169–178 (in German).

Bauer, F., 1972. Weitere Erfahrungen beim Uraninnachweis mit Aktivkohle. *Geol. Jahresbericht* C2, 19–27, Hannover (in German).

Brandecker, H., Maurin, V. and Zötl, J., 1965. Hydrogeologische Untersuchungen and baugeologische Erfahrungen beim Bau des Biessbach-Speichers (Steinernes Meer). *Steir. Beitr. Hydrogeol.*, 67–111 (in German).

Behrens, H., 1982. Verfahren zum qualitativen und quantitativen Nachweis von nebeneinander vorliegenden Fluoreszenztracern. *Beitr. Geol. Schweiz–Hydrol.*, 28I, 39–50 (in German).

Behrens, H. and Demuth, N., 1990. Lichtmessung in Gewässern über die Photolyse von Fluoreszentztracer. *Jahresber. 1989, Inst. Hydrol., GSF-HY* 1/90: 216–233 (in German).

Behrens, H. and Teichman, G., 1982. Neue Ergebnisse über den Lichteinfluss auf Fluoreszenzstracer. *Beitr. Geol. Schweiz–Hydrol.*, 28(I), 64–77 (in German).

Benischke, R., 1987. *Fluorescent tracers in hydrology*. Post-Graduate training course on groundwater tracing techniques, Institute for Geothermics and Hydrogeology, Research Centre Joanneum, Graz, Austria.

Bowman, R.S. and Gibbons, J.F., 1992. Difluorobenzoates as non-reactive tracers in soil and ground water. *Ground Water*, 30, 8–14.

Bronswijk, J.J., Hamminga, W. and Oostindie, K., 1995. Field-scale solute transport in a heavy clay soil. *Water Resour. Res.*, 31(3), 517–526.

Darland, G. and Brock, T.D., 1971. *Bacillus acidocaldarius* sp. nov., an acidophilic thermophilic spore-forming bacterium. *J. Gen. Microbiol.*, 67, 9–15.

Davis, S.N., Campbell, D.J., Bentley, H.W. and Flynn, T.J., 1985. *Ground Water Tracers*. Worthington, OH, USA: Natl. Water Well Ass., 200 pp.

Dechant, M., 1967. Die Färbung der Lycopodiumsporen. *Steir. Beitr. Hydrogeol.*, 241–247 (in German).

Dechant, M. and Hacker, P., 1988. Neue Entwicklung in der Methode des Sporennachweises. *Proc. 5th Int. Symp. Underground Water Tracing, Athens, IGME*, 149–155 (in German).

Dole, R.B., 1906. Use of fluorescein in the study of underground waters. *U.S. Geological Survey Water Supply and Irrigation. Paper 160, 73–85.*

Dunn, B. and Vaupel, D.E., 1965. Effects of sample and fluorometer-compartment temperatures on fluorometer readings. *Geol. Surv. Res., Prof. Pap.* 525-D, 225–227.

Dvgw, 1983. *Forum Mikroorganismen und Viren in Grundwasserleitern, München 1983.* DVGW-Schriftenreihe Wasser 35, 169 pp. (in German).

Everts, C.J., Kanwar, R.S., Alexander, Jr., E.C. and Alexander, S.C., 1989. Comparison of tracer mobilities under laboratory and field conditions. *J. Environ. Qual.*, 18, 491–498.

Fabiol, R. and Pouzancre, H., 1981. *Evolution de la fluorescéine en milieu thermal jusqu'a 200°C.* Bull. BRGM (2) III, 3 1980/81, Orléans, 249 pp. (in French).

Feuerstein, D.L. and Selleck, R.E., 1963. Fluorescent tracers for dispersion measurements. *J. Sanitary Eng. Div. Trans. Am. Soc. Civil Engineers*, SA4, 89, 1–21.

Florkowski, T. and Wojtas, J., 1979. Model Experiment for tritium balance of surface water reservoir including molecular exchange. *Nukleonika*, 24, 1247–1252.

Foster, T.H., 1951. *Fluoreszenz organischer Verbindungen.* Vandenhoek & Ruprecht, Göttingen, 312 pp. (in German).

Gardner, G.D. and Gray, R.E., 1976. Tracing subsurface flow in karst regions using artificially coloured spores. *Assoc. Engr. Geologists Bull.*, 13, 177–197.

Gioda, A., Le Barbe, L. and Delfieu, G., 1989. *Traçage à la rhodamine B.* Note technique, Laboratoire d'Hydrologie Dèpartment Eaux Continentales, ORSTOM, Montpellier, France (in French).

Heumann, K.G. and Seewald, H., 1985. Gleichzeitige Bestimmung von Cl^-, Br^- und I^- in Wässern durch massenspektrometrische Isotopenverdünnings analyse. *Frezenius Z. anal. Chem.*, 320, 493–497 (in German).

Ilgenfritz, E.M., Blanchard, F.A., Messelink, R.L. and Panigrahi, B.K., 1988. Mobility and effects in liner clay of fluorobenzene tracer and leachate. *Ground Water*, 26(1), 22–30.

International Atomic Energy Agency, 1981. *Low-level tritium measurement*, IAEA, Vienna, TECDOC-246, 205 pp.

International Atomic Energy Agency, 1982. Basic safety standards for radiation protection, 1982 edition, *Safety Series No. 9*, IAEA, Vienna.

International Atomic Energy Agency, 1988. Principles for the exemption of radiation sources and practices from regulatory control. *Safety Series No. 89*, IAEA, Vienna.

International commission on radiological protection (ICRP), 1977. *Recommendations of the ICRP. Publication n° 26*. Pergamon Press, Oxford, New York.

International commission on radiological protection (ICRP), 1992. *Annual limits of intake of radionuclides by workers based on the 1990 Recommendations. Publication n° 61.* Pergamon Press, Oxford, New York.

Jabro, J.D., Lotse, E.G., Fritton, D.D. and Baker, D.E., 1994. Estimation of preferential movement of bromide tracer under field conditions. *J. Hydrol.*, 156, 61–71.

Käss, W., 1964. Die unmittelbare Bestimung von Uranin-Spuren bei Färbversuchen. *Steir. Beitr. Hydrogeol., N.F.*, 15–16, 37–65 (in German).

Käss, W., 1965. Erfahrungen bei Färbversuche mit Uranin. *Steir. Beitr. Hydrogeol., N.F.* 17, 21–65 (in German).

Käss, W., 1976. 100 Jahre Uranin! *Pap. 3rd Int. Symp. Underground Water Tracing (3. SUWT) 1*, 113–122.

Käss, W., 1982. Fluoreszierende Sporen als Markierungsmittel. *Beitr. Geol. Schweiz–Hydrol.*, 29(I), 131–134 (in German).

Käss, W. (ed.), 1998. *Tracing technique in geohydrology*. A.A. Balkema, Rotterdam/ Brookfield.

Käss, W. and Reichert, B., 1988. Tracing of karst water with fluorescent spores. *Proc. 5th Int. Symp. Underground Water Tracing, Athens, IGME, 157–165.*

Käss, W., Ritter, R. and Sacré, C., 1983. Lebensdauer und Transport von Bakterien in typischen Grundwasserleitern — Oberreinische Schotterebene. *DVGW-Schriftenreihe Wasser*, 35, 127–138 (in German).

Laane, R.W.P.M., Manuels, M.W. and Staal, W., 1984. A procedure for enriching and cleaning up rhodamine B and rhodamine WT in natural waters, using a Sep-pack C18 cartridge. *Water Res.*, 18, 163–165.

Leap, D., 1992. Apparent relative retardation of tritium and bromide in dolomite. *Ground Water*, 30, 549–560.

Lukas, G., 1959. Die Triftung von Bakterien. Karsthydrologische Untersuchungen in Toten Gebirge mit besonderer Berücksichtigung der versorgungswasserwirtschaftlischen Belange in Tauplitzgebiet. *Österr. Wasserwirtschaft*, 16, 112–123 (in German).

Marichal, M. and Benoît, R., 1961. Dosage de la rhodamine B dans les eaux naturelles. *Chim. Anal.*, (2), 70–72 (in French).

Mattson, W., 1929. The laws of soil colloidal behaviour I. *Soil Science*, 27–28, 71–87.

Maurin, V. and Zötl, J., 1964. Karsthydrologische Untersuchungen im Totem Gebirge. *Österr. Wasserwirtsch.*, 16, 112–123 (in German).

Menhold, C.B., 1984. *Criteria for a De Minimis Level.* US Health Physics Society, New Orleans, LA., U.S.A.

Molinari, J., 1969. *Les traceurs salins et fluorescents en hidrologie: aspects techniques et économiques de leur utilisation.* Centre d'Etudes Nucleaires de Grénoble. Commissariat a l' Energie Atomique, 36 pp. (in French).

Molz, F.J., Melville, J.G., Given, O., Crocker, R.D. and Matteson, K.T., 1985. Design and performance of single-well tracer tests at the mobile site. *Water Resour. Res.*, 21, 1497–1502.

Moser, H. and Sagl, H., 1967. Die Direktmessung hydrologischer Farbtracer im Gelände. *Steir. Beitr. Hydrogeol.*, 179–183 (in German).

Pang, L., Close, M. and Nooman, M., 1998. Rhodamine WT and *Bacillus subtilis* transport through an alluvial gravel aquifer. *Ground Water*, 36(1), 138–144.

Payne, B.R., 1970. Water balance of Lake Chala and its relation to groundwater form tritium and stable isotope data. *J. Hydrol.*, 11, 47–58.

Perlega, W., 1976. Der Nachweis von Fluoreszenfarbstoffen mittels Aktivkohle. *Pap. 3rd Int. Symp. Groundwater Tracing (3. SUWT)*, 195–201 (in German).

Pfuhl, E., 1897. Ueber die Verschleppung von Bakterien durch das Grundwasser. *Z. Hyg. Infektionskr.*, 25, 549–554 (in German).

Porro, I., Wierenga, P.J. and Hills, R.G., 1993. Solute transport through large uniform and layered soil columns. *Water Resour. Res.*, 29, 1321–1330.

Pschyrembel, W., 1964. *Klinisches Wörterbuch*, pp. 154–158. W. De Gruyter und Co, Berlin, 980 pp. (in German).

Ptak, T. and Schmid, G., 1996. Dual-tracer transport experiments in a physically and chemically heterogeneous porous aquifer: Effective transport parameters and spatial variability. *J. Hydrol.*, 183, 117–138.

Richter, G., 1944. Zur Kenzeichnung unterirdisch fliessender Wasser—Untersuchungen an der Range und anderen Karstquellen bei Warstein (Westfalen). *Abh. Reichsamt Bodenforsch., N.F.* 299–336 (in German).

Rodda, J.C., 1977. *Institute of Hydrology Research Report 1974–6*. Institute of Hydrology (Wallingford, England), pp. 54–57.

Sabatini, D.A. and Al Austin, T., 1991. Characteristics of rhodamine WT and fluorescein as adsorbing ground-water tracers. *Ground Water*, 29, 341–349.

Sattar, S.A., Synek, E.J., Westwood, J.E.N. and Neals, P., 1972. Hazard inherent in microbial tracers; reduction of risk by the use of *Bacillus stearothermophilus* spores in aerobiology. *Appl. Microbiol.*, 23, 1053–1059.

Schröter, J., 1987. In G. Matthess, M. Isenbeck and J. Schröter (Eds.), *Reinigungswirkung des Untergrundes gegenüber Schadstoff-Gruppen in Trinkwasserschutzgebiete — Teil B: Felduntersuchungen*. Geol. - Paläontol. Inst. U. Mus. D. - Christian-Albrecht Univ., Kiel, 72+9 pp.

Seaman, J.C., Bertsch, P.M. and Miller, W.P., 1995. Ionic tracer movement through highly weathered sediments. *J. Contam. Hydrol.*, 20, 127–143.

Shontton, E. and Habeeb, A.S.F.A., 1955. The stimulation of fluorescein in dilute solutions. *J. Pharm. Pharmacol.*, 7, 456–462.

Smart, P.L., 1982. A review of the toxicity of 12 fluorescent dyes used for water tracing. *Beitr. Geol. Schweiz–Hydrol.*, 28, 101–112.

Smart, P.L., 1984. A review of the toxicity of 12 fluorescent dyes used for water tracing. *Nat. Speleol. Soc. Bull.*, 46, 21–33.

Smart, P.L. and Laidlaw, M.C., 1977. An evaluation of some fluorescent dyes for water tracing. *Water Resour. Res.*, 13, 15–33.

Smart, P.L. and Smith, D.I., 1976. Water tracing in tropical regions, the use of fluorometric techniques in Jamaica. *J. Hydrol.*, 30, 179–195.

Thompson, G.M., Hayes, J.M. and Davis, S.N., 1974. Fluorocarbon tracers in Hydrology. *Geophys. Res. Lett.*, 1, 177–180.

Uba, 1985. *Lebensdauer von Bakterien un Viren in Grundwasserleitern—Zusammengefasster Abschlussber*. Herausg. vom Unweltbundesamt, Materialien 2/85, Berlin, 108 pp. (in German).

Wagner, O.R., 1977. The use of tracers in diagnosing interwell reservoir heterogeneities. *J. Petrol. Technol.*, 1410–1416.

Watson, A.J., Liddicoat, M.I. and Ledwell, J.R., 1987. Perfluordecalin and sulfur hexafluoride as purposeful marine tracer: some deployment and analysis techniques. *Deep-Sea Res.*, 34, 19–31.

Wilson, J.F., Jr., 1968. *Fluorimetric procedures for dye tracing*. Techniques of Water-Resources Investigations US Geol. Surv., 3, Washington DC, 31 pp.

Wilson, R.D. and Mackay, D.M., 1993. The use of sulfur hexafluoride as a conservative tracer in saturated sandy media. *Ground Water*, 31, 719–724.

Wilson, R.D. and Mackay, D.M., 1996. SF_6 as a conservative tracer in saturated media with high intragranular porosity or high organic carbon content. *Ground Water*, 34, 241–249.

Wood, W.W. and Ehrlich, G.G., 1978. Use of baker's yeast to trace microbial movement in ground water. *Ground Water*, 16, 398–403.

Zojer, H. and Zötl, J., 1975. Hydrogeologische Untersuchungen im Bereich Grossrutschung des Gradenbachtales bei Döllach/Kärnten. *Steir. Beitr. Hydrogeol.*, 27, 65–84 (in German).

Zötl, J., 1957. Hydrologische Untersuchung is östlichen Dachsteingebiet. *Beitr. alpinen Karsforsch.*, 9, 187–205 (in German).

CHAPTER 8

Localisation and measurement of flows in boreholes

8.1 INTRODUCTION

The identification and measurement of flow patterns in boreholes using artificial tracers is an essential tool in investigating the nature of leaks from reservoirs and lakes. The repair works carried out to reduce or stop such leaks generally consist of reducing the permeability of the geological materials through which groundwater flows. To successfully and cost effectively achieve such a decrease in permeability, it is necessary to identify and determine the exact groundwater flow path that connects the reservoir with the downstream springs. The most direct way of achieving this is by conducting tracer experiments. The techniques, which allow the identification, localisation and measurement of the magnitude of the flows intercepted by boreholes, are described in this chapter. The measurement of vertical profiles of temperature and electrical conductivity, as well as the permeability tests described in Chapter 3 and the geophysical well-logging techniques described in Chapter 4, are complementary techniques.

As will be shown below, the techniques used for the localisation and measurement of flows in boreholes are diverse. In all cases, they require that water can flow freely across the borehole side-walls.

Water flow in a given borehole may have either a horizontal or a vertical dominant component. The techniques for the precise measurement of these flows are different and will be demonstrated later. A general finding of leakage studies close to dam sites is the presence of very high hydraulic gradients, derived from the high head existing between the water level in the reservoir and the downstream position of the springs or emerging waters. As a result, very rapid water flows are possible, complicating precise measurements of flow velocity. However, for the investigation of dam leakage, it is generally not necessary to measure water flows with such high precision. In most cases, it is more important to pinpoint the location where such flows occur and to determine the flow path across the geological formations and its continuity, rather than to quantify absolute values of the flow velocity.

Water flowing through boreholes hydraulically connected to the reservoir will obviously show strong variations with the water level fluctuation in the reservoir.

For this reason, it is very important to repeat the flow tests for different water levels in the reservoir. In many cases, flows will occur in some stretches of a borehole that are activated only when the water level in the reservoir has reached a particular level.

8.2 APPROPRIATE TRACERS

The tracer techniques described in this chapter can be carried out with the following materials:

1) *A saturated solution of table salt*. In this case, the tracer is detected with a conductivity meter equipped with a cable of sufficient length. For obvious reasons, this tracer is not appropriate where high salinity waters are present (for example, electrical conductivity higher than about 5 mS/cm). Such waters do not very frequently occur close to reservoirs. If they do, performing flow tests in boreholes is not very relevant because this groundwater is likely to be stagnant, providing only a very small contribution to the spring flow. This is often the case, even in instances where the phreatic or piezometric level in the boreholes fluctuates in response to that of the reservoir.
2) *The radionuclide iodine-131*. In this case, the tracer detection is carried out by using measuring probes fitted with a Geiger-Müller or a scintillation detector with a NaI(Tl) crystal. The authors have in the past used this tracer routinely, but, at present, its application is limited to very deep (more than 150 m) boreholes or boreholes tapping groundwater with high salinity values. In this case, a natural gamma probe of a commercially available well-logging unit is used to measure and record gamma ray profiles with good precision.
3) Some authors have suggested the use of the dye tracer known as *nigrosin*. This tracer has a black colour, and its concentration can be measured in situ using an infrared turbidity meter. These available measuring systems show a logarithmic response and allow the measurement of concentrations up to 200 mg/l. Clearly, the presence of suspended sediments, which mainly occurs in instances of highly turbulent flow, interferes with the detection of the dye.
4) Recently, portable fluorometers fitted with a fibre optic-based sensor have been developed (see Chapter 7). These fluorometers could be used to identify and measure vertical profiles of *fluorescent tracers* in boreholes. Therefore, these submersible probes can, in principle, be used for the tracer tests described in this chapter.

It must be pointed out that the tracers indicated in points 2, 3 and 4 above allow the tests to be performed with negligible changes in the mass concentration of water, that is, the mass of total salt injected is minimal. Therefore, no significant change of water density will be produced during the labelling process. This is not the case when table salt is used as tracer. Small density currents can be created with this tracer, as shown below.

8.3 FILTRATION VELOCITY

The flow rate Q through a cross section S of an aquifer with a porosity p is related to the velocity v_t of an ideal tracer in water, as expressed by the following equation:

$$Q = v_t \cdot p \cdot S \tag{8.1}$$

The product $v_t \cdot p$ is called *filtration velocity*, v_f or Darcian velocity. This is defined by Darcy's equation $v_f = K \cdot i$, where K is the permeability (given, for example, in m/day) and i is the hydraulic gradient (dimensionless parameter).

8.4 POINT DILUTION TECHNIQUE FOR THE MEASUREMENT OF HORIZONTAL GROUNDWATER FLOW

8.4.1 Principle of the method

In the point dilution method, the horizontal component of the groundwater flow velocity is obtained from the decrease in the concentration of the tracer injected into the borehole due to the flow perpendicular to its axis, as a function of time.

The principle of the method was first described by Kocherin in 1916. Common salt was used at that time as a tracer. The method was later further developed by a number of researchers (Borowczyk, 1965; Borowczyk et al., 1967; Drost, 1983, 1986; Drost et al., 1968a, b; Guizerix et al., 1963; Moser and Neumaier, 1970).

In principle, the method is based on the following considerations:

Let us suppose that a certain amount of tracer is injected in a defined stretch of the water column of a borehole of length h and diameter d, confined by two watertight packers, producing an initial concentration C_0. Suppose also that the following conditions are fulfilled:

1) The groundwater flow regime at the borehole site is in a steady state.
2) In the labelled cylinder of the borehole of volume $V = \pi d^2 h/4$, a homogeneous distribution of the tracer is permanently observed, that is, the tracer concentration at any given moment in all points within this volume is the same. This implies good and instantaneous mixing of water when injection occurs in the volume V.
3) The tracer escapes from volume V only because of the horizontal flow existing in the aquifer. This implies the absence of a vertical flow in the borehole, as well as the absence of significant losses of tracer due to osmotic diffusion

(i.e., diffusion of the tracer because of the concentration gradient existing between the fluid in the borehole and in the aquifer).

4) The detector used for the measurement of the concentration of the injected tracer should not detect the tracer already escaped from the volume V of the borehole. In other words, the fraction of the tracer already located outside the borehole should not affect the detector response.

Under these conditions, the tracer concentration in the volume V decreases as a function of time in accordance with the following differential equation:

$$\frac{dC}{dt} = -\frac{C}{V}\frac{dV}{dt} = Q\frac{C}{V} \qquad (8.2)$$

The term $dV/dt = Q$ represents the flow rate of the water circulating through the section $S = h \cdot d$, where S is the borehole cross section, h is the height and d is the borehole diameter.

The integration of equation (8.2) gives:

$$C_t = C_0 \, exp\left(\frac{Q}{V}t\right) \quad \text{and therefore:} \quad Q = \frac{V}{t}\, ln\, \frac{C_0}{C_t} \qquad (8.3)$$

where C_0 is the initial tracer concentration in volume V and C_t is the concentration after time t.

Taking into account that the flow rate Q is given by the product $v_s \cdot S$, where v_s is the flow circulating inside the borehole through the section S, and that the volume V is given by the expression $V = \pi d^2 h/4$, the following equation is obtained:

$$V_S = \frac{\pi \cdot d}{4t}\, ln\, \frac{C_0}{C_t} \qquad (8.4)$$

On the other hand, the water velocity inside the borehole v_s, is related to the filtration or Darcian velocity v_f by the equation:

$$v_S = \alpha \cdot v_f \qquad (8.5)$$

where α is a coefficient needed for the correction of the hydrodynamic disturbance of the natural groundwater flow produced by the presence of the borehole.

Due to the high permeability existing inside the borehole, the water velocity in the volume V is higher than the velocity in the aquifer. The coefficient α represents the ratio between the flow rate Q circulating through the inner section of the borehole and the flow rate circulating through the same section of the aquifer Q_a, that is to say:

$$\alpha = \frac{Q_S}{Q_a} \qquad (8.6)$$

Taking into account equation (8.6), and taking for the ratio C_0/C_t a value of 10, the following expression is obtained:

$$v_f = \frac{\pi \cdot d}{4\alpha t_{1/10}} ln\, 10 = \frac{1.81d}{\alpha t_{1/10}} \tag{8.7}$$

where $t_{1/10}$ is the time needed for the tracer concentration in the volume V to be reduced to 1/10 of its initial value, that is, the time needed to fulfil the condition $C_0/C_t = 10$.

Equation (8.7) allows the determination of v_f as a function of $t_{1/10}$ in a practical and simple way. The value of $t_{1/10}$ is determined experimentally by measuring the variation of the tracer concentration as a function of time using an appropriate detector located inside the volume V.

8.4.2 Coefficient of hydrodynamic disturbance

According to equation (8.6), the coefficient of hydrodynamic disturbance α gives the ratio between the flow rates that occur through an equal cross-section of the borehole and the aquifer. The difference between the two flow rates is a consequence of the hydrodynamic disturbance produced by the borehole, which is illustrated by the flow lines shown in Fig. 8.1, for cases where a gravel pack around the borehole is installed or absent.

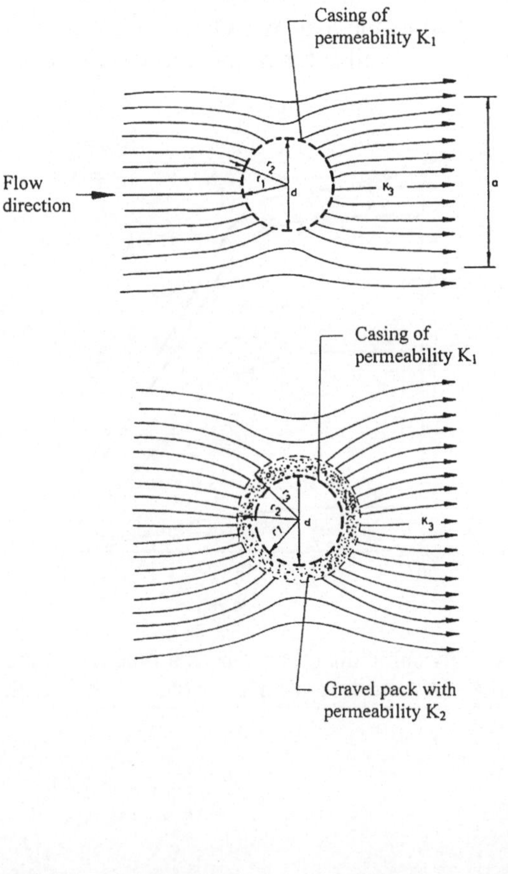

Fig. 8.1. Hydrodynamic disturbance produced by a borehole in an aquifer with parallel flow lines, both without gravel pack and installed gravel pack.

In the first case (absence of a gravel pack), the α coefficient is given by the ratio a/d, where a is defined by the flow lines that are tangent to the borehole casing.

When a gravel pack is present around the casing, the value of α is given by the following equation (Halevy et al., 1967):

$$\alpha = \frac{8}{\left(1+\frac{K_3}{K_2}\right)\left[1+\frac{r_1^2}{r_2^2}+\frac{K_2}{K_1}\left(1-\frac{r_1^2}{r_2^2}\right)\right]+\left(1-\frac{K_3}{K_2}\right)\left[\frac{r_1^2}{r_3^2}+\frac{r_2^2}{r_3^2}-\frac{K_2}{K_1}\left(\frac{r_1^2}{r_3^2}-\frac{r_2^2}{r_3^2}\right)\right]} \tag{8.8}$$

In the absence of the gravel pack, $r_2 = r_3$ and $K_2 = K_3$. The previous equation is simplified to:

$$\alpha = \frac{4}{1+\frac{r_1^2}{r_2^2}+\frac{K_3}{K_1}\left(1-\frac{r_1^2}{r_2^2}\right)} \tag{8.9}$$

In the absence of casing, $r_1 = r_2$ and $\alpha = 2$. Usually, the radii r_1 and r_2 are very similar, and, therefore, the ratio r_1/r_2 is very close to 1. In addition, if the permeability of the casing is much higher than the permeability of the aquifer, that is to say, $K_1 \gg K_3$, the value of α is close to 2. Equations (8.8) and (8.9) are valid only for laminar flow. For turbulent flow, the values of α are somewhat lower than those given by these equations (Halevy et al., 1967).

The variation of α as a function of the ratio K_3/K_2 for different values of the ratio K_2/K_1 given by equation (8.8), as well as the inverse, for specific values of r_1/r_2 and r_2/r_3, are plotted in Fig. 8.2. In a similar form, the variation of α as

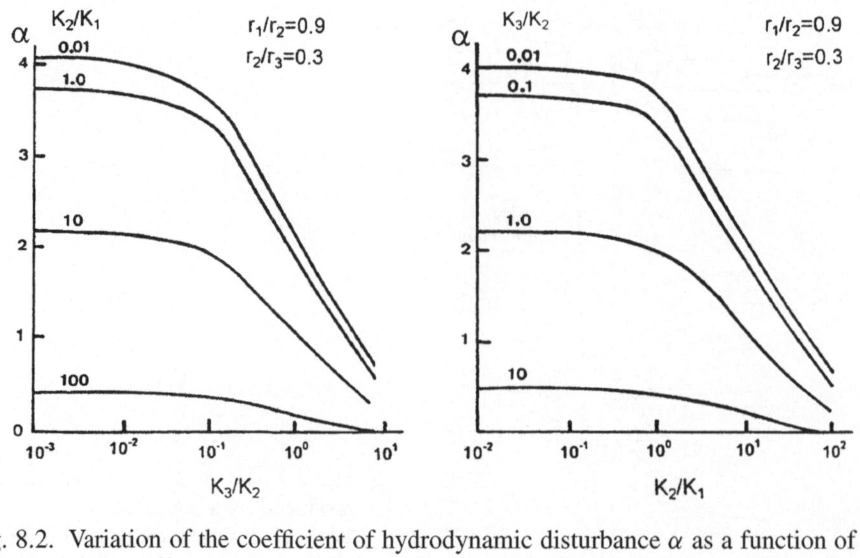

Fig. 8.2. Variation of the coefficient of hydrodynamic disturbance α as a function of the ratio K_3/K_2 for different values of the ratio K_2/K_1 and vice versa. Case of a borehole with a gravel pack (after Halevy et al., 1967).

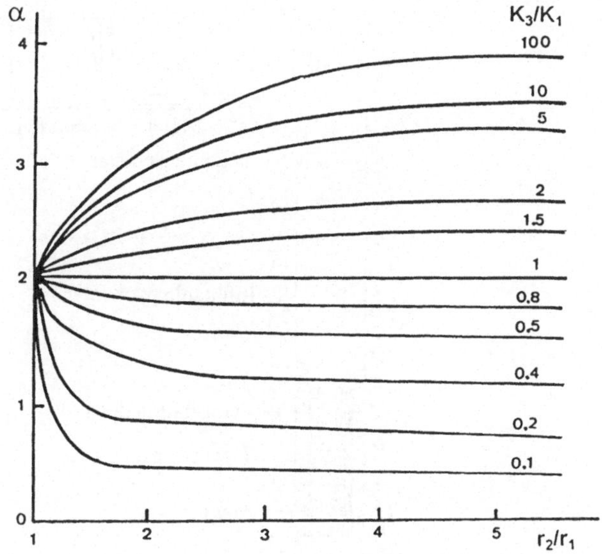

Fig. 8.3. Variation of the coefficient of hydrodynamic disturbance α as a function of the ratio r_2/r_1 for different values of the ratio K_3/K_1. Case of a borehole without a gravel pack (after Halevy et al., 1967).

a function of the ratio r_2/r_1 for different values of the ratio K_3/K_1 is plotted in Fig. 8.3, using equation (8.9).

The α coefficient can be determined in laboratory tests by reproducing the field conditions in an appropriate permeameter. The flow tests are performed, in this case, in such a way that the Darcian velocity v_f is known. Therefore, the coefficient α is computed using equation (8.7) by using the value of time $t_{1/10}$ obtained experimentally. The analysis of the curves presented in Fig. 8.2 demonstrates that to keep small the uncertainty associated with the value of α, it is necessary that the condition $K_1 > K_2 > 10K_3$ be satisfied. In the same way, these curves show that where a gravel pack is absent, the conditions are more favourable when the permeabilities K_1 and K_2 are similar. The value of K_1 can be calculated using the method described by Drost et al. (1968a, b). In practice, it is recommended that casing with a perforated surface area larger than 1% be used (see Annex 4).

8.4.3 Experimental methodology

Different measuring systems have been developed for the determination of the groundwater flow by means of the previously described point dilution method. Some of these systems are commercially available (Drost, 1983, 1986). All these devices are fitted with two packers for the definition of the volume V that will be labelled with the selected tracer. An additional aim of these packers is to eliminate any interference derived from vertical flows that may exist in the borehole. Such vertical flows occur where the borehole connects aquifers or permeable, saturated horizons with different piezometric heads or where the borehole penetrates a confined aquifer with a permeable upper stratum.

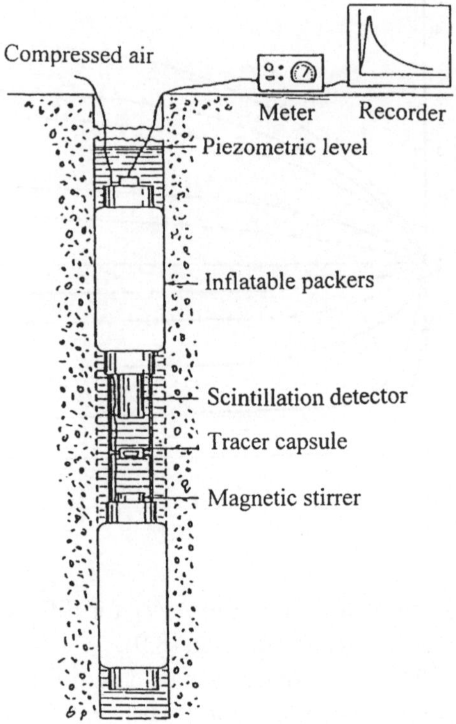

Fig. 8.4. Illustration of the basic configuration for the point dilution method used for the measurement of groundwater flows.

The basic configuration for these measuring devices is illustrated in Fig. 8.4. All instruments developed for this purpose use a radioactive material as tracer, generally, ^{131}I or ^{82}Br. Usually, the packers consist of inflatable rubber tubes. The tracer can be injected by means of different methods, such as the following:

1) Using a small pump driven by an electrical motor.
2) Injection made from the ground through a plastic tube of small diameter.
3) Using a small glass ampoule filled with the tracer. The ampoule is broken by means of an electrical impulse (see Fig. 8.4), etc.

A shaking device located in the volume V improves the precision of the technique because it helps in satisfying the exponential conditions defined by equation (8.5), which requires a permanent constant tracer concentration in the measuring volume. The method is applied by following these steps:

1) The measuring device is lowered to the depth where groundwater velocity has to be determined.
2) Packers are inflated.
3) Tracer is injected in the measuring volume.
4) The dilution curve (concentration versus time) is measured over a sufficiently long period to determine the parameter $t_{1/10}$ required by equation (8.7).

If the measuring procedure is repeated at different depths, a flow velocity profile is obtained that allows the determination of the "velocity transmittancy" T_v, which is given by the expression:

$$T_v = v_i \cdot h_i \tag{8.10}$$

where h_i is the thickness of the section for which the velocity v_i has been obtained. The value of T_v divided by the local hydraulic gradient gives the aquifer transmissibility.

A highly sophisticated measuring system has been developed by Drost (1986), allowing the measurement of both the horizontal flow component and the flow direction. For the flow direction, the measuring system uses a revolving collimator that allows directional measurements of the tracer distribution around the borehole to be carried out, therefore indicating the direction in which the tracer is escaping. A compass provides the orientation of the measuring probe and, therefore, the direction of the groundwater flow. The magnetic north direction is determined by means of an optical system fitted to the lower end of the revolving collimator. A small lamp and a photocell allow the identification of the position of the compass needle on the basis of the reflected light. The eventual presence of a vertical flow through the gravel pack is detected by using two detectors located inside the probe at positions above and below the measuring volume, where the radioactive tracer is injected. The velocity of the vertical flow is determined from the time elapsed between the tracer injection and the maximum activity peak detected by one of these detectors.

8.4.4 Limitations

The point dilution method has several limitations that complicate its practical use and strongly reduce its acceptance as a routine technique. The most important ones are the following:

1) The method has a strong point character, so that only in exceptional cases does it provide representative values of the groundwater flow velocity for a given aquifer. Most aquifers are characterised by a high degree of anisotropy. As a result, groundwater flow generally shows large spatial variations. These changes are particularly important with depth due to the particular structure or stratification in the aquifer. Therefore, to obtain representative values of the flow velocity, many measurements are necessary.
2) The use of packers does not guarantee the absence of vertical flows. Vertical flow may still exist in the space between the casing and surroundings and, more particularly, through the gravel pack. Such vertical flow interferes with the point dilution technique.

3) The adoption of a radioactive material as tracer is very inconvenient. Despite the very low activity requirements, the use of such material requires special authorisation in most countries. Even worse, in some countries, the prevailing legislation forbids the use of such materials. This problem can be overcome simply by using a saline tracer. Due to the recent development of fibre optic fluorometers, the use of a fluorescent tracer for this purpose seems also to be possible.
4) The technique can often only be used in boreholes drilled specifically for investigation purposes. This is because of the requirements with respect to the inner diameter of the borehole and the permeability of the casing.
5) The technique does not operate properly in aquifers on fractured rocks due to the two following reasons: (i) groundwater flow is concentrated exclusively in the fractures of the rocks, and therefore, very strong variations of the flow velocity with depth will be observed, and no representative velocity values can be obtained; (ii) the α coefficient is not defined in a unique way because the groundwater flow pattern shown in Fig. 8.1 is not applicable. Such flow patterns are controlled by the fracture size, shape and orientation. In most cases, a value $\alpha = 1$ will account for the prevailing flow pattern.

These and other limitations drastically reduce the likelihood of the future adoption of this technique. The technique can provide valuable results for the investigation of leaks from dams in only a few instances. Nevertheless, an understanding of the theoretical principles of the point dilution method will assist in interpreting the results obtained from using the method based on the labelling of the entire water column, which is described in the following section.

8.5 METHOD BASED ON THE LABELLING OF THE ENTIRE WATER COLUMN

8.5.1 Experimental methodology

The method described in this section is simple and allows the rapid determination of qualitative and, under favourable conditions, semiquantitative information on the movement of water along or across the borehole. The technique has been developed by the senior author of this book and has been applied for more than 35 years in investigating more than 5,000 boreholes (Baonza et al., 1970).

The method is based on the labelling of the entire water column in the borehole, using one of the tracers mentioned in Section 8.2. The labelling procedure is illustrated in Fig. 8.5A. A plastic tube of about 8 to 10 mm inner diameter is inserted to the bottom of the borehole. A weight fitted to its lower end assists in the insertion process.

The tube is open at both ends. A volume of tracer solution equal to the inner volume of the plastic tube, measured from the phreatic level to the bottom of the

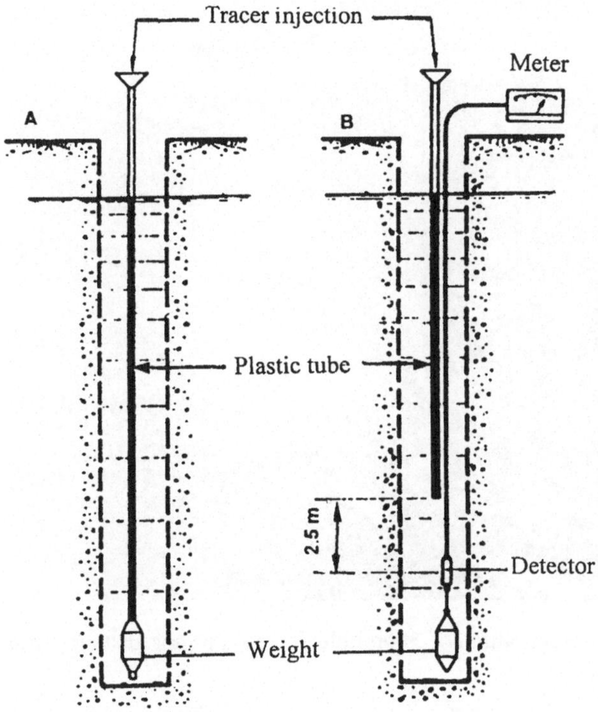

Fig. 8.5. Illustration of the techniques used by the authors for the labelling of the whole water column of a borehole with a tracer. (A) Labelling of the entire water column at one time (case of slow flows). (B) Labelling in successive stretches (case of very rapid flows).

borehole, is injected through the upper end via a funnel (see Fig. 8.6). The tube is then extracted slowly at a more or less steady pace, allowing the tracer to be uniformly distributed throughout the water column.

Once the tracer has been injected, successive vertical profiles of tracer concentration are measured by moving the detector up and down the water column. The measurements can be made in discrete steps by raising or lowering the detector, for example, in 1 or 2 m intervals, depending on the total length of the water column to be scanned and on the observed concentration variations in the profiles. Figure 5.2 shows the device used by the authors for this purpose, which is fitted with a depth meter. An alternative solution is to mark with adhesive tape the cable of the measuring probe metre by metre (Fig. 8.7). If a radioactive tracer is used, the measurement can be performed using, for example, the natural gamma probe of a commercially available well-logging system. A continuous profile in digital form is then obtained.

The frequency of measurement of the concentration profiles has to be chosen in accordance with the rate of the tracer dilution. Usually, at the beginning of the experiment, short intervals are adopted (e.g., one profile every 5 or 10 min). After 1 or 2 h, the frequency is increased as a function of the rate of the tracer dilution. It must be pointed out that it is very important to prolong the measurement of concentration profiles throughout a period of time long enough until most of the

Fig. 8.6. Labelling of the entire water column of a borehole drilled at a short distance from the Sabaneta dam, Dominican Republic.

Fig. 8.7. Measurement of tracer concentration profiles in a labelled borehole in the vicinity of the Giribaile dam.

injected tracer has escaped from the borehole. The reason for such strategy is that, very frequently, the first two or three (or more) concentration profiles are not only governed by the local groundwater flow, but they can be strongly influenced by the initial tracer distribution in the annular space between the casing and the terrain. This is especially true when a low permeability casing is used. More information on this point is given below.

For the interpretation of the results, the concentration values obtained in each profile are plotted against depth in the same graph. Guidelines to help in this interpretation will be given in the next sections.

The method illustrated in Fig. 8.5B is used only when the groundwater flow velocity is so high that most of the tracer escapes from the borehole during the period of time needed to extract the injection line and to insert the measuring probe. This is often experienced in dam leakage studies because of the high prevailing hydraulic gradients. In such cases, both the injection line and the measuring probe are introduced in the boreholes and tied together with adhesive tape. The lower end of the injection line is then positioned at a certain distance above the detector (e.g., 2.5 m), as indicated in Fig. 8.5B. The injection line is then filled with tracer solution and the measuring line (injection line and detector) is then raised 5 m. In doing so, a stretch of the water column of similar length is labelled and the detector remains located in the central part of the labelled stretch. Keeping the injection/measuring line fixed at this position, the change in concentration with time is determined. The tracer will rapidly escape from the borehole because of the rapid groundwater flow. Once the measurement interval has elapsed, the injection/measuring line is raised a further 5 m, and the measurement of the tracer dilution is repeated. The procedure is repeated until the whole section affected by the rapid flow is covered.

Obviously, for the interpretation of the results obtained using the previous method, it is necessary to take into account whether or not a vertical flow exists along the water column.

8.5.2 Borehole requirements

Obviously, the method described in the previous section requires that water can freely flow across borehole side-walls. Ideally, the use of boreholes without inner casing is preferable. However, such boreholes can exist only in particular cases where exclusively consolidated rocks are intercepted. In most cases, the character of the geological formation requires that the holes be cased. Then, the casing has to be perforated with slots over the full length of the saturated zone. For the determination of this length, the variation of the phreatic or piezometric level in the formation resulting from the variation of the water level of the reservoir needs to be taken into account. Therefore, the best approach is to perforate the full length of the casing below the piezometric level reached at the maximum level of the reservoir. The influence of the casing on groundwater flow will depend on its relative permeability with respect to the permeability of the rock formation.

This influence will be negligible for a casing permeability higher than about five times the permeability of the aquifer.

To maintain the integrity of the casing, it is preferable that the slots be cut parallel to the casing axis or slanted with a small angle. An appropriate size for the slots is 5 cm length and 1.5 mm width. To obtain sufficient flow though the casing, the surface area of slots should be between 0.5% and 3% of the casing area (see Annex 4). Plastic casing with a surface area of slots of about 10% is commercially available. The use of such casing is highly recommended for the investigation of leaks.

An additional problem is the partial or total siltation of boreholes due to the penetration of fine sediment particles through the casing slots. Very frequently, the accessible depth of existing boreholes decreases drastically with time due to this effect. The solution based on the use of a gravel pack hardly provides good results in cases of small diameter boreholes. Usually, these are drilled using a tool of a diameter between 6 and 10 cm, and then a casing of 5 cm diameter is placed. Under these conditions, it is very difficult or impossible to fill with gravel the annular space between the casing and the terrain. The best solution is to cover the slots with a permeable cloth specially produced to this purpose. However, in any case, this annular space can create difficulties for the interpretation of the tracer test because the measured concentration profiles can be affected by the condition of the annular ring. It may remain empty or be filled with gravel or with materials detached from the borehole side-walls.

The ideal inner diameter of the boreholes is between 75 and 100 mm. Nevertheless, all tests described in this and the next sections can be carried out in boreholes of 50 mm diameter. For obvious reasons, casing perforated at the bottom 2 or 3 m only is not appropriate. In some cases, such boreholes may nevertheless provide valuable information.

8.5.3 Tracer injection techniques

Tracer injection is usually made by gravity. In this case, before injecting the tracer solution, running water is injected through the funnel to prime the injection line. The air contained in the inner space of this line can create a big bubble that hinders the admission of water. This problem is solved by pulling out between 5 and 10 m of the injection line without stopping the water injection (no air should enter through the funnel). Due to the suction created inside the plastic tube during its extraction, the admission of water starts again. When no more water is admitted, the line is rapidly inserted again to the bottom of the borehole, always keeping the funnel full of water. The air bubble still remaining in the plastic tube will be discharged through its lower end. Once water is flowing freely, the tracer solution is added. To avoid an undesirable dilution of the tracer solution with the funnel water, the injection is started only at the precise moment when this funnel becomes almost empty.

An alternative way to overcome this priming problem is to fill the injection line with water before it is inserted into the borehole. In this case, additional water has to be injected during the insertion process to maintain the water flow through the injection line. Once the lower end of this line reaches the bottom, the water injection is stopped and the injection of tracer solution started.

Of course, tracer injection can also be made using a peristaltic pump. The volume of tracer solution to be injected, according to the calculation made previously, is placed in a container and the pumping is started. Once the container becomes empty, the pump has to be immediately switched out and the injection line disconnected from the pump to allow the tracer solution inside this line to equilibrate by gravity with the borehole piezometric level.

The tracer distribution along the water column during the extraction of the injection line is characterised by the following features:

1) To guarantee that the entire length of the water column is labelled, a volume of tracer solution slightly higher than the inner volume of the injection tube between the piezometric level and the borehole bottom is injected. Obviously, this excess of volume is released at the bottom, producing at this point a tracer concentration higher than the average concentration of the water column.
2) The previous situation also occurs if air bubbles are mixed with the tracer solution during the injection, for example, due to the formation of a vortex in the funnel. To avoid this mixing, one should ensure that the injection is made with the funnel as full as possible with tracer solution.
3) Due to the friction of the tracer solution, the level inside the injection line during the extraction of the plastic tube is slightly higher than the piezometric level in the borehole. Therefore, if this extraction is interrupted, a section of the borehole with higher tracer concentration is created at the point where the end of the injection line was placed at this moment.
4) Due also to this friction, when the operator starts pulling up the injection line, no or a very small amount of tracer solution is released through the lower end of this line until a certain level inside it is reached. As a result, a lower tracer concentration will be produced in the deeper stretch of the water column above the bottom of the borehole. The length of this lower concentration stretch depends on the extraction velocity of the injection line. Usually, it is less than 2 or 3 m. The same shortage of tracer occurs when the extraction of the injection line is renewed after an eventual interruption. This effect can be minimised if the injection line is extracted very slowly.
5) Due to the higher level of the tracer solution inside the injection line with respect to the piezometric level, a higher tracer concentration can be created at the top of the water column.

Let us suppose, on the other hand, that a saturated solution of common salt is used as a tracer. The solubility of common salt in water is about 350 mg/l. The electrical conductivity of such a solution is about 467 mS/cm [see equation (5.2)]. If the inner diameter of the plastic tube is 10 mm, the cross sectional area is 0.785 cm^2. If a borehole is used with an inner diameter of 6 cm, that is, a cross-section of

28.3 cm^2, the tracer dilution produced inside the borehole is equal to a factor of $28.3/0.785 = 36$. Therefore, the electrical conductivity of the borehole water will result in a conductivity increase of $467/36 = 13$ mS/cm. This conductivity increase is adequate to perform the test if the initial conductivity of the borehole water is lower than about 5 mS/cm.

However, if a borehole or well with an inner diameter of, for example, 50 cm is used (very infrequent in dam leakage studies), the dilution factor will be $50^2/1^2 = 2500$ and the conductivity increase will amount to only 198 μS/cm. This increase will be insufficient for the experiment. Clearly, in such instances, the problem can be solved by using an injection tube of higher inner diameter (for the previous example, an injection tube of about 8 cm inner diameter would be required). The handling of such an injection line is, however, very cumbersome, and the solution adopted by the authors, in such instances, is to carry out the tracer injection using a small peristaltic pump. The volume of saturated salt solution necessary to attain the required conductivity is first calculated. This volume is then placed in an appropriate plastic bottle or container. The injection is done by means of the peristaltic pump moving up and down the lower end of the injection line (from the piezometric level to the bottom of the borehole) until the total quantity of solution has been injected. The injection is made with a low flow rate over sufficient time to allow several sweepings of the water column. If done in this way, the tracer distribution will be more or less uniform, although this condition is not absolutely necessary for the test.

Obviously, the problem mentioned in the previous paragraphs does not exist if a radioactive material is used. There is then no limitation regarding the concentration of the solution to be injected. This can be increased to achieve the appropriate concentration inside the borehole, independent of the diameter of the borehole and the diameter of the injection line.

In cases when the borehole is very deep (above 150 m) and the phreatic or piezometric level is also very deep, the injection and measuring lines may be too heavy to manually carry out the test. In such cases, it becomes necessary to employ a drum driven by a motor to mechanise the insertion and extraction both of the injection and measuring lines. The senior author has constructed such a device. The principle of it is illustrated in Fig. 8.8. The velocity of the motor can be continuously regulated, and its turning can be switched in both directions to extract or to insert, respectively, the injection/measuring line. The device is fitted with a brake that allows the operator to keep these lines in a fixed position. A depth meter continuously provides indication of the position of the measuring probe or the lower end of the injection line.

8.5.4 Interpretation of results

The method based on labelling the entire water column determines the horizontal velocity of the groundwater flow only when no vertical flow component occurs within the borehole. Because such vertical flows are common, only in a few cases

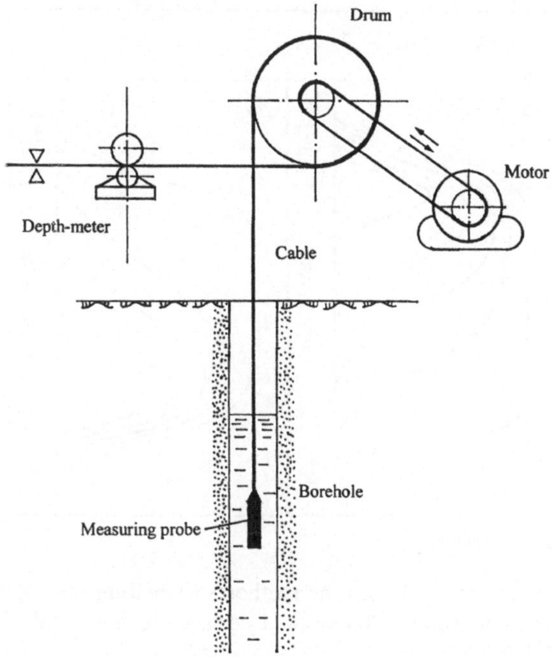

Fig. 8.8. Illustration of the device used by the authors for tracer injection and measurement of concentration profiles in cases of very deep boreholes and piezometric water tables.

can valuable data on the horizontal flow be obtained. This usually occurs in shallow boreholes where the length of the water column is short. In deep boreholes, vertical flow is a common feature, and it is frequently observed in consolidated rock aquifers where the borehole penetrates two or more fractured or karstified zones. Nevertheless, even where this happens, the technique provides information on the position of the permeable horizons intersected by the borehole, as well as the positions where water flows into the borehole and escapes from it.

The technique is highly appropriate for the investigation of leaks from reservoirs and lakes. Precise quantitative data on the groundwater flow is usually not required, and the objective pursued in applying the technique is to identify the flow pattern in the rock formation through which the leaks take place.

Figure 8.9 shows typical concentration profiles obtained with this technique. In the case of Fig. 8.9A, a horizontal flow exists in the central section of the water column. The successive concentration profiles 1 to 5 show how the tracer concentration decreases with time. From these observations, the velocity of the horizontal component of the groundwater flow can be derived using equation (8.7).

In the case shown in Fig. 8.9B, a vertical descending flow takes place in the borehole. The ingress of water takes place in the section marked with the character I, while the outlet is located at the stretch marked with the character O. The successive profiles show how the tracer concentration progressively decreases between the two sections in which flow occurs. The tracer losses produced above and below section I are due to the turbulence associated with the flow of water inside the borehole.

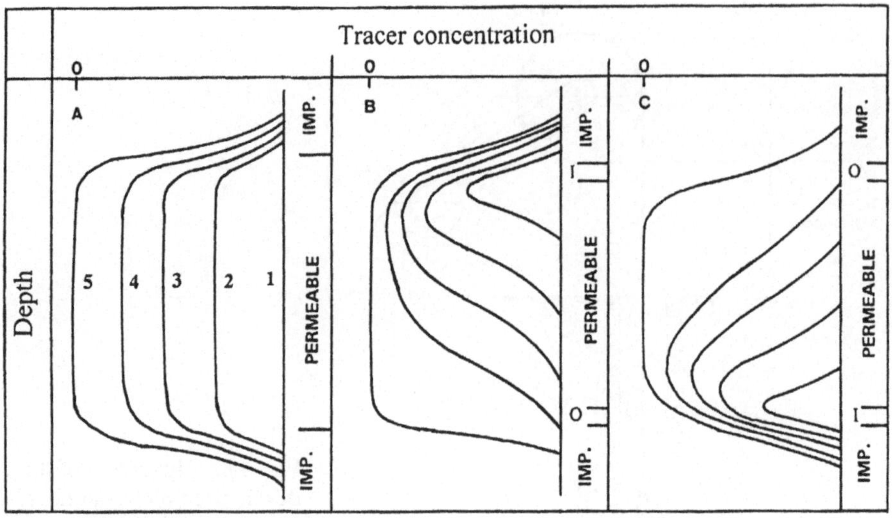

IMP. = IMPERMEABLE; I = INFLOW; O = OUTFLOW

Fig. 8.9. Typical concentration profiles obtained with the method of labelling the entire water column of a borehole: (A) horizontal flow; (B) vertical descending flow, and (C) vertical ascending flow.

When a saturated table salt solution is used as tracer, the salinity of the borehole water, and therefore its density, increases as a result of the tracer injection. Therefore, when the labelled borehole water in the section between I and O is replaced by groundwater from the aquifer, water with a higher salinity (density) is left at the top of section I. The higher density of the water parcel located above point I causes the parcel to sink to lower levels (because of its higher density), allowing the tracer to disappear also from the upper section. In this case, the flow velocities are much lower than in the section mainly affected by vertical flow. The reverse is true for the section of the borehole near the outlet (O), where the water column is stable because the parcel of water showing higher density is situated below the lower salinity water.

The case presented in Fig. 8.9C is similar to the one in Fig. 8.9B, but with vertical ascending flow. In both instances, the velocity of the vertical flow can approximately be determined from the movement of successive fronts of tracer inside the borehole. This is done by measuring the distance between points of equal concentration of two consecutive fronts, as well as the time elapsed.

The electrical conductivity and temperature profiles measured before the tracer injection are of great help in interpreting the results. The same applies for the interpretation of vertical flow tests that are described below. The presence of a vertical flow results in the homogenisation of the electrical conductivity and temperature values along the section affected by such a flow. Horizontal flows usually result in changes of both parameters in the profiles. On the other hand, it is common that in stretches where only very low or null flow velocities occur, groundwater shows a much higher conductivity than in the sections with high flow velocity. Temperature of water can be higher or lower, depending on local conditions.

8.5.5 Case studies

As mentioned before, more than 5,000 boreholes have been studied by the authors using the previously described technique. The interpretation of the results obtained in the experiment is not always easy. The main difficulties are related to the way the borehole has been constructed. The basic parameters that are required for a correct interpretation of the results include: (1) the inner diameter of borehole (drilling diameter), (2) location and types of screens inserted, (3) inner and outer casing diameter, (4) size and distribution of the slots or holes perforating the casing, and (5) data pertaining to the gravel pack eventually present.

For obvious reasons, apart from the borehole parameters, information on the geological formations or strata intercepted will also be very valuable for the correct interpretation of the results. The same can be said for the observations made during the drilling of the borehole, such as variation of water admission of the borehole, changes of the static piezometric level and presence of cavities in the case of consolidated rocks. These are usually detected through the falling down of the drilling pipeline.

To help the reader in the interpretation of the results obtained using this technique, a set of real cases was selected that cover most typical situations and conditions that can be found in practice.

Example no. 1 (Fig. 8.10) This case study corresponds to a borehole where no vertical flow occurs located in the River Gurguéia valley, northeast Brazil. The tracer used for the test was ^{131}I. The profiles obtained in the test are similar to the theoretical ones shown in Fig. 8.9A. A pure horizontal flow was found in a layer of alluvial sand located between 20 and 43 m depth. This layer was covered by an impervious clay layer and underlain by an also impervious marl formation. The variation with time of the average concentration corresponding to the alluvial sand is plotted in the lower diagram of Fig. 8.10, in a logarithmic scale. It clearly shows that the exponential law defined by equation (8.5) is adequately satisfied. The slope of the straight line gives a value of the parameter $t_{1/10}$ equal to 33 h. With this value, knowing that the borehole inner diameter was 46 cm and using $\alpha = 2$, equation (8.7) provides a Darcian velocity of the groundwater flow equal to 30.3 cm/day.

Example no. 2 (Fig. 8.11) This example corresponds to a borehole located at the dam site of the Bellus reservoir, Spain. The borehole is tapping, below the piezometric level, a calcarenite rock formation. The initial conductivity profile (Fig. 8.11A) showed the presence at the top of the water column of a water layer of

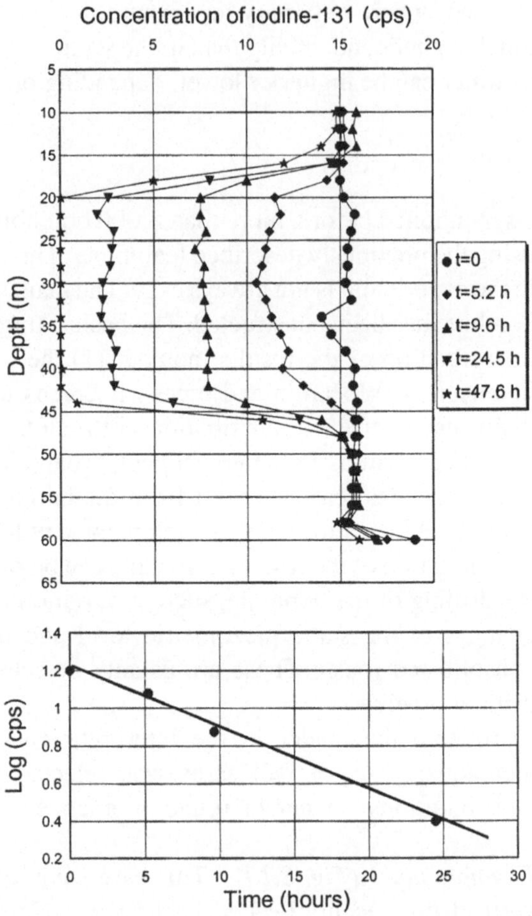

Fig. 8.10. Results obtained by applying the method based on the labelling of the entire water column in a borehole with horizontal flow located in a permeable stratum of alluvial materials. The borehole was located at the River Gurgueia valley, northeast Brazil.

about 10 m thick with an electrical conductivity higher than the water underneath. This situation creates an unstable water column due to the higher density of the water located at higher elevation. Therefore, it represents a transitory situation that can prevail only if a permanent water flow across the borehole exists. Both the chemical and isotopic contents (stable isotopes deuterium and oxygen-18) showed that groundwater in the upper horizons was derived from the neighbouring reservoir, while the deep groundwater represented locally recharged groundwater. Reservoir water showed an electrical conductivity of 1350 μS/cm because of the pollutants transported by the river, whereas local groundwater showed conductivity values between 500 and 600 μS/cm. Local groundwater was saturated in calcite, with very low concentrations of Cl^-, $SO_4^=$ and Na^+ ions, contrary to the observed characteristics of the reservoir waters.

The results obtained by the labelling test of the entire water column (Fig. 8.11B) showed a rapid sweeping out of the tracer from an upper layer of 5–10 m thickness.

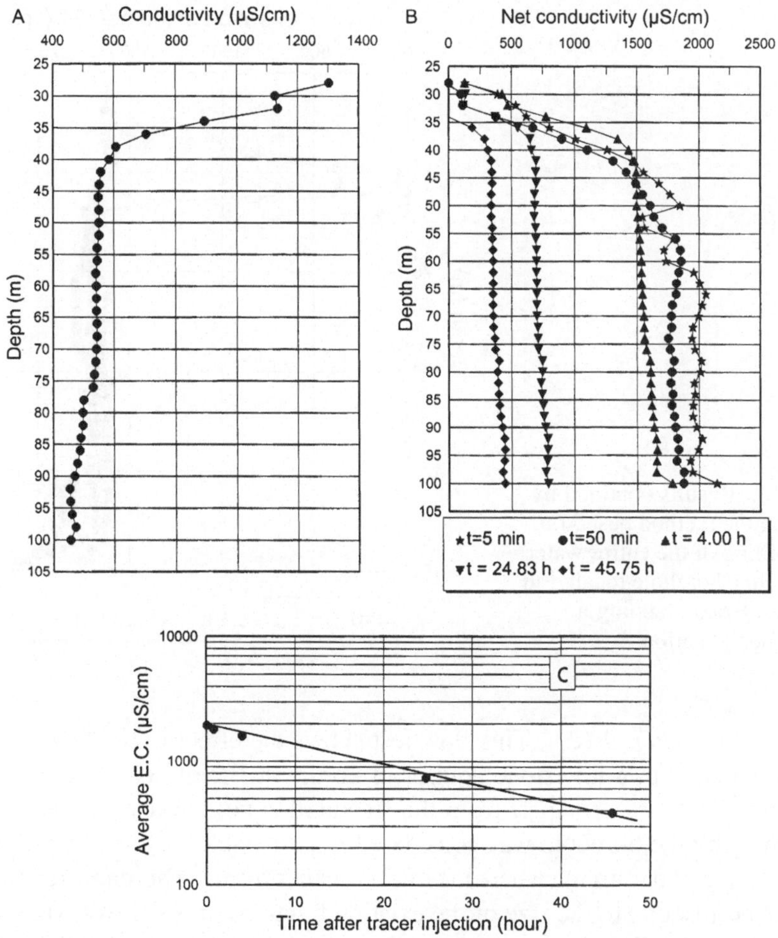

Fig. 8.11. Results obtained by applying the method based on the labelling of the entire water column in a borehole with vertical descending flow at the upper layer and horizontal flow below 40 m. A calcarenite rock formation was intercepted. The borehole was located in the vicinity of the Bellús dam, Spain. The initial conductivity profile shows that the vertical descending flow was generated by water escaping from the reservoir.

The form of the electrical conductivity profiles in this layer indicates that the reservoir water was entering the borehole at higher levels than the piezometric level (probably side-walls dripping), creating a descending vertical flow. This water was escaping from the borehole through the upper stretch above 50 m.

Below this depth, only horizontal flow was present. The average conductivity of the water column at the deeper stretch decreased with time according to the straight line shown in Fig. 8.11C (half-log plot). A value of the parameter $t_{1/10}$ equal to 60 h was obtained, which provided an estimated Darcian flow velocity of about 2.4 cm/d.

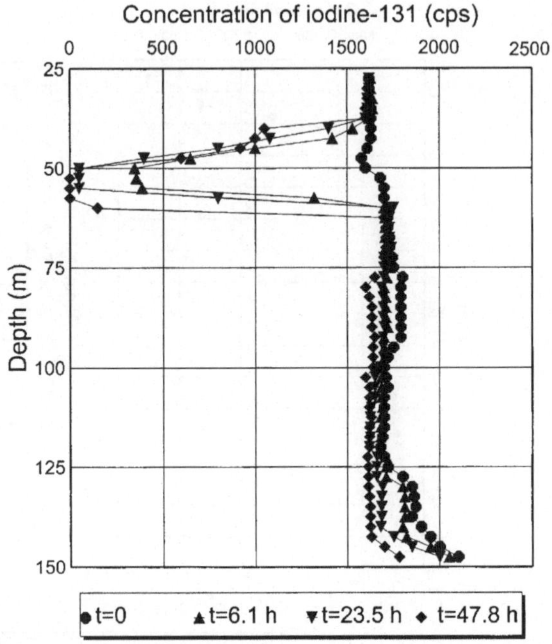

Fig. 8.12. Results obtained by applying the method based on the labelling of the entire water column in a borehole located in northeast Brazil, tapping a limestone formation.

Example no. 3 (Fig. 8.12) This flow test shows the effect in the flow pattern of a fractured zone in a limestone aquifer in northeast Brazil. The test was carried out using ^{131}I as tracer. Apart from the fractured zone, the rest of the formation intercepted by the borehole was considered impermeable. It must be pointed out that the length of the stretch where the tracer escapes from the borehole (about 10 m) does not correspond to the size of the existing fracture. This size was surely much shorter than 10 m. The reason for defining such a wide stretch is that the sweeping out of the tracer is affected by turbulence and by the flow pattern created inside the borehole by the water entering through the permeable stretch.

If table salt solution had been used as a tracer instead of ^{131}I, the tracer in the water column above the fractured stretch would also have been swept out after several hours due to the density effect. The flow through the permeable stretch would have removed the tracer and produced a layer of density lower than the one of the overlying water. The radioactive tracer allowed the labelling without any significant change of density (carrier-free tracer), especially if the tracer is diluted with the same water present in the borehole.

Frequently, when dealing with consolidated rock formations, many boreholes intercept rock cavities with no or very low flow. In such cases, the flow test shows a higher tracer dilution in the position of the cavity, providing a profile similar to the one shown in Fig. 8.12. However, in this case, tracer concentration does not reach the zero value. It takes a more or less constant value above zero after a certain period of time, which depends on the average cross-section (or water

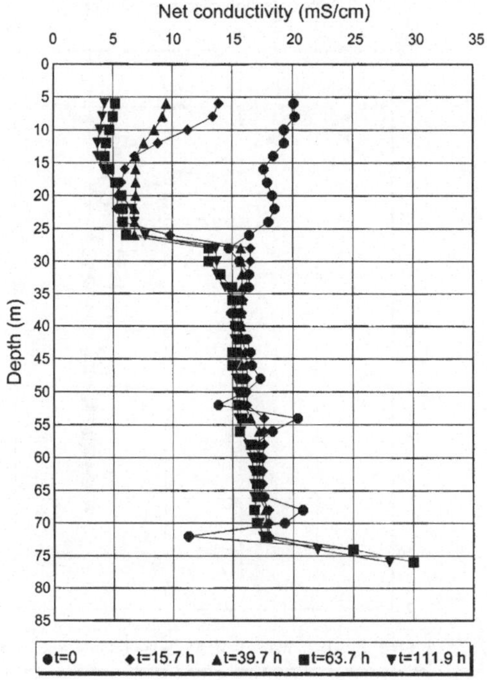

Fig. 8.13. Results obtained by applying the method based on the labelling of the entire water column in a borehole located in the landslide zone of Benameji, Spain, tapping an impervious clay formation.

volume) of the cavity. If table salt solution is used as a tracer in such a case, the profile attains after a prolonged period of time a uniform concentration, even in the stretch corresponding to the cavity, due to the flow produced along the water column by the generated density gradient.

Example no. 4 (Fig. 8.13) This case study corresponds to a borehole tapping an impervious clay formation of the landslide area of Benamejí, Spain. A rapid tracer dilution was observed during the first 16 hours at the upper stretch up to a depth of 26 m. However, the tracer concentration in this stretch, as well as in the lower stretch, remained constant for 111.9 hours, indicating the extremely low permeability of the intercepted rock formation. The initial tracer dilution of the upper stretch is simply explained by the higher diameter of the bore. The mixing of the tracer injected inside the casing with groundwater from the circular annulus between the casing and the terrain took place slowly due to the low density of the slots perforated in it.

Example no. 5 (Fig. 8.14) This example corresponds also to the same limestone formation cited in example number 3. The test shows the presence of a karstified stretch close to the water table. The first three concentration profiles show clearly that the piezometric level is located slightly above the fractured zone. In other words, groundwater is not supplied by a perched aquifer. When this situation

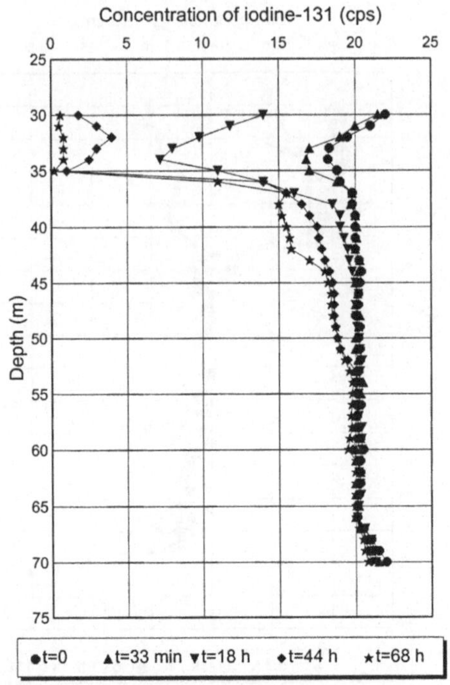

Fig. 8.14. Results obtained by applying the method based on the labelling of the entire water column in a borehole located in northeast Brazil, tapping a limestone formation.

happens, the tracer losses start at the top of the water column and descend gradually with time, as shown in Fig. 8.11B. A vertical descending flow is then generated that ends at a point corresponding to the bottom of the deepest permeable layer.

The last three profiles of Fig. 8.14 show that a small vertical descending flow may exist below 35 m depth.

Example no. 6 (Fig. 8.15) This example corresponds to a borehole located at a landslide zone near Malaga, Spain, which was tapping detritic materials. The casing was formed by nonperforated inclinometric pipeline, so that water could enter inside the borehole only through the joints of the 3-m length pipe pieces. The concentration profiles of Fig. 8.15 show that water was entering at the depth of 3, 9 and 15 m, which effectively corresponds to joints of the inclinometric pipelines. The results demonstrated that a high velocity groundwater flow existed in the zone. Obviously, due the low permeability of the casing, the calculation of this velocity was not possible because of the uncertainty in the value of the α coefficient in equation (8.7). In any case, a very low value of this coefficient would have to be used. The borehole was intercepting impervious materials below a depth of 24 m, as shown by the concentration profiles.

Example no. 7 (Fig. 8.16) This borehole was located at a dike site in the landfill of Cetransa, Valladolid, Spain. The intercepted formations were colluvial

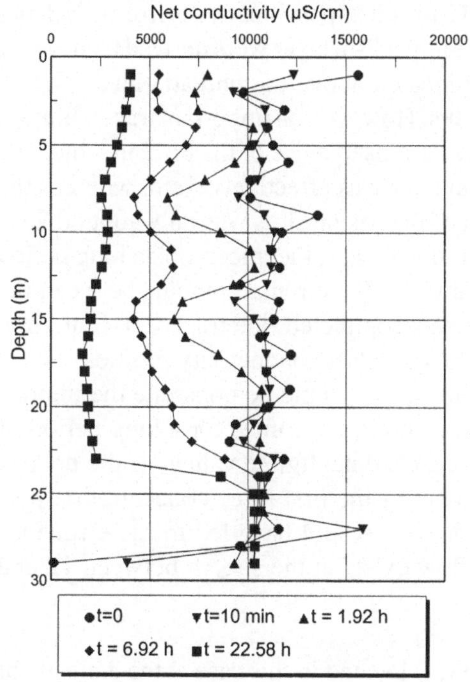

Fig. 8.15. Results obtained by applying the method based on the labelling of the entire water column in a borehole located in a landslide area near Malaga, Spain. The casing was formed by an inclinometric pipeline, which was permeable only in the joints at 3-m intervals. A rapid horizontal flow was detected.

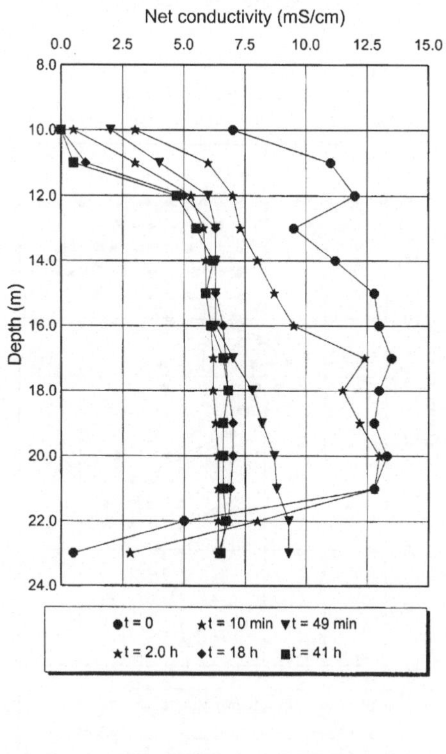

Fig. 8.16. Results obtained by applying the method based on the labelling of the entire water column in a borehole located in the Cetransa landfill, Valladolid, Spain, where a vertical descending flow was identified in the upper layer. The initial tracer losses observed below 14 m were due to the slow dilution of the tracer with the water contained in the circular annulus between the casing and the terrain.

half-permeable materials up to a depth of 12 m and impervious marls from this depth to the bottom. Tracer losses observed in the upper 2-m layer were derived from water inflow in the form of side-wall dripping located above the phreatic level. Below 12 m depth, the marl formation is impervious. However, an important tracer dilution was observed during the first 2 hours. It was a consequence of the form in which the borehole was constructed. The casing diameter was effectively 5 cm, whereas the drilling diameter was 10 cm. The fraction of the casing slots was only about 0.25% of the total casing surface. Under these circumstances, the tracer took a long period of time to be mixed with the water contained in the circular annulus between the casing and the terrain. Once this mixing was completed, the tracer concentration remained constant, as can be seen from the three last conductivity profiles.

These results and those presented in the next example demonstrate the importance of extending the measurement of concentration profiles for a long period of time or until the tracer concentration has reached negligible values. If the present test had been stopped after the measurement of the first three concentration profiles, the interpretation of the available profiles would have led to the erroneous conclusion that an important horizontal flow exists in the stretch between 12 and 22 m deep.

Example no. 8 (Fig. 8.17) The borehole was located in this case at the dam site of the reservoir Santa Liestra, Spain. As in the previous case, tracer losses observed

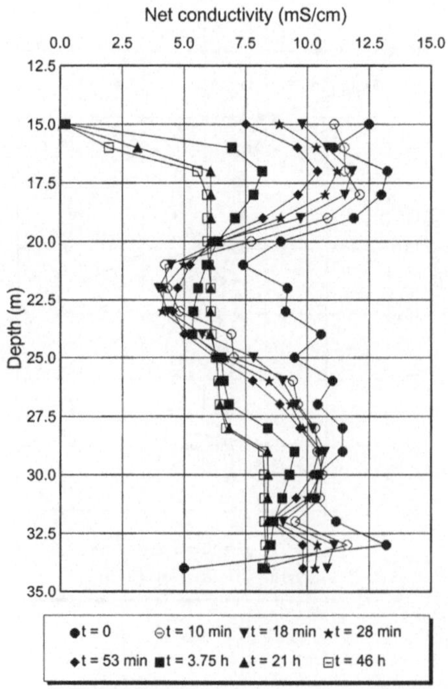

Fig. 8.17. Results obtained by applying the method based on the labelling of the entire water column in a borehole located in the Santa Liestra dam site, Spain. A vertical descending flow was identified in the upper layer. The initial tracer losses observed below 17.5 m depth were due to the slow dilution of the tracer with the water contained in the circular annulus between the casing and the terrain.

at the upper 2 m were produced by the ingress of water above the phreatic level. Below a depth of 17 m, the intercepted material was impervious clay. The tracer dilution observed during the first 3.75 hours was a consequence of the entry of the tracer in the annular spaces located between the casing and the terrain, which in this case showed an irregular distribution. The two last profiles measured 21 and 46 hours after the injection showed no variation of the tracer concentration.

Example no. 9 (Fig. 8.18) Figure 8.18 shows the signature of a vertical descending flow in a borehole drilled at a landslide zone in Benameji, Spain. The ingress of water occurred through a sandstone stratum located at an average depth of about 28 m and the outflow through a similar stratum located at about 63 m. The rest of the rock materials intercepted by the borehole were impervious clay. The velocity of the vertical descending flow was calculated from the position of the tracer fronts corresponding to two consecutive profiles, as illustrated in the figure, and the time elapsed between them. A value of 0.40 m/h was obtained for the stretch between 32 and 51 m and 0.33 m/h for the stretch between 51 and 67 m. This lower value is because the vertical flow ends at a depth of 67 m.

Example no. 10 (Fig. 8.19) This example corresponds to a borehole located at the left side of the future reservoir near the town of Santa Liestra, Spain. A thick clayey formation outcrops in the area, with clear indications of being affected by sliding in the past. The risk of further sliding processes may increase once the reservoir starts

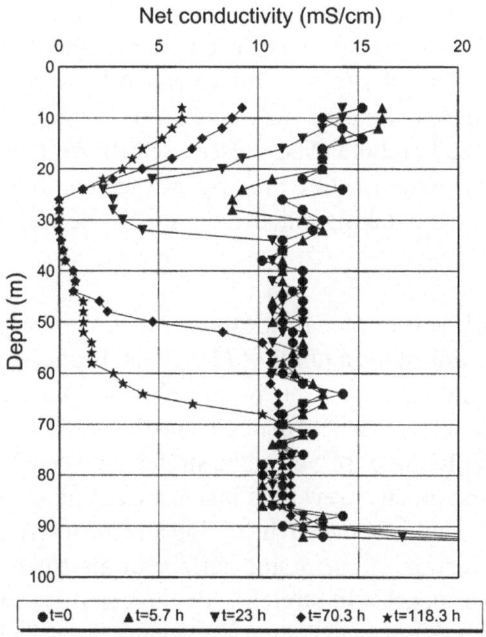

Fig. 8.18. Results obtained by applying the method based on the labelling of the entire water column in a borehole drilled at a landslide zone in Benameji, Spain. A vertical descending flow was identified connecting two sandstone permeable layers.

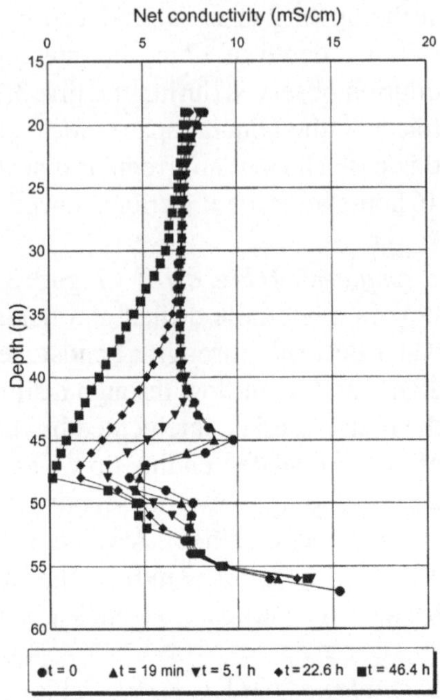

Fig. 8.19. Results obtained by applying the method based on the labelling of the entire water column in a borehole drilled at a landslide zone in the left bank of the future Santa Liestra reservoir, Spain. The intercepted rock consisted of impervious clayey materials. A fracture was detected at 48 m depth, probably related to a landslide process.

impounding water. The flow test illustrated in Fig. 8.19 showed the presence of a vertical ascending flow produced by water entering at a depth of 48 m. This water inflow is due to the presence of a narrow fracture in the rock formation, which could be related to sliding processes affecting the area in the past. The tritium content of the incoming water (7 T.U.) in 1999 matches the typical values found for recent precipitation in northern Spain. This suggests that the rock fracture is connected to the surface at a short distance from the borehole. The presence of the vertical flow indicates that the rock materials above the fracture are impervious (confined water). The outflow of the water was probably taking place at the top of the water column near the piezometric level.

Example no. 11 (Fig. 8.20) This borehole was located inside the perimetral gallery in the Algar dam, Spain. It was tapping a limestone aquifer of Muschelkalk age. As in the previous case, a vertical ascending flow was present. Water inflow was taking place along a fracture in the rock formation located at 90 m depth and the outflow at a stretch located above 20 m depth. Tracer losses observed at the stretch between 25 and 70 m are explained by the lower water velocity prevailing inside the casing with respect to the velocity along the circular annulus between the casing and the terrain. This flow is a consequence of the low fraction of perforated surface of the casing (approximately one slot of 1.5 × 40 mm per 30 cm stretch). The injected tracer inside the casing was continuously being swept out by the rapid external vertical

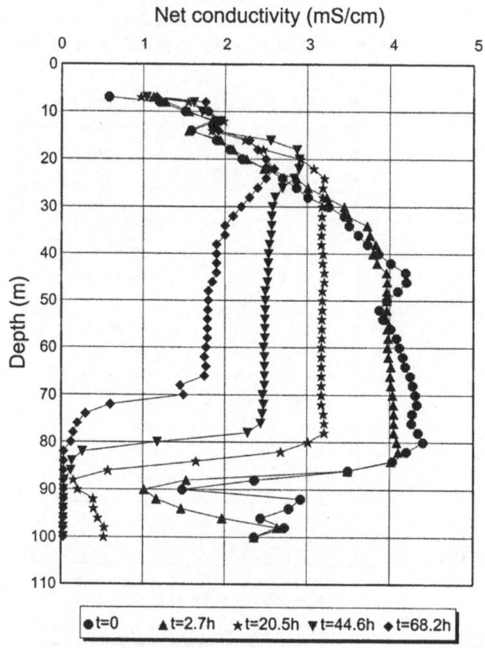

Fig. 8.20. Results obtained by applying the method based on the labelling of the entire water column in a borehole drilled near the Algar dam site, Spain. A vertical ascending flow was generated by a fracture in the limestone formation located at 90 m depth. Tracer losses between 30 and 70 m depth were due to the sweeping out of the tracer contained inside the casing by groundwater flowing along the circular annulus between the casing and the terrain.

flow. In fact, the stretch between 20 and about 85 m corresponded to impervious materials.

The above-mentioned difference in the flow velocity inside the casing and in the circular annulus can lead to situations in which a vertical flow can be interpreted erroneously as a more or less homogeneous horizontal flow. The successive profiles may show an almost uniform decrease of the tracer concentration with time due to the sweeping out of the tracer along the whole water column. Of course, this problem will not exist if a high permeability casing is placed in the borehole.

Example no. 12 (Fig. 8.21) This borehole was drilled close to the Giribaile dam site, Spain, and was tapping a limestone formation. The lower tracer concentrations found in the initial profiles below a depth of 65 m were, in principle, suggesting the presence of a vertical ascending flow. But, this was not the case. The lack of tracer at this deeper stretch was derived from a bad injection. When the injection line was inserted in the borehole, the lower end of the weight used to facilitate this operation was blocked by mud deposited near the bottom of the borehole. Therefore, a very small amount of tracer was injected in the lower end of the borehole during the extraction of the injection line, until the blockage produced by the mud disappeared. Due to the density effect created by the saline tracer and, especially, due to the prevailing vertical descending flow, the conductivity profile became homogeneous at this deeper stretch once several hours elapsed after injection. This problem occurs frequently when the tracer outflow inside the injection line takes place through an

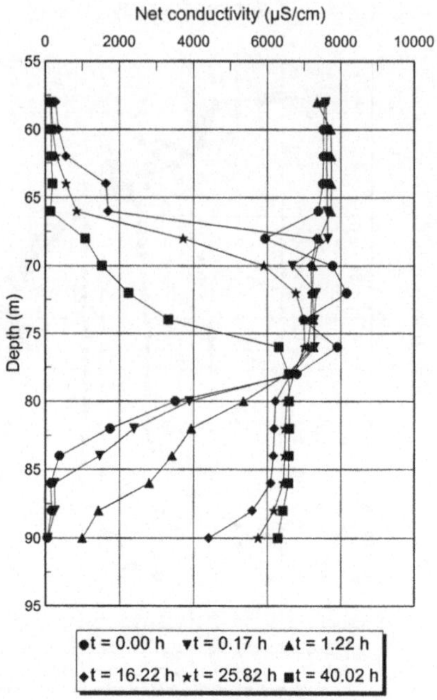

Fig. 8.21. Results obtained by using the method based on the labelling of the entire water column in a borehole drilled close to the Giribaile dam site, Spain. The initial lack of tracer at the bottom stretch was due to the blockage of the lower end of the injection line.

opening located at its lowest end. This potential problem can easily be solved if this opening is placed some tens of centimetres above the lowest end by drilling a small hole in the plastic tube.

8.5.6 Advantages of the method

Important advantages of the method based on the labelling of the entire water column are the following:

1) The method is versatile because it can be used in any well or borehole. However, for the interpretation of the results, it is necessary to know the main characteristics of the casing.
2) Operations are simple, even in the case of deep boreholes. Apart from the measuring system, all others pieces of equipment and components can be obtained off the shelf.
3) The technique provides information on the flow pattern existing in the entire water column of the borehole, as opposed to the point dilution method, which provides the flow velocity only at a fixed depth. This is extremely important in cases of consolidated rocks, where a concentrated flow is usually present.
4) The technique can be applied to all different geological settings, including consolidated rocks and multilayer aquifers where strong changes of permeability with depth usually are found.

8.6 MEASUREMENT OF VERTICAL FLOWS

8.6.1 Generalities

The presence of vertical flows is a very frequent process in boreholes. These flows are generated in the following instances:

1) When the borehole penetrates two aquifers or permeable horizons with different piezometric heads. Vertical flow is established with groundwater flowing from the aquifer with the highest piezometric head to the aquifer with a lower head. Therefore, flow can either be ascending or descending.
2) When the borehole connects a confined aquifer with an upper permeable saturated or unsaturated horizon or stratum, a vertical ascending flow is created.
3) When water from a perched aquifer enters the borehole, a vertical descending flow usually is established with groundwater flowing from the top of the water column to the lower end of the permeable stretch.
4) Where flow lines within an aquifer are oblique to the borehole axis. The vertical flows created in such cases are usually very low.

In investigating dam leakage, vertical flows are often encountered in the boreholes located in the zone affected by leaks. Boreholes located close to the springs often present vertical ascending flows, indicating that water from the reservoir moves toward the discharge point in the ground. The contrary is valid for boreholes located close to the reservoir. Due to the usually high gradient of the hydraulic heads between water in the reservoir and the level in boreholes, vertical flow velocities can be extremely high. When a borehole is located at an elevation lower than the water level in the reservoir, water frequently emerges from the borehole, becoming artesian. If this is not the case, this is usually interpreted as a bad hydraulic connection between the borehole and the reservoir.

Vertical descending flows in the leakage zone may occur in the following two instances, amongst others:

1) When the leakage at the reservoir is taking place at relatively high elevations and the escaping water flows through a shallow aquifer. In such a case, a deep borehole may connect the flow fed by the reservoir with a deeper aquifer, whose piezometric level is lower than the water level in the reservoir. Such an example was found in Lake Laja, Chile, as will be shown in Chapter 12.
2) When groundwater not related to the reservoir, with a piezometric level higher than the water level in the reservoir, is present in the zone of leakage. In this instance, if the borehole intercepts both water flows, a vertical descending flow is generated inside the borehole. Groundwater from the aquifer is mixed with groundwater derived from the reservoir and follows the same pathway through the geological formations located downstream. This was the case found at the Colbún reservoir, Chile, as described in Chapter 12.

It must be pointed out that when a vertical flow exists inside a borehole, the next important action should be the measurement of the flow velocity variation as a function of depth. Any increase in the flow velocity will be a consequence of water inflow into the borehole, in the same way that any decrease will be interpreted as water outflow. A constant value of the flow velocity along a certain stretch of the borehole means that we are dealing with impervious materials.

8.6.2 Measurement techniques using tracers

8.6.2.1 *Previous tests*

Different techniques have been developed by the senior author for the measurement of vertical flows using tracers. These have been routinely used for many years. As a general rule, the initial measurement to be carried out in a borehole is the profiling of temperature and electrical conductivity. A constant value of these parameters in a specific stretch of the water column is, in principle, a hint about the possible existence of a vertical flow, although this is not always the case.

The next action should be the performance of the test based on the labelling of the entire water column. The results obtained from these tests normally will provide sufficient information to clarify whether a vertical flow takes place, and in an affirmative case, to identify the flow direction. In the same way, the labelling of the entire water column will provide the approximate magnitude of the vertical flow and will identify the length of the borehole section affected. The information thus obtained can be used to select the most appropriate method to precisely measure the vertical flow. The methods used for this purpose are described in the following sections.

8.6.2.2 *Migration of the tracer cloud*

If the vertical flow is sufficiently slow, a simple method is used. The method is based on the point injection of a tracer at a depth close to the point where the vertical flow starts. The tracer injection is carried out by means of the method illustrated in Fig. 8.22. A plastic tube is introduced into the borehole, with the lower end located at the depth where the injection is to be made. The tracer solution is injected through the top, via a funnel, followed by the injection of sufficient clean water to replace the volume of water stored inside the plastic tube between the piezometric level and the injection point. For example, if a saturated table salt solution is used as a tracer, a volume of this solution between 250 ml and 2 to 3 litres is sufficient, depending on the borehole diameter. The injection is gravity fed or is done using a peristaltic pump, following the indications made in Section 8.5.3.

Immediately after the tracer injection, the determination of successive vertical profiles of tracer concentration is started. The aim is to define the movement of the tracer cloud with time. Obviously, the frequency of these profiles will depend on the flow velocity, which will be obtained from the migration of the tracer cloud.

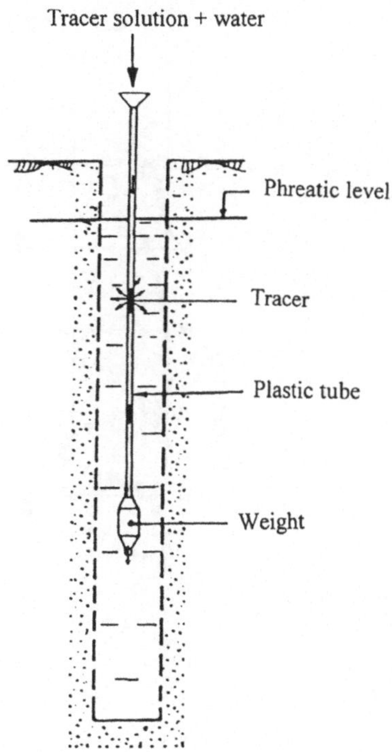

Fig. 8.22. Illustration of the point injection technique for the measurement of vertical flow in a borehole by applying the method based on the migration of the tracer cloud.

When this cloud arrives at the zone where groundwater escapes, the velocity decreases and, at the same time, the total amount of tracer decreases. This decrease can be determined by integrating the curves showing the variation of the concentration versus depth, for a constant cross-section of the borehole.

Figures 8.23 to 8.25 show examples of tests performed using the above-described technique. The first case corresponds to a borehole located at the River Gurguéia valley, northeast Brazil. A prevailing vertical ascending flow was identified. Groundwater entered the borehole through a fracture located at 98 m depth and escaped through a sandstone layer located above 68 m depth. A confining layer of shells and argillite was present in the stretch between 68 and 98 m. The variation of the vertical flow velocity is also shown in the figure. It can be seen that, as expected, the flow velocity in this confining layer showed a constant value. Both the flow velocity and the surface covered by the successive tracer cloud curves decreased within the sandstone layer due to the water outflow.

The case shown in Fig. 8.24 corresponds to a borehole where both the technique based on the labelling of the entire water column and the method described in this section were applied. The borehole was tapping a limestone formation in northeast Brazil. A vertical ascending flow was identified; groundwater inflow was taking place through a fractured stretch located close to the bottom (50 m depth), and the outflow was through another fracture situated close to the piezometric level

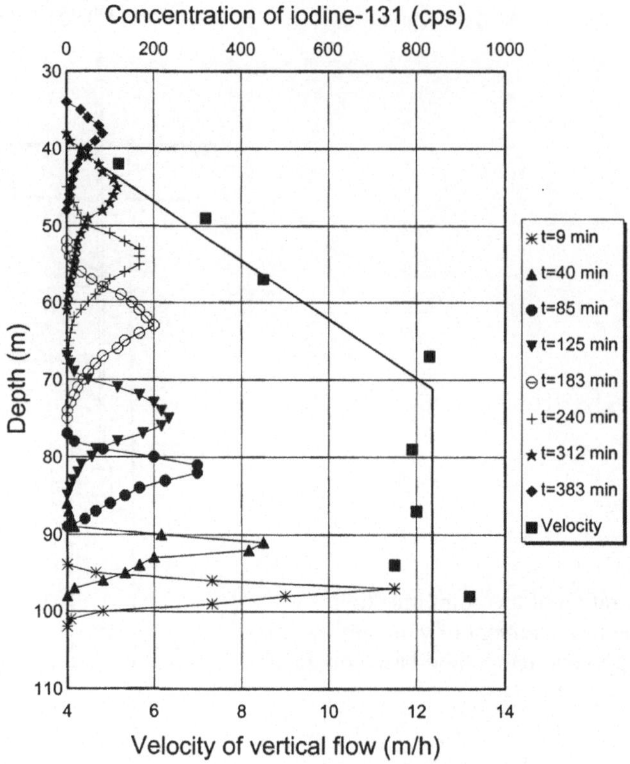

Fig. 8.23. Results obtained by a test performed applying the technique illustrated in Fig. 8.22 for the measurement of vertical flows. A point injection of tracer (^{131}I) was made at a depth close to 100 m. The tracer outflow occurs through a pervious stratum located above a depth of 68 m.

(14 m depth). The results showed that both techniques provided similar values of the vertical ascending flow. Effectively, the first mentioned technique provided approximately the following velocity values: $V_{40\text{–}36.5} = 0.74$ m/min, $V_{36.5\text{–}30} = 0.72$ m/min, $V_{30\text{–}23} = 0.87$ m/min and $V_{23\text{–}16} = 0.87$ m/min (the subscripts give the depth intervals). The second technique gave the values: $V_{47\text{–}44} = 0.75$ m/min, $V_{44\text{–}40} = 0.95$ m/min, $V_{40\text{–}36.5} = 0.87$ m/min, $V_{36.6\text{–}22} = 0.86$ m/min, $V_{22\text{–}17} = 0.90$ m/min, $V_{17\text{–}15.2} = 0.3$ m/min. The lower velocity value found in the last case is a consequence of the water outflow. The rock formation between the two fractures is impervious, as shown by the constant value of the vertical flow.

The previous cases correspond to normal situations. On the contrary, the case shown in Fig. 8.25 corresponds to a borehole in which the velocity of the vertical ascending flow in the circular annulus between the casing and the terrain was much higher than the velocity existing inside the casing. This was a consequence of the low fraction of surface of the casing covered by the slots. Due to this fact,

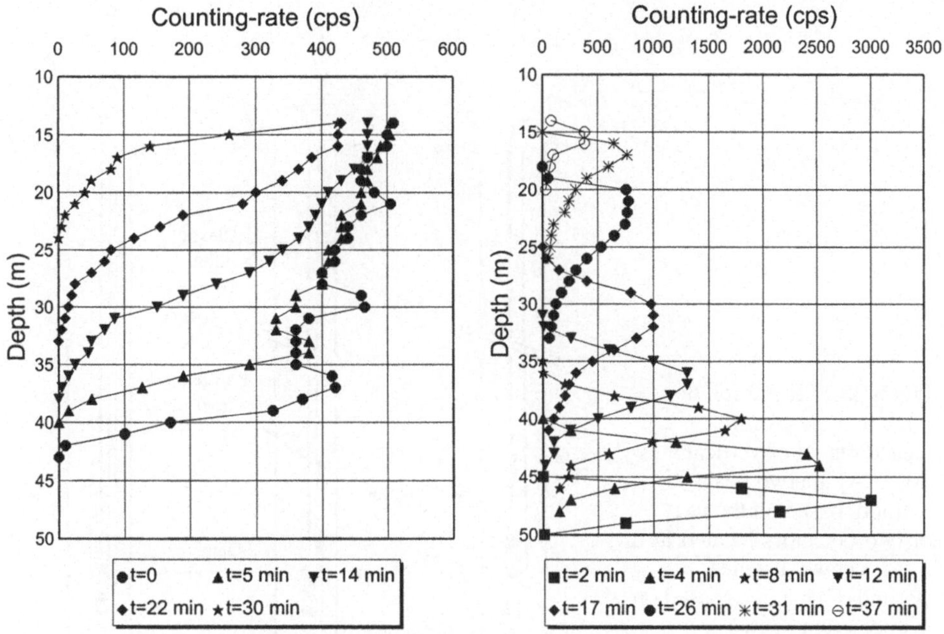

Fig. 8.24. Comparison of the results obtained using the technique illustrated in Fig. 8.22 and the one based on the labelling of the entire water column in a borehole with a vertical ascending flow.

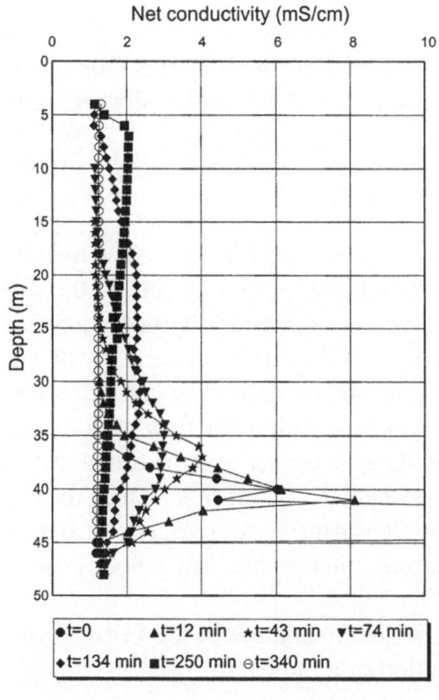

Fig. 8.25. Results obtained in a test performed by means of the technique illustrated in Fig. 8.22 for the measurement of vertical flow. The observed high tracer dispersion was due to the higher velocity of the water circulating through the circular annulus between the casing and the terrain with respect to flow velocity inside the casing.

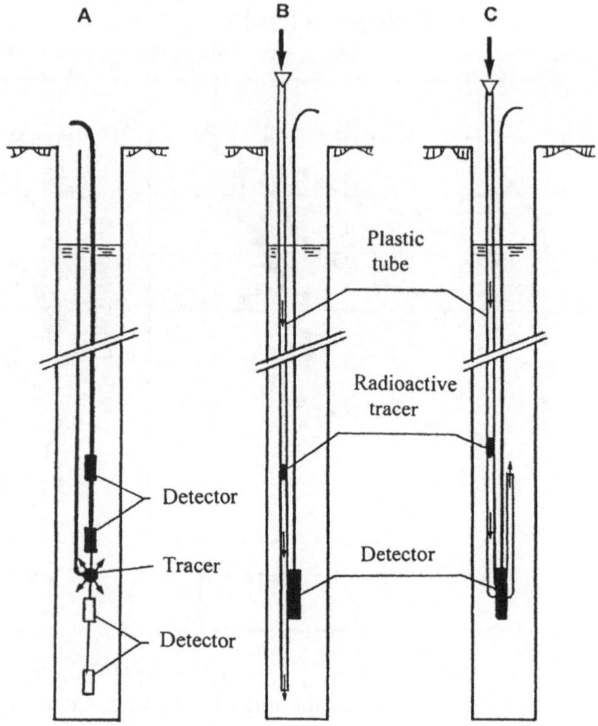

Fig. 8.26. Illustration of three methods for the measurement of vertical flow: (A) the two peak method, based on the use of two detectors located at a fixed, known distance; (B) and (C) methods based on the detection of two concentration peaks using a single detector and a radioactive tracer.

the tracer injected at 41 m depth was rapidly dispersed in the upward direction and after 2 hours was distributed along almost the entire water column. Obviously, under these circumstances, only qualitative information of the vertical flow velocity can be obtained.

8.6.2.3 *The two peak method*

For the two peak method, two detectors are located at a fixed distance. The method is illustrated in Fig. 8.26A. Tracer is injected at a point located just below the two detectors, if vertical ascending flow occurs, and above them, if the flow direction is descending. The injection is made instantaneously from the surface or by means of a small peristaltic pump that is operated for several seconds. The injected tracer results in two concentration peaks crossing the two detectors. The flow velocity can then be calculated from the distance between these detectors and the time interval elapsed between the two peaks, which is experimentally determined. Once the test is completed, new tests are performed with the measuring system placed at different depths. This method provides quantitative information on the flow velocity along the water column.

The above method provides precise data on the magnitude of the vertical flow and can be applied using either a saline solution or a radioactive tracer. The only disadvantage is the need to introduce two detectors into the borehole together with the tracer injection device. This may be difficult, particularly in small diameter boreholes. For obvious reasons, the method is more appropriate in cases of high and intermediate values of the flow velocity, for example, higher than about 10 cm/min. In case of lower velocities, the method described in the previous section is recommended.

A simpler alternative used by the senior author requires only one detector, as illustrated in Figs. 8.26B and 8.26C. The method can be applied only when a radioactive tracer is used. The tracer is injected in the form of a single pulse from the surface using an injection line following the technique described in Section 8.6.2.2 (Fig. 8.22). A small volume of the radioactive solution is first injected, followed by a water volume sufficiently large to transport this solution to the lower end of the injection line. A first, short duration peak of radioactivity is recorded when the tracer passes the detector during its descending motion inside the injection line. A second peak of much lower intensity and higher amplitude is recorded when the tracer again reaches the detector position, transported in this case by the vertical flow existing in the borehole.

In the situation depicted in Fig. 8.26B, the vertical flow is ascending, while in Fig. 8.26C, the flow is descending. In the latter case, a loop has to be made in the injection line to force the tracer to pass first through the position where the detector is located. Obviously, the method is affected by the small error produced by the time required for the tracer to travel from the position of the detector to the end of the injection line. This time is usually very short, and the error is negligible. If necessary, the error can be evaluated and the vertical flow velocity corrected by determining the velocity of the tracer inside the injection line. This can be calculated from the time required to inject a predetermined volume of water.

Once the measurement of the flow velocity has been accomplished, the measuring line is moved to a different depth in the borehole and a new measurement is carried out. The variation of the flow velocity along the water column is then obtained.

8.6.2.4 *Movable measuring line*

The method of the movable measuring line is illustrated in Fig. 8.27. A tracer injection line and a detector, firmly tied to each other with adhesive tape, are inserted into the borehole. The distance between the lower end of the injection line and the detector is initially kept fixed (e.g., 2 m). The lower end of the injection line is placed above the detector if the vertical flow is descending and below if the flow is ascending. Generally, the measuring line is introduced to the end of the borehole or, better, to the depth at which the vertical flow starts, as determined by the method based on the labelling of the entire water column. The lower end of the injection line is fitted with a terminal rod perforated with a small diameter hole (about 5 mm). The mission of this rod is to keep low the losses of tracer solution when the measuring line is maintained in a fixed position.

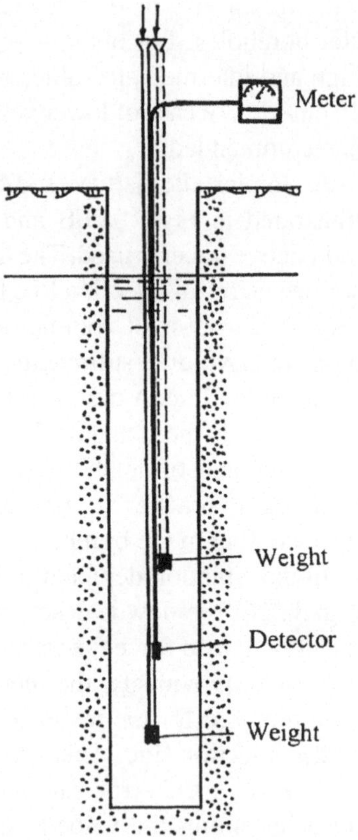

Fig. 8.27. Method for the measurement of vertical flows based on a multiple tracer injection and by moving up the measuring line in successive steps.

Once the measuring line (injection tube plus detector) has been placed at the correct position, the injection tube is filled with tracer solution from the top. A tracer solution volume slightly higher than the inner volume of the plastic tube, measured from the piezometric level to the lower end of this tube, is injected. The line is kept at rest for several minutes. This is done to allow the excess volume of tracer that has escaped through the lower end of the tube to be distributed or to migrate in accordance with the flow existing at this point of the borehole. The measuring line is then rapidly raised by, for example, 2 m. With this action, the volume of tracer solution stored in an equal length of the plastic tube is released through the lower end of the tube. Because of the inertia of the water flow inside the tube, most of the tracer will be released once the measuring line reaches its new position. The injected tracer will then move inside the borehole in accordance with the prevailing vertical flow direction. The time elapsed from the moment the measuring line has reached its new position to the arrival time of the tracer (concentration peak) at the position of the detector is determined. The flow velocity is then calculated by dividing the distance of 2 m by the elapsed time.

The test is repeated by raising the line to a new position, for example, 2 m upward, until a position is reached where no vertical flow is detected. When the vertical flow is ascending, the tracer injected at a certain depth in the borehole can be found 2 m above when the measuring line is raised to conduct the next test. If this happens, a longer time should be allowed to elapse between two successive tests to allow the tracer cloud to travel outside the new position of the detector.

The method described is not precise because the tracer injection is not instantaneous. Tracer will be spread along the 2-m water column when the line is raised. However, if the line is raised rapidly (2 or 3 seconds are sufficient), most of the tracer volume would have escaped from the lower end of the plastic tube by the time the line reaches its new position. Note that the mean tracer injection time is not, therefore, well-defined. However, it must be taken into account that in most cases, we are more interested in obtaining relative values of the vertical flow velocity than absolute values.

8.6.2.5 *Point injection with a peristaltic pump*

This technique is illustrated in Fig. 8.28. A line consisting of the tracer injection tube and the detector (for example, the electrode of the conductivity meter) is used for the measurement. A small size electromagnetic valve is fitted at the end of the injection tube. The injection tube, the detector cable and the conductor of the electromagnetic valve are firmly tied with adhesive tape. The injection tube is

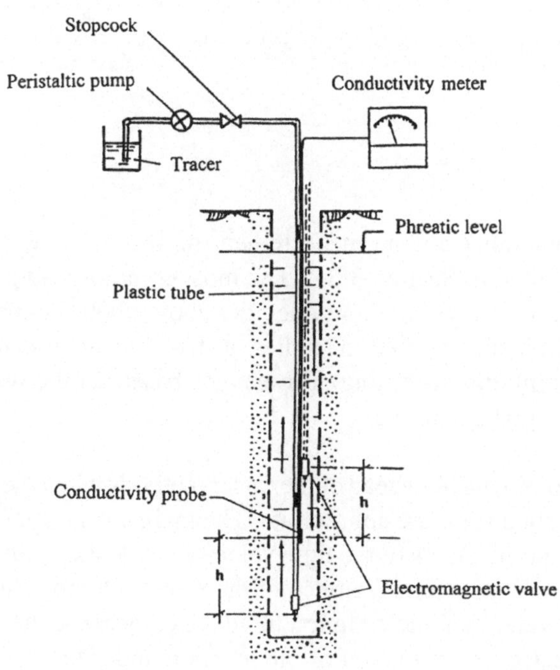

Fig. 8.28. Method for the measurement of vertical flows based on the point injection of the tracer using a peristaltic pump and, if necessary, an electromagnetic valve.

filled with tracer solution. Then, both the stopcock and the electromagnetic valve are closed and the measuring line is inserted to the bottom of the borehole, or better, to the point where the vertical flow starts. Once this line is in the required position, the stopcock is opened and the peristaltic pump is operated for a short period of time (e.g., 5 seconds). The tracer injected through the lower end of the line migrates along the water column according to the prevailing vertical flow and reaches the position of the detector. The time interval elapsed between the injection time (mean pumping period) and the arrival time of the tracer to the detector is measured. The flow velocity is calculated from this time and from the known distance between the end of the injection line and the detector. The electromagnetic valve guarantees that the tracer injection ends at a precise and well-known time. The measuring line is then raised a distance equal to the distance h between the detector and the end of the injection tube, and the next test is performed. The procedure is repeated until the entire stretch affected by vertical flow has been covered.

When the piezometric level in a borehole is deeper than a few metres, the peristaltic pump will not be required because the injection can be made by gravity. The tracer injection is then made simply by opening both the stopcock and the electromagnetic valve for a short period of time (several seconds). Even under favourable circumstances, the tracer injection can be done without the electromagnetic valve simply by opening the stopcock. In this case, a metallic rod perforated with a small diameter borehole is fitted to the end of the injection line to better define the time interval of the injection and to reduce the leakage of tracer after closing the stopcock.

Of course, other different kinds of tracer injection systems can also be developed. For example, the use of small pumps submerged in the borehole together with the detector can provide satisfactory results.

8.6.3 Use of measuring gauges

Different types of gauges are available for the measurement of vertical flow in boreholes. They have been described in Section 3.5.2. The most common gauges use propellers (current meters) and can measure flow velocities above about 2 cm/s. Such high flow velocities are often encountered in studies of dam leakage due to the high hydraulic gradients commonly prevailing. The gauges based on the use of thermal and electromagnetic probes can work with flow velocities as low as 0.5 cm/s (see Section 3.5.2).

The use of these gauges is recommended instead of the tracer methods when they are available and the measuring conditions are appropriate. The main advantage of such gauges is that they provide point values of the flow velocity, while the values provided by the tracer methods refer to a certain stretch of the water column. This is very important, especially in studies conducted in fractured rocks, because these measuring gauges allow the precise determination of the location and size of the cracks and cavities through which the water flows. In any case, a better definition of the flow velocity along the water column will be obtained using these gauges.

8.6.4 Interpretation of the results

The interpretation of the results from tests conducted to measure vertical flows is, in principle, simple and is carried out by plotting the variation of the flow velocity versus depth in the same way as shown in Fig. 8.23. In the stretch of the water column where water enters the borehole, the vertical flow velocity component increases. This velocity remains steady in impervious stretches. In stretches where water escapes from the borehole, flow velocity decreases.

Figure 8.29 shows two examples of flow tests performed with the technique described in Section 8.6.2.4. The length of each segment (2 m) is defined by the distance from the injection point to the position of the detector. In the case of Fig. 8.29A, a vertical descending flow was present due to the ingress of water in the borehole from a perched aquifer. The outflow of groundwater is located at the upper layer between the piezometric level and 16 m depth. Below this depth, the materials intercepted by the borehole are impervious (marl layer).

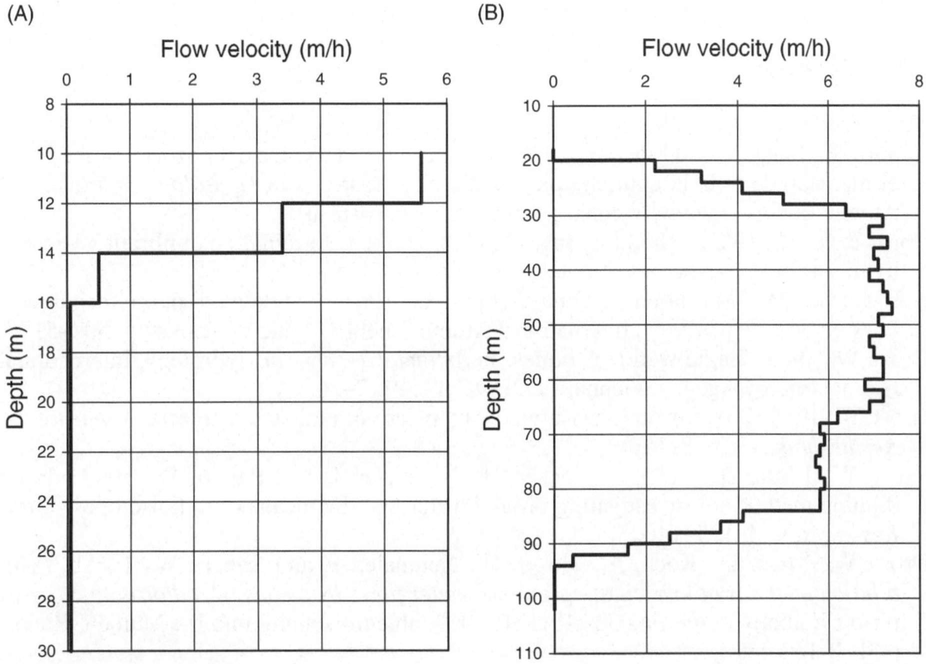

Fig. 8.29. Vertical profiles of flow velocity measured in two different boreholes using the method illustrated in Fig. 8.28.

In the case of Fig. 8.29B, a vertical ascending flow was present. Groundwater was entering the borehole through two fractured stretches located, respectively, at 64–70 and 84–94 m depth. Groundwater outflow was taking place between about 20 and 30 m depth. The vertical ascending flow created an increase of the piezometric level above the top of the upper permeable stretch of about 4 m.

The techniques described in Sections 8.6.2.3, 8.6.2.4 and 8.6.2.5 are usually applied in cases where the vertical flow velocity is relatively high. Therefore, the flow regime will be, in most cases, turbulent. Under these circumstances, the difference of the flow velocity inside the casing and the velocity in the circular annulus between the casing and the terrain tends to be small. However, this difference eventually has to be taken into account for the correct interpretation of the field data.

Finally, it is important to keep in mind the effects of the measuring lines on the vertical flow velocity. Usually, this velocity or, more precisely, the flow rate along the water column, will be higher in the case of a free borehole. The measuring probes, as well as the injection lines, create a loss of hydraulic charge due to friction and therefore, a decrease of the flow rate. Nevertheless, in dam leakage studies, this effect is not very important because the interest is in finding relative differences of the flow existing along the water column.

REFERENCES

Baonza, E., Plata, A. and Piles, E., 1970. Aplicación de la técnica del pozo único mediante el marcado de toda la columna piezométrica. In *Isotopes in Hydrology*, International Atomic Energy Agency, Vienna, pp. 695–711 (in Spanish).

Borowczyk, M., 1965. Laboratory investigations on the determination of filtration velocity by means of radioisotopes. *Atomkernenergie*, 10, 51–56.

Borowczyk, M., Mairhofer, J. and Zuber, A., 1967. Single-well pulse techniques. In *Isotopes in Hydrology*, International Atomic Energy Agency, Vienna, pp. 507–517.

Drost, W., 1983. Single well techniques. In *Artificial Tracers in Hydrology*, International Atomic Energy Agency, Vienna, TECDOC n° 291, 7–16.

Drost, W., 1986. Groundwater measurement by means of radioactive tracers. *Atomenergie, Kerntechnik*, 48(2), 81–86.

Drost, W., Klotz, D., Koch, A., Moser, H., Neumaier, F. and Rauert, W., 1968a. Point dilution methods of investigating ground water flow by means of radioisotopes. *Water Resour. Res.*, 4, 125–146.

Drost, W., Klotz, D., Koch, A., Moser, H., Neumaier, F. and Rauert, W., 1968b. *Point dilution methods of investigating ground-water flows by means of radioisotopes*. Institut für Radiohydrometrie, Gesellschaft für Strahlenforschung m.b.H., Münich, Report GSF-R-15 1968.

Guizerix, J., Grandclèment, G., Gaillard, B. and Ruby, P., 1963. Appareil par la mesure des vitesses relatives des eaux souterraines par la méthode de dilution ponctuelle. In *Isotopes in Hydrology*, International Atomic Energy Agency, Vienna, pp. 25–35 (in French).

Halevy, E., Moser, H., Zellhofer, O. and Zuber, A., 1967. Borehole dilution techniques: A critical review. In *Isotopes in Hydrology*, International Atomic Energy Agency, Vienna, pp. 531–564.
Kocherin, O.J., 1916. Materiely po vodnym izyskanijam Krium, idrometriceskoij. *Izd. Otd. Zem. Ulcsenij*, vyp. 1, Simferopol, 1916 (in Russian).
Moser, H. and Neumaier, F., 1970. *Nukleare Methoden in der Hydrologie des Bauwesens*. Konferenz über die Strahlungs- und Isotopenanwendung in Bauwesen, Brussels (in German).

CHAPTER 9

Interconnection experiments

9.1 INTRODUCTION

Interconnection experiments constitute one of the essential stages in investigations related to dam leakage. These experiments are frequently undertaken to obtain an irrefutable proof of the hydraulic connection between two water bodies, generally indicated or supported by previous investigations using one or several of the techniques described in the previous chapters. Briefly, the experiments consist of the injection in a water body of a determined amount of an appropriate tracer and the subsequent determination of its arrival at boreholes or springs located downstream.

Depending on the nature of the experiment, tracer injection is carried out inside the reservoir or in boreholes located in the vicinity of the dam site. Similarly, the detection of the arrival of the tracer is done in a discharging spring or in a borehole. In any case, the experiment requires careful planning. Obviously, the first aspect to check is whether the hydraulic connection to be investigated is possible from the point of view of the chemical and isotopic contents of the involved waters. It is necessary to prove in advance that both chemistry and isotope signatures indicate that the expected interconnection and the transit time involved are possible and reasonable, that is, whether the contents of some parameters are identical or very similar at the injection point (reservoir or borehole) and at the detection point (borehole or springs). However, for this preliminary evaluation, the possibility should be taken into account that reservoir water or groundwater at the injection point can undergo some kind of mixing with water/groundwater of a different origin and, therefore, have a different chemical and isotopic composition. In the case of the chemical composition, the possible existence of certain chemical changes between the injection point and the detection point should also be considered. For example, if the reservoir water shows a concentration of the Cl^- ion of 85 mg/l and groundwater sampled from a spring or a borehole where the tracer is expected to be detected contains only 20 mg/l, it is clear that the groundwater can not derive from the reservoir. It can be just a minor fraction, at best. The same argument can be applied if the reservoir water has a δ^{18}O-value, for example, equal to $-8.5‰$, while groundwater sampled from a spring or a borehole shows a value of $-6.5‰$. Therefore, the determination of the chemical and isotopic composition of the involved waters is strongly recommended at the time of planning any interconnection experiment. Of course, the piezometric levels at the involved boreholes and the elevation of the springs relative to the water level at the reservoir have also to be taken into account to define the possibility of success of any interconnection experiment. Such studies are necessary but not sufficient. Other additional studies must be performed before the interconnection experiment is carried out. These are described below.

9.2 THEORETICAL CONSIDERATIONS

9.2.1 Flow equations

Interconnection experiments with a point injection of a certain amount of tracer generally produce bell-shaped breakthrough curves at the control points (springs or boreholes) where the tracer passage is detected. The shape of the breakthrough curves depends on the dispersivity of the rock formation through which groundwater flows, as well as on the flow pattern prevailing in the vicinity of the injection point. Each individual fraction of groundwater labelled with a tracer usually flows underground through very different pathways and with different flow velocities. This is true even in the case of groundwater flowing through a single fissure. Flow velocity in the central part of the fissure is faster than on the borders, near the rock walls. These processes control the magnitude of the tracer dispersion. In practice, the most important factor governing the dispersion process is the lack of homogeneity of the rock formation through which groundwater flows. This is particularly true in the case of consolidated fractured rocks, where a fraction of groundwater flows at faster velocities through the cavities and large open pores or fissures, while another fraction flows slowly through narrow pores or fissures.

Depending on the properties of the geological medium, in the case of a point injection, the breakthrough curve may show a Gaussian distribution. This can be narrow, with high and well-defined passage, or be broad and flat, depending on the dispersive characteristics of the rock formation affected by the flow. Strictly speaking, a one-dimensional flow can be produced only when groundwater flow takes place along some kind of a pipeline. In the same way, a two-dimensional flow requires the presence of two impermeable plates. These circumstances are never found in natural geological formations, where a highly complex flow pattern is usually found.

The combination of advection and dispersion processes is described by the following differential equation:

$$\frac{dC}{dt} = D_x \frac{d^2C}{dx^2} + D_y \frac{d^2C}{dy^2} + D_v \frac{d^2C}{dz^2} - v_a \frac{dC}{dx} \qquad (9.1)$$

where C is the solute concentration, t is time, v_a is the mean effective flow velocity of groundwater and D_x, D_y and D_z the dispersion coefficients in the directions x, y and z, respectively.

For porous homogeneous aquifers, the three-dimensional solution of the advection-dispersion equation is possible. But this solution cannot be applied for the evaluation of tracer tests because of the many unknown parameters. However, most of the field tests conducted using tracers can be considered a one-dimensional flow (Sauty, 1980), although the injection and sampling procedures may have a significant influence on the breakthrough curve obtained. An overview of this topic is given by Kreft and Zuber (1978).

In the case of fissured and karst aquifers, transport processes present obvious differences with respect to porous aquifers. Apart from the advection and dispersion processes, the dispersive characteristics of the rock formation usually play a predominant role. The breakthrough curves obtained in these tests can be evaluated by means of a multidispersion model developed by Maloszewski (1992, 1994). In this model, the experimental breakthrough curves are considered to be a superposition of several individual curves.

In the case of one-dimensional transport, the differential equation is the following:

$$\frac{dC}{dt} + v_a \frac{dC}{dx} = D \frac{d^2C}{dx^2} \qquad (9.2)$$

The solution for this equation can be written in the following form:

$$C(x,t) = \frac{m}{S \cdot n_e \sqrt{4\pi \cdot D \cdot t}} \, exp\left[-\frac{(x - v_a \cdot t)^2}{4D \cdot t} \right] \qquad (9.3)$$

where $C(x, t)$ is the tracer concentration, which obviously depends on the time t elapsed since injection and the distance x to the injection point, m is the mass of the injected tracer, S is the cross-section of the rock formation involved in the flow, n_e is the effective porosity of this formation and D is the linear dispersion coefficient along the axis x.

The first half of the equation, before the exponential term, represents the maximum tracer concentration of the breakthrough curve. The function $C(x)$ is symmetric but the function $C(t)$ is not. The variance σ_x^2 of the function $C(x)$ is given by the expression:

$$\sigma_x^2 = \int x^2 C(x)\, dx = 2Dt \tag{9.4}$$

Therefore, the standard deviation σ_x is equal to $\sqrt{2Dt}$.

For the function $C(t)$, the variance σ_t^2 is:

$$\sigma_t^2 = \int C(t)(t - t_a)^2\, dt = \frac{2D}{v_a}\left[\frac{x}{v_a} + \frac{4D}{v_a^2}\right] \tag{9.5}$$

where t_a is the average transit time corresponding to v_a.

The same solution can also be expressed using the well-known Peclet number, *Pe*, given by the following expression:

$$Pe = \frac{x}{\alpha_l} = \frac{v_a \cdot x}{D} \tag{9.6}$$

where α_l is the dispersivity, which is characteristic of the rock formation. The Peclet number is dimensionless. The dispersivity or dispersion length α_l represents the lack of homogeneity of the rock formation. Its value is not constant under natural conditions and increases with the length of the flow path.

9.2.2 Point injection

When a point injection of tracer is performed, a bell-shaped breakthrough curve is usually obtained. This curve provides the following values of the flow velocity (see Fig. 9.1):

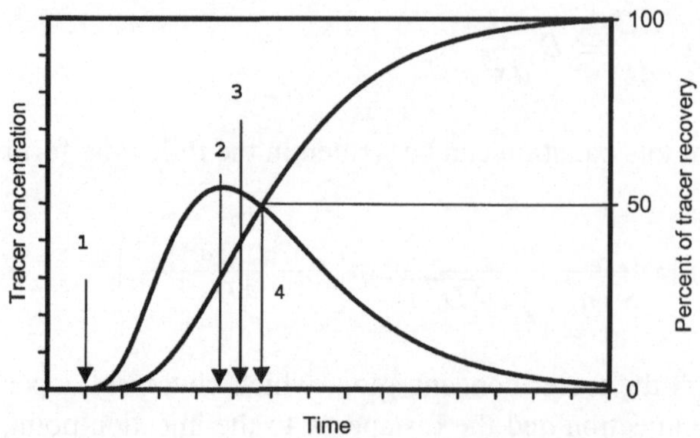

Fig. 9.1. Typical breakthrough curve obtained in interconnection experiments in the case of a point injection of the tracer. The time t_1 corresponds to the maximal flow velocity (first arrival of tracer). The time t_2 corresponds to the dominant flow velocity (peak of the breakthrough curve). The time t_3 corresponds to the real effective flow velocity. The time t_4 corresponds to the mean flow velocity (50% of tracer recovered).

1) Velocity that corresponds to the first appearance of the tracer, which sometimes is called maximal velocity.
2) The dominant velocity derived from the time corresponding to the maximal concentration.
3) The mean flow velocity derived from the time on the breakthrough curve at which 50% of the recovered tracer amount has passed the measuring station.

However, none of these flow velocity values represents the real effective flow velocity. The time corresponding to this flow velocity occurs after the passage of the maximum concentration and before the time of 50% passage, as indicated in Fig. 9.1. This time cannot be obtained with high precision from the analysis of breakthrough curves. A good approximation is to calculate this flow velocity as the average value of the times corresponding, respectively, to the maximal concentration and to the 50% of recovered tracer.

The tracer transit time obtained from the breakthrough curve measured at the detection point, together with the flow rate of the discharge, can be used to estimate the water volume stored in the rock formation involved in the flow between the injection and the detection points. This volume can be computed as the product of the flow rate and the mean effective transit time. The quotient of this volume and the distance between these two points provides the average value of the cross-section of the flow path.

The mean residence time, t_r, of the tracer in the rock formation between the injection and detection points can be derived by integrating the breakthrough curve, according to the following equation:

$$t_r = \frac{\int_0^\infty t \cdot C(t)\, dt}{\int_0^\infty C(t)\, dt} \tag{9.7}$$

The integration can also be manually computed by plotting separately the product $C \cdot t$ versus the time t and C versus t. The quotient of the areas below these two curves will give the mean residence time. This calculation can be done independently of the shape of the breakthrough curve.

For the Gaussian distribution given by equation (9.3), the mean residence time is given by the expression:

$$t_r = \frac{2D}{v_a^2} + \frac{x}{v_a} \tag{9.8}$$

9.2.3 Continuous injection

When a continuous injection of tracer is performed and the conditions for one-dimensional flow are applicable, the tracer arrival at the detection or sampling point occurs according to the following equation:

$$C(x,t) = \frac{C_0}{2} \cdot erfc\left(\frac{x - v_a \cdot t}{2\sqrt{D \cdot t}}\right) \tag{9.9}$$

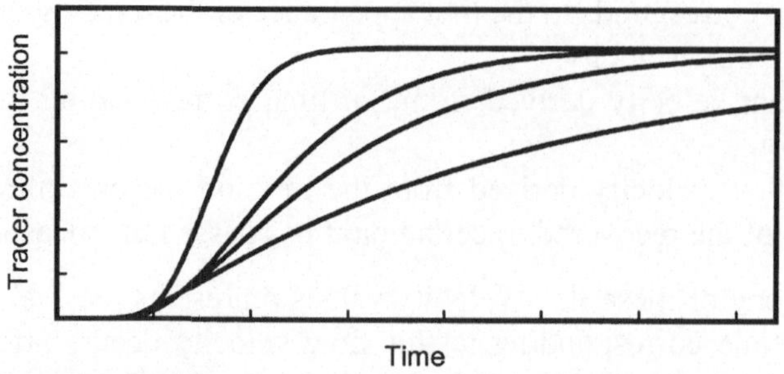

Fig. 9.2. Typical breakthrough curves obtained in interconnection experiments in the case of a continuous tracer injection at a constant flow rate for different values of the longitudinal dispersion coefficients.

where C_0 is the concentration of the injected tracer solution. All other variables are the same as indicated in previous equations. The complementary error function (*erfc*) is related to the Gaussian function, and its value can be found in mathematical tables of most handbooks on statistics, hydraulic modelling or hydrology.

The breakthrough curves determined at the measuring station may have the shape shown in Fig. 9.2 for different relative values of the dispersion coefficient. This is also approximately the type of breakthrough curves that are generally obtained in springs or boreholes when the labelling of the whole water volume of the reservoir is performed.

9.2.4 The longitudinal dispersion coefficient

The longitudinal dispersion coefficient of natural flows can vary over several orders of magnitude. Its value can be expressed as a function of the flow velocity v and the molecular diffusion coefficient D_m, by means of the following equation:

$$D = D_0 \cdot v + D_m \tag{9.10}$$

where D_0 is the dispersivity of the aquifer and has the dimension of a length. According to Lenda and Zuber (1970), the value of D_0 varies from a few centimetres to a few tens of centimetres for sandy aquifers, from 10 cm to 1 m for gravel formations and from 2 to 300 m for fissured rocks.

For very slow groundwater flow velocities in porous media (less than 1 cm per day), the value of D is approximately equal to $0.7D_m$. Typical values of D_m are between 6×10^{-10} m^2/s and 7×10^{-9} m^2/s (Custodio and Llamas, 1976). For faster flow velocities, the coefficient D_m can be neglected in comparison with $D_0 \cdot v$. The value of the longitudinal dispersion coefficient D can be computed by means of the following empirical equation (De Josseling de Jong, 1958):

$$D = \frac{1}{6} v \cdot l(\lambda + 0.75 - \ln \gamma) \tag{9.11}$$

where l is the length of the individual channel, given approximately by the product of the porosity and the mean grain size, v is the real flow velocity, $ln\,\gamma$ is the Euler constant (equal to 0.577) and λ is a parameter that increases with the distance along the flow direction. Its value can be derived from the following expression:

$$N = \frac{e^{2\lambda}}{2(\lambda - 0.75 - ln\,\gamma)} \tag{9.12}$$

where N is the number of channels affected by the flow along the aquifer, whose minimum value is $N_{min} = x_o/l$ (x_o = travelled distance) and the most probable value is given by $N = 1.5x_o/l$.

In the case of fissured and karstified rocks, flow velocity of water can reach very high values. The senior author has found a groundwater flow velocity of about 12 km per day in a Brazilian karst aquifer. For such an aquifer, the molecular diffusion coefficient D_m can be neglected. Values of D_0 as large as 300 m have been reported (Plata, 1986), and the longitudinal dispersion coefficient can be as high as 0.1 m^2/s.

9.3 GUIDELINES FOR CONDUCTING INTERCONNECTION EXPERIMENTS

9.3.1 Tracer injection

9.3.1.1 *Injection inside the reservoir*

As mentioned, interconnection experiments with a tracer injected inside the reservoir require a previous investigation oriented toward the identification of the most appropriate injection point where water infiltrates and the type of injection method to be used. Obviously, the initial point for this investigation is detailed information on the geological characteristics of the bottom formations. The techniques to be applied for the localisation of seepage or infiltration areas will be described in the next chapter.

A critical point to be studied before undertaking the tracer injection in the reservoir is the possible stratification of the water body. As explained in Chapter 2, this stratification can be either thermal or due to salinity variations. If the reservoir is stratified, the vertical mixing of the tracer is produced with great difficulty. Therefore, if a complete distribution of the tracer in the water body is desired, the best solution is to perform the tracing experiment at a time when no stratification exists in the reservoir. This occurs usually, in the case of temperate climates, during the winter season. However, lakes located in tropical and high latitude areas may present stratification during the whole year. An alternative method to obtain a relatively good distribution of the tracer in the whole reservoir consists of performing a multipoint injection. The tracer is injected at different depths using a pump or by means of the method described in Chapter 8 for the labelling of the entire water column of boreholes. Obviously, this problem does not exist in the case of very shallow reservoirs (water column less than about 10 m). Here, the current generated by local winds will produce the mixing of the tracer in the whole water volume.

One very illustrative case study of the problems created by the stratification in a lake is the experiment carried out by the senior author in 1976 in the Contreras reservoir, Spain. To investigate the nature of the leaks affecting the reservoir, 1 Ci of ^{46}Sc in the form of EDTA complex was injected at a certain depth of the lake. A point injection was made without having previously investigated the stratification conditions of the water body. A few hours after the injection, the tracer became distributed across a thin layer of about 1 m thickness and an extent that covered the whole reservoir width (about 250 m). Obviously, the tracer was not detected in the downstream springs due to the lack of vertical mixing.

Frequently, even in case of stratified lakes, good mixing of the tracer can be achieved within the hypolimnion or the epilimnion when the vertical gradient inside these two layers is small. If this is the case, both layers can be labelled separately. Therefore, it is important to measure the temperature of the spring (or borehole) water to identify in which of the two layers the emerging water infiltrates. As explained, the comparison of the temperature of the reservoir water and the emerging spring may provide information on the elevation at which the emerging water is being infiltrated. In Chapter 11, two case studies are described that illustrate how temperature of water can help in the identification of the seepage area inside the reservoir.

9.3.1.2 *Injection in a borehole*

Obviously, if the tracer injection is to be performed inside a borehole, the chemical and isotopic compositions of groundwater sampled in this borehole have to be compared with the contents found in groundwater collected from the spring to determine whether the interconnection between these two water types is possible. Additionally, it is absolutely necessary to verify that the borehole intercepts a groundwater flow with a velocity sufficiently rapid to transport the tracer towards the spring. Frequently, interconnection experiments are carried out where the tracer is injected in boreholes tapping exclusively stagnant water. This was the case observed by the senior author in boreholes at the Joumine (Tunisia) and Takna (Cyprus) reservoirs. Fluorescent tracers were injected in both cases for interconnection experiments. The injected tracers were detected in two boreholes 6 years and 6 months, respectively, after the tracer injection.

In conclusion, before injecting any tracer in a borehole for an interconnection experiment, it is essential to verify that the borehole intercepts a significant groundwater flow. This can be achieved by using the tracer techniques described in Chapter 8 based on the labelling of the entire water column of the borehole. The identification of the type of flow existing in the borehole will also help for the definition of the most appropriate method for the tracer injection. For example, if a vertical flow exists inside the borehole, the tracer should be injected at the precise depth of the water column where the flow was identified, so that the tracer leaves the borehole through the same strata where groundwater flows. If the borehole intercepts a pure horizontal flow, the tracer should be injected at the depth where the highest flow velocity has been detected.

If both the vertical and horizontal components of the flow are very low, after the tracer injection, a significant volume of water should be injected inside the borehole to facilitate the incorporation of the tracer to the groundwater flow toward the discharging spring. Sometimes, a great water volume needs to be injected, for example, 10 or 20 m^3. In fact, the greater the injected water volume, the better the initial migration of the tracer will be. However, it is important to take into account that the injection of a great volume of water does not always solve the problem. For example, let us suppose that a vertical descending flow exists inside the borehole starting at a depth of 20 m and finishing at 80 m. If the tracer is injected at a depth of 90 m, it will remain trapped at this depth even after injection of a high water volume.

Obviously, if no significant flow exists inside the borehole, this borehole is not appropriate for the interconnection experiment. A different borehole has to be chosen. In fact, when the injection has to be carried out in a borehole, the groundwater flow intercepted by all existing boreholes has to be determined in order to select the most appropriate one for the interconnection experiment. The chemical and isotopic composition of the borehole waters will also assist in this selection.

9.3.2 Experimental methodology

As mentioned before, the most appropriate tracers for interconnection experiments involving a large volume of water are the following: uranine, bromide ion, SF_6, ^{131}I and tritium. In the case of fractured or karstified rocks, the lycopodium spores can be added to this list. The experiments should be carried out in a quantitative way to get the maximum possible information about the flow patterns and the corresponding hydraulic parameters. To achieve this aim, the amount of injected tracer has to be known with sufficient precision so that complete and precise breakthrough curves are obtained. Usually, the tracer to be injected is dissolved or diluted in a known volume of water, for example, 25 l. A small aliquot (for example, several millilitres) of this volume can be taken for the calibration of the measuring system. Normally, the tracer concentration of this aliquot is very high and has to be further diluted to get a measurable tracer concentration. If this is the case, the dilution factor has to be known. Once the measuring system has been calibrated, both the amount of injected tracer and the tracer concentration determined at the emerging spring can be expressed in the same units provided by the measuring system, for example, fluorescence units, in the case of fluorescent tracers, counts per second, in the case of a radioactive tracer, or milligram per litre (ppm) in the case of a chemical tracer, etc. This means that the absolute value of the amount of tracer injected is not required. Evidently, the experiment can be conducted working only with relative values. Obviously, the measuring conditions for the calibration sample (injected solution) and for the samples collected at the spring have to be similar (same measuring efficiency).

The injection method depends on the particular characteristics of each test. The tracer is usually injected using a plastic tube, the end of which is located at the point where the injection is to be made. The tracer solution can be injected by gravity or by using a small pump.

However, when the injection is to be made in the body of the reservoir, it may be convenient to spread the tracer in a wide zone. This spreading of the tracer can be necessary when the location of the leakage point is not known or when the seepage is taking place over a wide area in the bottom. The tracer can be injected along a line close to the bottom or along several lines to cover the whole area suspected of being affected by seepage.

In the case of small reservoirs, it may be interesting to label the whole volume of the reservoir. If no stratification exists, the tracer can then be injected at the surface of the reservoir using a peristaltic pump. The tracer solution is then continuously injected at a constant flow rate through the plastic tube, the end of which is placed in the track of the outboard motor of a boat. The turbulence created by the propeller of the motor facilitates its mixing. The senior author has successfully used this method for the labelling with tritium of the whole Montevideo bay to investigate the distribution of the residence time of water.

Guizerix et al. (1967) and Molinari (1976) have described an alternative method for the labelling of the whole water volume that is recommended for reservoirs with a storage capacity lower than about 10×10^6 m^3. The method is based on the use of an injection device fitted with several sprinklers that is dragged by a boat. The injection is carried out using a constant rate pump.

In some cases, it may be advantageous to carry out the simultaneous labelling of different zones of the reservoir using different tracers. It may be of interest when these zones are clearly defined by straits, branches or other natural or artificial structures that render difficult the horizontal mixing of the injected tracers. Such a labelling was carried out by the authors at the Ulldecona reservoir (Spain). Five kilograms of uranine were injected in the zone between the dam and the cofferdam, which at that time was not fully covered by water. Three kilograms of rhodamine B were injected at the zone between the mentioned cofferdam and an old bridge located several hundreds metres upstream. Finally, two curies of tritium were injected upstream of this bridge. As expected, a partial mixing of the injected tracers was slowly produced over time. However, the comparison of the respective breakthrough curves at the emerging waters should provide information about the zone where the reservoir water was being infiltrated. In the case of the Ulldecona reservoir, only tritium was detected in the downstream springs.

Once the tracer has been injected in the water body, a sampling program has to be initiated, which should include all potential water points that may be affected by the underground flow (springs and boreholes). If both springs and boreholes are available, the control of the tracer recovery should not be restricted to the springs. The passage of the tracer cloud through the borehole also can provide valuable information on the pathway of the tracer and the relative flow velocities. If the tracer can be detected on site, vertical profiles of concentration should be measured from time to time to identify the depth at which the tracer is migrating. If the tracer cannot be measured on site, water samples should be collected and measured at different depths for the same purpose. This kind of control was done by the senior author during the investigations performed at Maguaca and Sabaneta reservoirs in Dominican Republic using, respectively, ^{131}I and uranine as tracers. The permeable strata responsible for the leakage could be identified at different boreholes located between the injection point and the springs.

If the number of potential sampling points is too large, a selection of the most representative points can be made, taking into account (1) their location with respect to the anticipated flow pattern (elevation, distance to dam, etc.), (2) geological nature and structure of the rock formations, (3) measured flow rates in boreholes and springs (springs with negligible flow rate can be neglected) and especially, (4) isotopic and chemical composition of the involved water types.

The sampling frequency is very important for the success of the experiment. Usually, when the tracing experiment is started, the transit time of groundwater between the injection and sampling points is not known. In establishing the sampling program, one has to take into account that nature is usually very heterogeneous and can bring many surprises, sometimes unimaginable. This is particularly the case when dealing with fractured rocks. For this reason, the initial estimations of the transit time of water based on assumed global hydraulic parameters very frequently fail. It must be taken into account that these experiments are carried out in zones with pronounced hydraulic gradients, which can generate very rapid groundwater flows. As a general rule, it is recommended to start with a sampling program highly compressed in time. As time elapses with no arrival of tracer at the sampling points, the sampling intervals are progressively increased. For example, a time interval between successive samples equal to 5% or 10% of the time elapsed after the injection will be reasonable in most cases. The dispersion of the tracer cloud along its pathway inside the rock formation usually increases with time. Therefore, long transit times are associated with high amplitude breakthrough curves. The utilisation of automatic water samplers, such as the device shown in Fig. 9.3, is recommended. These devices allow the collection of 24 water samples with a volume up to 1 l, according to a previously established sampling program. Combined samples can also be collected, in which the 1-l sample is completed by mixing fractions taken at two or three different time intervals.

The tracer concentration in the water samples collected from springs or boreholes has to be determined in absolute or relative units. In the latter case, the relative tracer concentration with respect to a standard solution representing an aliquot of the injected mass of tracer is determined.

Fig. 9.3. Automatic sampler used for interconnection experiments.

To obtain the tracer recovery at each spring, complete breakthrough curves have to be measured. In most cases, the sampling can be stopped once the tracer concentration has decreased to 5% or 10% of the maximal concentration measured at the peak of the breakthrough curve. The tail of this curve can be extrapolated by plotting the logarithm of the concentration versus time. A more or less straight line will be obtained. In the case of a spring with a constant flow rate Q, the recovered mass M of the tracer is obtained by means of the following equation:

$$M = Q \int C\,dt = Q \cdot C_m \cdot t_S \tag{9.13}$$

where C is the concentration at time t, C_m is the average concentration, and t_S is the sampling time.

If the flow rate Q is not kept constant during the duration of the experiment, the mass of recovered tracer can be calculated by adding the products $Q_i \cdot (C_m)_i \cdot (t_S)_i$ obtained for different periods of time in which the flow rate could be considered constant.

The mass of recovered tracer also provides information on the volume of the cavities of the geological formation involved in the interconnection experiment. In principle, this volume is approximately given by the product $Q \cdot t_a$, with t_a being the average transit time of tracer between the injection and the sampling points.

This time is determined from the breakthrough curve following the indications given in Section 9.2.

In some cases, the spring or springs represent only a fraction of the total leakage because the rest of the groundwater has not been accounted for, since it continues flowing downstream without emerging at the surface. In such cases, the total leakage rate can be computed using equation (9.13) written in the following form:

$$Q = \frac{M}{\int C\,dt} = \frac{M}{C_m \cdot t_S} \tag{9.14}$$

The utilisation of this equation requires that the following conditions are fulfilled (see Annex 2):

1) The total amount of the injected tracer is incorporated to the groundwater flow that feeds the spring (or springs).
2) Good mixing of the tracer in this groundwater flow is achieved upstream of the point where it branches into different pathways.

9.3.3 Tracer amount

Obviously, for conducting interconnection experiments, a decision on the amount of tracer to be injected must be made. Several empirical equations have been suggested for the calculation of the amount of tracer to be used. A rough calculation can be made using equation (9.14). Let us suppose, for example, than the total flow rate of the existing springs amounts to 600 l/s and the distance to the injection point is 2 km. The duration or passage time of the whole breakthrough curve (for example, 90% of tracer recovery) is estimated to be about 3 days. The water volume that has emerged through the springs during this time is about 1.55×10^5 m^3. Uranine is going to be used as the tracer. For this tracer, an average concentration of 10 ppb (or mg/m^3) is considered acceptable. Therefore, the tracer amount to be injected is 1.55×10^6 mg or 1.55 kg.

The previous calculation can be considered valid when the tracer is injected inside a borehole because, in such a case, all the tracer will be incorporated into the underground flow in a period of time relatively short compared with the duration of the breakthrough curve. However, if the tracer is injected in the reservoir water, it is important to take into account the previous dilution undergone by the tracer before it is incorporated into the underground flow feeding the springs. Obviously, this previous dilution can drastically influence the duration of the breakthrough curve. An even more unfavourable case occurs when a significant fraction of the tracer is not incorporated to the underground flow circulating towards the spring because it is swept away from the infiltration zone by the natural currents prevailing in the reservoir. In such a case, part of the tracer will be lost, and accordingly, a larger amount of tracer needs to be injected.

If the tracer is distributed in a large water volume of the reservoir, for example, using the multipoint injection technique, the flow rate of the springs can play a secondary role. The amount of tracer to be injected can then be calculated as the product of the total water volume to be labelled and the required average tracer concentration.

Obviously, the duration of the breakthrough curve can vary strongly depending on the type of rock formation (granular or consolidated medium), permeability distribution in the pathways of water, dispersivity of the aquifer, etc. Only rough estimations can be made in advance. Several empirical equations have been suggested for this purpose. One of these equations, valid for uranine and rhodamine B, is the following (UNESCO, 1973–1983):

$$M = K((Q \cdot L)/v) + 20\,V \quad \text{(kg)} \tag{9.15}$$

where K is a coefficient of the aquifer ($K = 500$ for porous aquifers and $K = 3{,}000$ for fissured aquifers), Q is the discharge of the springs given in m^3/s, L is the distance between the injection and the sampling points (km), v is the estimated groundwater flow velocity (m/day) and V is the volume of the traced groundwater (m^3).

The following empirical equation, also valid for the above-mentioned tracers, has been suggested by Leibundgut (1974):

$$M = 0.0018Q \cdot t \cdot C_{max} \cdot A \cdot S \quad \text{(kg)} \tag{9.16}$$

where t is the estimated passage time of the breakthrough curve, C_{max} is the estimated peak concentration (μg/l), A is the adsorption coefficient (higher than 1) and S is a factor higher or lower than 1, depending on the particular conditions of the tracing test.

9.4 CASE STUDIES

9.4.1 Canelles reservoir, Spain

The Canelles reservoir is located on the Noguera-Ribagorzana River in northeast Spain. The arch dam of 120 m high was constructed on a limestone formation. A number of springs appeared on the left abutment during the first filling (see Fig. 9.4), which rendered impossible the complete filling of the reservoir. The leaks occurred along a line of caves located on a highly karstified limestone formation (Plata and Del Val Cob, 1972). The study was carried out during the years 1962 and 1963 and was based exclusively on the performance of 12 interconnection experiments between boreholes and springs. ^{82}Br was used as a tracer in this case. The modern techniques described in this book were not available to the senior author at that time. All tracer injections performed in boreholes located in the limestone formation gave positive results; that is to say, the injected tracer was detected at the springs. On

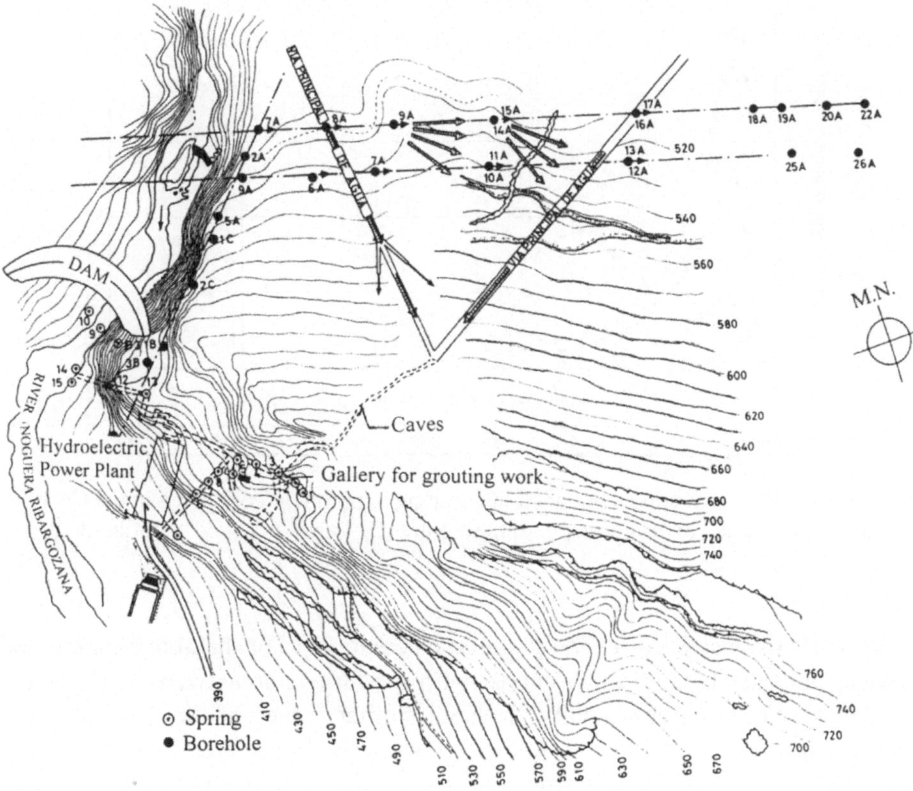

Fig. 9.4. Illustration of the leaks studied in the Canelles reservoir, Spain. The arrows indicate the direction of the main leakage flow.

the contrary, the tracer was not detected when it was injected outside the limestone formation.

The results obtained in one of these experiments are shown in Fig. 9.5. Breakthrough curves were measured in eight different springs. Typical bell-shaped breakthrough curves with relatively short tails were obtained. It can be easily seen that the areas of the breakthrough curves are different. The smaller areas are explained by the confluence of two different water flows, one of them labelled with the injected tracer mixed with a groundwater component that was not labelled.

The total flow rate of the visible springs with the reservoir filled to about one-third of its maximum level was about 250 l/s. However, the water balance carried out by comparing the flow rate at the ingress point and the spring flow rate indicated that the leakage had to be much higher. Most of the leakage was discharging at the bottom of the Santana reservoir, the tail of which was located at a very close distance to the Canelles dam (several hundreds of metres). The total leakage was determined by the integration of the breakthrough curves obtained in the interconnection experiments.

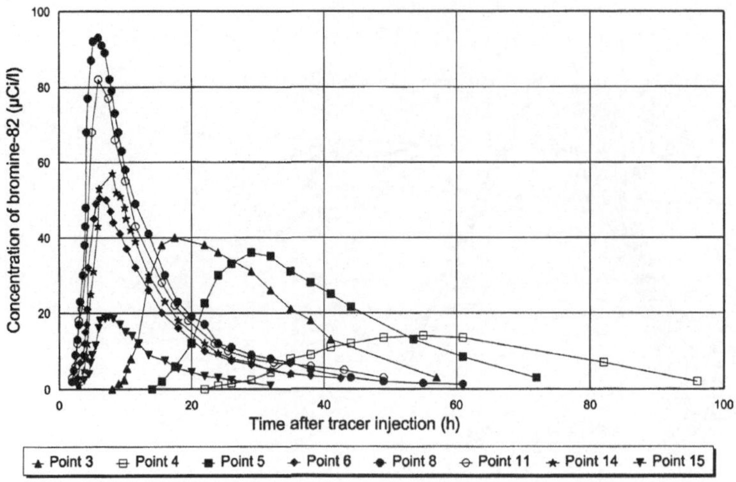

Fig. 9.5. Breakthrough curves obtained in a tracing experiment conducted in the Canelles reservoir, Spain. The curves corresponding to points 3, 4 and 8 showed a similar area.

Equation (9.14) was used for the required calculations. For the above-mentioned reasons, only the breakthrough curves with the largest areas were used for the calculation. These curves usually provided similar values of the flow rate. For example, in Fig. 9.5, the curves 3, 5 and 8 showed similar areas and, therefore, provided similar values of the flow rate. A total flow rate of the leakage close to 3 m^3/s was obtained. This value matched satisfactorily the calculation made by the water balance.

9.4.2 Lake Laja, Chile

The investigation carried out by the senior author at Lake Laja in Chile is described in a more detailed form in Chapter 12. Here, the results obtained in a tracing experiment are described, where 20 Ci of tritium in the form of tritiated water were injected.

As indicated in Chapter 12, the leakage from the lake is taking place through volcanic materials of a natural dike acting as a dam, which formed the lake in the past. A flow rate between 25 and 40 m^3/s, depending on the water level, escapes from the lake. The prevailing flow rate of the leak during the experiment was on the order of 26 m^3/s, according to the information provided by the electrical power plant that uses the leakage water for power generation. Groundwater emerges along a series of numerous springs located about 3.5 km from the lake.

The tracer was injected at a point of the lake where water was clearly escaping through a horizon formed by recent lava. The results obtained are shown in Fig. 9.6. The first arrival of the tracer was detected about 5 days after the injection. The

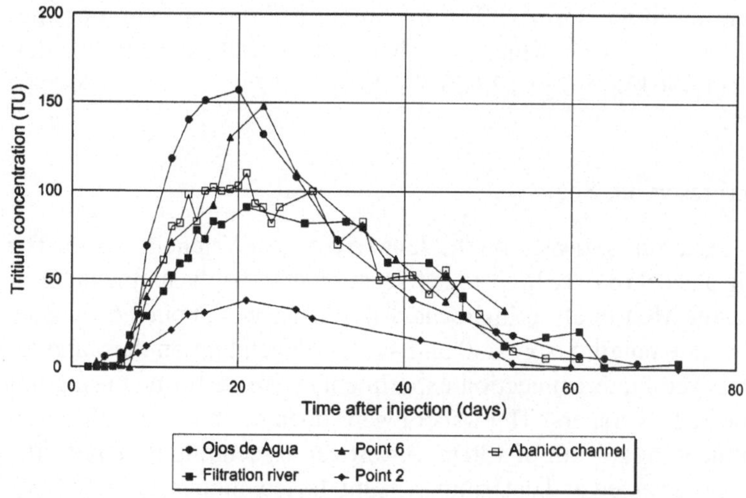

Fig. 9.6. Breakthrough curves obtained in a tracing experiment performed in Lake Laja, Chile.

maximum concentration was obtained about 20 days after the injection, and the tails of the breakthrough curves extended for more than 2 months. The lower areas of some breakthrough curves are explained, in this case, by the mixing of the labelled water from the lake with groundwater recharged in the intermediate basin, as indicated in Section 5.3.3.

The area of the two largest breakthrough curves corresponds approximately to a rectangle with a concentration of 60 T.U. ($0.189\,\mu Ci/m^3$) passing over about 45 days (3.89×10^6 s), that is to say to $0.735\ Ci \cdot s \cdot m^{-3}$. The injected activity (20 Ci) divided by this value gives a flow rate of $27.2\,m^3/s$, which is similar to the value provided by the electrical power plant. This fact proves that the tracer recovery was almost quantitative (96% of tracer recovery).

On the other hand, the transit time corresponding to the average velocity of the tracer was close to 23 days (water velocity equal to 152 m/d). The volume of water discharged at the springs during this period of time was $51.7 \times 10^6\,m^3$. Supposing a porosity of the rock equal to 30%, this volume implies the existence of a volume of saturated rock formation of $172 \times 10^6\,m^3$ and an average cross-section of the groundwater flow between lake and springs equal to $49{,}000\,m^2$. Such a big storage capacity of the rock formation between Lake Laja and the springs is difficult to accept. In Chapter 12, an alternative hypothesis that explains these results is described.

It must be stressed here that such an interconnection experiment, with quantitative tracer recovery and involving the labelling of such a huge water volume, can only be performed using artificial tritium as the tracer. No groundwater pollution was derived from this experiment. The tracer was discharged to the sea a few days after emerging in the springs. The maximum tritium concentration measured at the springs was 160 T.U., which is far below the trivial concentration described in Chapter 7 (1,496 kBq/m^3 or 12,600 T.U.).

9.4.3 Tous reservoir, Spain

The Tous reservoir is located on the Júcar river, near Valencia, Spain. The leakage was taking place through a limestone formation of the right abutment. Groundwater emerges at the Mourterall spring located at a distance of about 1.6 km from the dam with a flow rate ranging from 250 and 400 l/s, depending on the water level in the reservoir. Several interconnection experiments were performed using uranine and rhodamine WT as tracers. The tracers were injected in a borehole located in the right abutment, at a depth of 130 m. A vertical descending flow prevailing in this borehole was detected and determined using the techniques described in Chapter 8 (Section 8.6.2.2). The outflow of groundwater inside the borehole was located at a fracture located at a depth of 130 m. This depth was about 50 m lower than the altitude of the bottom of the reservoir.

The water impounded in the reservoir contained a high tritium concentration (between 35 and 55 T.U.) due to the discharge of radioactive liquid wastes from a nuclear power plant located in the Tagus basin and transferred to the Júcar basin. The tritiated water reaches the Tous reservoir via the aqueduct that connects the Tagus river with the Júcar river. However, the tritium concentration at the Mourterall spring was about 5 T.U., proving that most of the emerging water is groundwater from the nearby aquifer and not related to the reservoir. One of the obtained breakthrough curves is shown in Fig. 9.7.

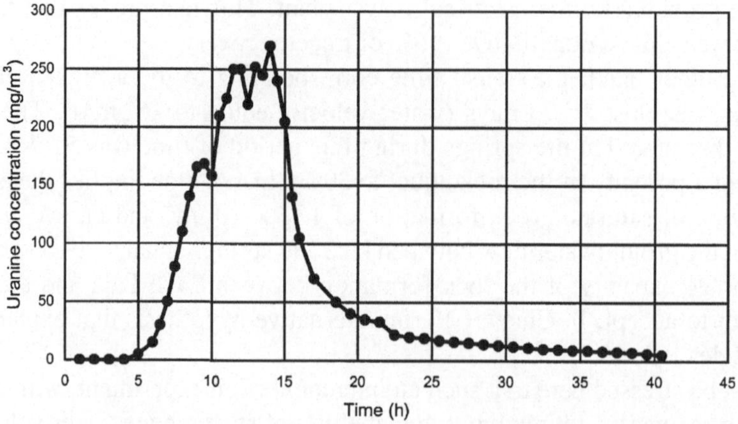

Fig. 9.7. Breakthrough curve obtained in a tracing experiment carried out in the vicinity of the Tous reservoir, Spain.

9.4.4 Joumine reservoir, Tunisia

At the Joumine reservoir in Tunisia, two tracing experiments were performed using, respectively, uranine and rhodamine WT as tracers. These tracers were injected in boreholes located a few tens of meters downstream of the dam at a distance of only 3 m from each other. The rock formation consists of karstified limestone. Tracer tests using the labelling technique of the entire water column described in Chapter 8 were performed in both boreholes. A high velocity flow was detected at the bottom of both boreholes, where the tracers were injected. Two springs were located at a distance of about 200 m from these boreholes. The breakthrough curves shown in Fig. 9.8 were obtained. The results show that the tail of the breakthrough curve was much higher in the case of rhodamine WT than for uranine, which suggests a strong interaction of the rhodamine tracer with the rock formations. The tracer recovery was only about 65% in the case of rhodamine WT and more than 95% for uranine. These results indicate, as expected, a more conservative behaviour of uranine with respect to rhodamine WT under these experimental conditions. Rhodamine B and WT are not recommended for quantitative groundwater tracing experiments.

9.4.5 PUNA reservoir, Africa

The dam of the PUNA reservoir (see footnote of Section 2.2.2) was constructed on dolomitic and limestone formations. A large number of leaks was identified during the first filling at both sides of the rock formation downstream of the dam. To

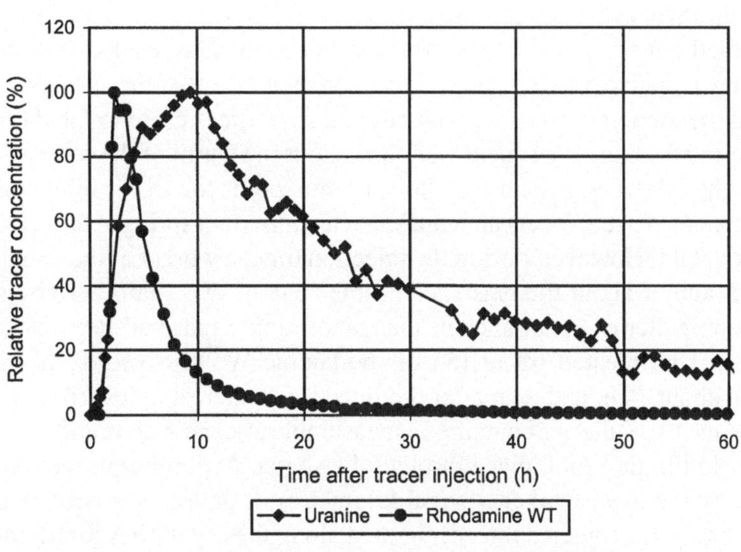

Fig. 9.8. Breakthrough curves obtained in a tracing experiment performed in the Joumine reservoir, Tunisia. The larger amplitude of the breakthrough curve corresponding to rhodamine WT suggests more pronounced interaction of the tracer with the rock formation.

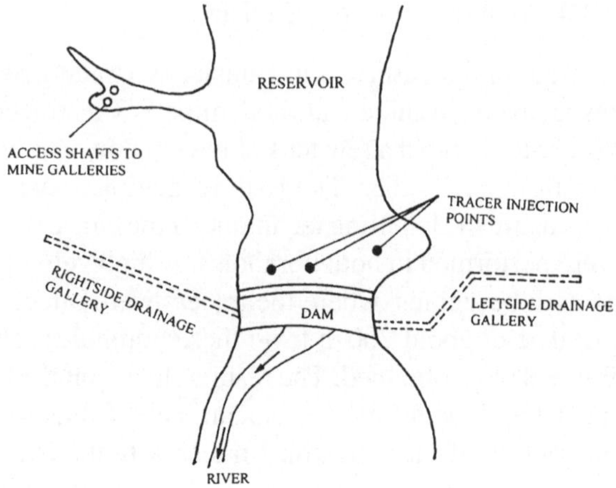

Fig. 9.9. Schematic view of the dam site of the PUNA reservoir, Africa.

reduce subsurface pressures, two drainage galleries were constructed, as illustrated in Fig. 9.9. At full reservoir level, the leaks were estimated to discharge about 3 m^3/s on the right side gallery and about 0.5 m^3/s on the left side gallery. Groundwater emerged at about 80 points on the right side gallery and about 50 points on the left side gallery. After a first grouting campaign carried out on the right side gallery, the total discharge of the leak was reduced to about 0.6 m^3/s.

Four interconnection experiments using fluorescent tracers (uranine and rhodamine WT) were performed by the senior author. In the first three experiments, the tracers were injected at points located close to the dam at distances between 100 and 200 m. Intensive sampling programs of the water flowing out of the galleries were carried out for several days after the injection. The results of these experiments were negative: the tracers were not detected in any of the springs. It became clear that the escaping water was not coming from areas close to the dam site.

The fourth tracing experiment was carried out in an arm of the reservoir located on the right side. The reason for choosing this point for tracer injection was, in principle, related to the recent appearance of an important spring at the opposite side of the mountain. However, during the injection time, a worker at the dam informed the senior author about the existence in this zone of two shafts for accessing the exploitation galleries of an ancient manganese mine that had been closed many years ago. The injected tracer (5 l of rhodamine WT) arrived at the right side gallery in about 12 h, and 24 h after the injection, water samples from most of the springs of the two galleries contained a measurable tracer concentration. Obviously, the tracer infiltration took place through the shafts. According to the information provided by the local worker, several kilometres of galleries existed in the zone. Most probably, the injected tracer migrated along these galleries to the zone of the dam site.

Due to the unpredictable character of these results, the designed sampling program in this case did not allow the measurement of complete breakthrough curves.

Therefore, the senior author recommended the repetition of the experiment under better sampling conditions. Unfortunately, however, the new experiments were not authorised by the staff responsible for the management of the dam. A new grouting campaign on the right side gallery was scheduled.

REFERENCES

Custodio, E. and Llamas, M.R., 1976. *Hidrología Subterránea*, Ediciones Omega, Barcelona (in Spanish).

De Josseling De Jong, G., 1958. Longitudinal and transverse diffusion in granular media. *Trans. Am. Geophys. Union*, 39, 524–536.

Guizerix, J., Molinari, J., Gaillard, B., Santos-Cottin, H., Mornas, A. and Corda, R., 1967. Localisation des fuites sur un grand réservoir à l'aide de traceurs radioactifs. In *Isotopes in Hydrology*, International Atomic Energy Agency, Vienna, pp. 587–598 (in French).

Kreft, A. and Zuber, A., 1978. On the physical meaning of the dispersion equation and its solution for different initial and boundary conditions. *Chem. Eng. Sci.*, 33, 1471–1480.

Lenda, A. and Zuber, A., 1970. Dispersion of tracers in underground water experiments. In *Isotope Hydrology*, International Atomic Energy Agency, Vienna, pp. 619–642.

Leibundgut, Ch., 1974. Fluoreszierende Markierfarbstoffe in der Hydrologie. *Mitt. Naturforsch. Ges. Bern, N.F.*, 31, 63–84 (in German).

Maloszewski, P., 1992. Mathematical modelling of tracer transport in different aquifers: Results from ATH test fields. In H. Hötzl and A. Werner (Eds.), *Tracer Hydrology. Proc. 6th Int. Symp. Water Tracing, 21–26 Sept. 1992, Karlsruhe*, Balkema, Rotterdam, pp. 25–30.

Maloszewski, P., 1994. *Mathematical modelling of tracer experiments in fissured aquifers. Freiburger Schriften zur Hydrologie* 2, Freiburg i. Br, 107 pp.

Molinari, J., 1976. Recherche et localisation de fuites sur retenues naturelles ou artificielles par techniques de traceurs. *Houille Blanche*, 3/4, 301–312 (in French).

Plata, A. and Del Val Cob, M., 1972. Estudio de las fugas de agua de un gran embalse con trazadores radiactivos. *Agua*, 70, pp. 68–76 (in Spanish).

Plata, A., 1986. Investigation of the origin of the water discharging at the Reocin mine using isotope techniques. In *Isotope Techniques in the study of the hydrology of fractured and fissured rocks*, International Atomic Energy Agency, Vienna, pp. 239–252.

Sauty, J.P., 1980. An analysis of hydrodispersive transfer in aquifers. *Water Resour. Res.*, 16, 145–158.

UNESCO, 1973 with Supplements 1975, 1977 and 1983. *Groundwater studies: an international guide for research and practice*. UNESCO, Paris, France.

CHAPTER 10

Tests inside the reservoir

10.1 INTRODUCTION

The tests performed inside the reservoir or lake related to leakage studies have as the main or unique objective the localisation of the areas where water losses are taking place. This is really a difficult task because of the point character of the available techniques and the large extent of the impounded areas in the reservoir where the leakage or seepage can occur. For this reason, before applying these techniques, and to reduce the area in which the techniques have to be applied, it is very important to explore exhaustively all the possibilities offered by the techniques based on the measurement of the natural tracers of water, such as temperature, stable isotopes of water, and chemical composition, as described in previous chapters. In the same way, the filtration law, that is, the variation of the leakage rate as a function of the water level in the reservoir can provide very valuable information on the topographic elevation at which the water losses are taking place. Of course, the geological setting of the bottom of the reservoir and the surrounding formations has to be considered as the starting point.

Obviously, the above-mentioned techniques must be complemented with a detailed study of the geological and hydrogeological characteristics of the materials outcropping at the bottom of the reservoir. For example, the identification of fractures or faulted zones, areas where dolines or diaclases are present, geological contacts, karstified strata, remains of alluvial material, etc. can lead to the identification of areas where the leakage or seepage of water is more likely.

In the same way, a detailed visual inspection of the reservoir bottom, taking advantage of eventual periods of partial emptying, may be used to identify specific areas where water infiltration may take place. The accumulation of plant debris at specific points or the absence of deposited sediments can provide additional evidence or indications of water outflow. Other marks suggesting water infiltration can be the presence of depressions and channels in recent sediment layers, as well as the grain size distribution of these sediments; the fraction of fine particles can be smaller in the infiltration zones due to their preferential washing out by the infiltrating water. This can be important in cases of spread seepage.

The number of studies performed at the bottom of the reservoir for the identification of preferential leakage or seepage zones is very limited. Even in most of the reported studies, the validity of the results obtained were not confirmed in an irrefutable way using other more direct techniques such as, for example, interconnection experiments using artificial tracers. The suggested techniques have been presented as new methods that demand an ulterior development that never took place. In other words, the use of most of the proposed techniques did not have the required continuity in time and never were applied in a complete form. The case studies described in Chapter 12, which were conducted by the senior author in the Colbún reservoir and Lake Laja, both in Chile, are probably the only known exceptions to the previous statement.

All techniques proposed for the localisation of areas of preferential leakage or seepage are based upon the water flows generated at the bottom of the water body. Apart from exceptional cases where huge water losses were concentrated in a small area of the bottom of the reservoir, the flows or currents generated by most leaks tend to be very small. Very frequently, these flows are much lower than the natural flows or currents produced by factors such as wind, temperature differences, etc. In spite of the fact that these natural water flows tend to be very small for depths lower than about 10 m, they constitute the main difficulty for the development of reliable techniques for the localisation of preferential leakage zones at the bottom of reservoirs.

Despite the previous statements, the development of simple and reliable devices for the rapid measurement of water flows, including their direction and velocity, produced at the bottom of the water body by the leakage, seems to be possible by means of recently developed methods. This is a pending task that most probably will be tackled in the near future. Most probably, the solution to this problem will not be related to further developments in the field of artificial tracers, but to techniques such as electromagnetic induction, ultrasonic-based methods, radar techniques, migration of thermal pulses, and others. The measuring system to be developed should allow the survey of large areas of the bottom of the reservoir in a relatively short period of time. However, the techniques based on the use of artificial tracers will play, in any case, an important role for the confirmation of the water flows detected with the measuring device to be developed. This confirmation would be done by means of infiltration tests using the techniques described in this chapter.

The techniques for the localisation of leakage or seepage zones at the bottom of reservoirs are very attractive because, if positive results are obtained, they can allow the suppression or reduction of the water losses through the impermeabilisation of the materials through which these are taking place. These impermeabilisation works will surely be less expensive than those to be carried out downstream in a zone close to the springs. This is because the volume and surface of rock formation involved in the leakage in the latter zone can be of much higher dimensions, due to the dispersion of the water flow. This was the case found in the PUNA reservoir (see Section 2.2.2), where most of the escaping water was infiltrating through two submerged shafts from an old manganese mine. Large sums were invested in grouting works carried out at a front of about 200 m of a limestone formation located at the dam site. The efficiency of these grouting works was very poor. Most of the leaks could have been eliminated simply by stopping up the flow at the mouths of the two shafts.

In the next sections, a summary is presented of the main techniques used or suggested for the localisation of preferential leakage or seepage zones at the bottom of lakes and reservoirs. The high diversity of these techniques does not allow their classification in well-defined groups.

10.2 USE OF FLOAT DROGUES

The use of float drogues for measuring water flows is very common in oceanography. The devices commonly used are schematically illustrated in Fig. 10.1. Two metallic or plastic plates mounted perpendicularly among them are hung from a buoy using a fine rope. The length of this rope is chosen in such a way that the drogue remains at the depth where the water flow needs to be measured or identified. The float drogue displaces according to the prevailing water flow. To minimise the effect of the flow existing in the upper layers, a very fine rope, for example a fishing line, and a small buoy are used. Preferably, the average density of the

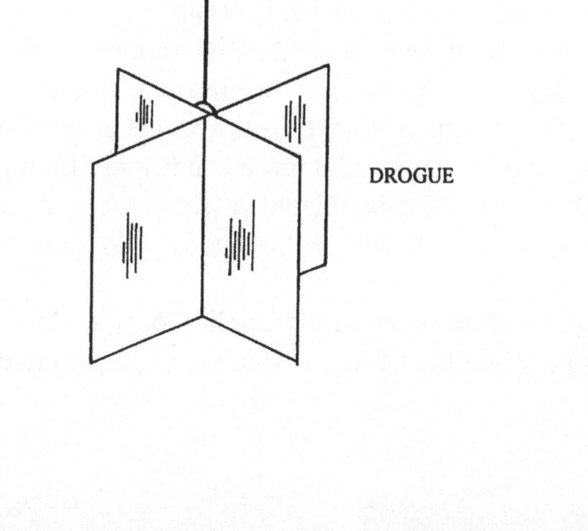

Fig. 10.1. Details of a float drogue for the measurement of water flows in lakes and the sea.

drogue should be only slightly higher than the water density. Obviously, the use of plates with a large surface increases the dragging force produced by the water flow, making the migration of the measuring device derived mainly from the velocity of the flows at the point where it is located. Plate surfaces between 0.5 and 1 m^2 are appropriate. In the case of oceanographic studies, the buoy is labelled with a flag and/or a radar reflector to facilitate its location after a long period of time has elapsed since its launching. In the case of small lakes or reservoirs, this may not be necessary.

The experimental technique consists of launching several float drogues in the area suspected of leakage with the plate device placed close to the bottom. Then, the migration of each float drogue is followed during a certain period of time. Depending on the accuracy requirements, global positioning systems (GPS) or, alternatively, theodolites can be used for the positioning of the float drogues. The accuracy of GPS reaches some tens of metres, depending on the latitude of the study area, while for theodolites it can be even finer. The best choice is the use of laser-based theodolites as a range and azimuth positioning system. The theodolite bounces a laser beam off of a prism located on the study boat. Then, it records the distance and azimuth with respect to the boat. These data are electronically processed and provide the x–y position of the boat relative to the base station (theodolite position). This information can be transmitted via radio to an on-boat computer, for example, every second.

After several hours or maybe days, the float drogue may remain in a fixed position of the lake, which may correspond to the site of a preferential point of infiltration. Obviously, the stop of the float drogue can also occur because the device has touched the bottom of the lake. This has to be checked. When it happens, the fishing line or rope on the device will not be in a vertical position. In any case, it is recommended to repeat the exercise several times to check that the stop of the drogue always occurs at the same point.

The technique can be particularly interesting in cases where the leakage is linked to fractures or cavities located at the lateral walls of the reservoir. In this case, float drogues are launched at different depths at several points close to the wall. Also interesting can be the case when a doline or sinkhole exists at the bottom of the reservoir. If the escaping water infiltrates at this point, a vortex may exist, and the float drogue will remain permanently fixed after reaching this point.

The technique can be successfully applied in cases of shallow lakes or reservoirs with a high leakage rate. This is the case of the Koka Lake in Ethiopia, where the average water depth is close to 7 m at maximum level, and a leakage rate of several tens of cubic metres per second is suspected to exist under these conditions. Taking into account that the bottom material consists of consolidated volcanic rocks, float drogues can be used to localise the zone of the lake where the investigated leakage is concentrated.

For obvious reasons, this technique can be successfully applied only in cases of very high leakage rates, for example larger than 1 m^3/s, which are concentrated in small areas. Only in this case, water flows of sufficient high flow rate may exist in the reservoir so as to govern the migration of the float drogue inside the reservoir.

10.3 TRACING OF THE RESERVOIR WATER

The localisation of preferential infiltration zones at the bottom of the reservoir can be undertaken by means of interconnection experiments between the reservoir and the springs located downstream. The method is based on the execution of tracer injections at different points of the reservoir located close to the bottom and the measurement of the eventual breakthrough curves at the existing springs or downstream boreholes. Each tracer injection provides information on the water leakage corresponding to a small area close to the injection point. The breakthrough curves will have different forms, depending whether the injection is made at a point close to the infiltration area or sinkhole, or far away from it. The following differences will be observed:

1) The mean transit time of the tracer between the reservoir and the measuring point will be shorter if the injection is made at a point close to the area where preferential infiltration occurs. Even if the injection is made at a point located very far away from the infiltration area, most probably the tracer will never be detected at the spring due to the high dilution produced inside the reservoir.
2) The amplitude of the measured breakthrough curves will be smaller if the injection is made close to the infiltration area. The reason for that is the dilution undergone by the tracer before its infiltration in the terrain. The differences between different breakthrough curves can be extreme. Obviously, the maximum tracer concentration of the breakthrough curve will be drastically affected by this initial dilution.
3) Very frequently, the fraction of tracer recovered at a specific spring will be higher when the tracer is injected at a point close to the infiltration area. If the tracer is injected far away from this area, the natural flows existing in the reservoir can transport all or part of the tracer in a different direction, and it never will be detected.

To accelerate the investigation, simultaneous injections of different tracers can be made at different points of the reservoir. The arrival of these tracers at the springs is determined at the same time and in the same water samples. Automatic water samplers that operate 24 hours a day can be used (see Fig. 9.3 below). According to the indications given in Chapter 7, the most appropriate tracers for these multitracing experiments are uranine, rhodamine B or WT, ^{131}I and tritium. An additional tracer can be the Br^- ion in the form of NH_4Br. However, to ensure the detection of the tracer, very high amounts of this product have to be injected, for example more than 50 or 100 kg, depending on the flow rate of the springs.

A multitracing experiment was carried out by the authors at the reservoir Ulldecona, Spain, using uranine, rhodamine B and tritium. Due to the high dilution undergone by these tracers, both in the reservoir and in the aquifer, only tritium could be detected in the springs located downstream of the dam. This tracer arrived at the springs about 6 months after the injection, and only a small fraction of the injected tritium activity was recovered.

Obviously, in a multitracing experiment, two or more tracers can emerge at the spring at the same time, and consequently, some interferences might occur with their separate detection. For the above-mentioned tracers, the only possible interference corresponds to the detection of admixtures formed by uranine with rhodamine and tritium with ^{131}I. In the first case, the individual concentration of the two fluorescent tracers can be precisely determined by measuring the water sample twice with the fluorometer. One measurement will be performed using the uranine filter setting and the second with the rhodamine filter. The fluorometer response for uranine using the rhodamine filter setting will be very small, and the same occurs in the inverse case. The measuring technique of admixtures of these tracers requires the calibration of the fluorometer for both of them, that is to say, using the settings for both filters. Let us call $(E_U)_U$ and $(E_U)_{RH}$, respectively, the fluorometer efficiency for uranine using the uranine and rhodamine filter settings, $(E_{RH})_U$ and $(E_{RH})_{RH}$ the efficiency for rhodamine using the same filter settings, and $(R_U)_S$ and $(R_{RH})_S$ the readings given by the fluorometer for a water sample in which both tracers are mixed, using the uranine and rhodamine filter settings, respectively. The uranine and rhodamine concentrations C_U and C_{RH} can be obtained by solving the following system of equations:

$$(R_U)_S = C_U(E_U)_U + C_{RH}(E_U)_{RH} \tag{10.1}$$

$$(R_{RH})_S = C_U(E_{RH})_U + C_{RH}(E_{RH})_{RH} \tag{10.2}$$

All parameters involved in this system of equations are experimentally determined. The solution of this system gives the respective tracer concentrations. The efficiency is computed as the ratio between the reading given by the fluorometer, expressed, for example, in milligrams per cubic metre (ppb), for a standard tracer solution and the known concentration of this solution. For example, using a filter fluorometer model Turner AU-10, the following efficiency values were obtained: $(E_U)_{RH} = 0.0078$ (0.78%), $(E_{RH})_U = 0.012$ (1.2%), $(E_{RH})_{RH} = 1$ (100%) and $(E_U)_U = 1$ (100%).

In the case of samples where both tritium and ^{131}I are present, the respective concentration of both radionuclides can be easily determined because tritium does not interfere at all in the measurement of ^{131}I since it is a pure beta emitter. Therefore, the ^{131}I concentration is directly obtained by measuring the water sample with a NaI(Tl) crystal scintillation detector. On the contrary, ^{131}I will interfere in the measurement of tritium by the liquid scintillation technique due to the beta particles emitted by that isotope. For this reason, the determination of the counting efficiency of ^{131}I using the liquid scintillation detector with the same channel setting used for tritium is necessary. Let us call $(E_I)_T$ this counting efficiency, given by the ratio between the readings of the liquid scintillation detector, expressed in counts per second, obtained for a standard ^{131}I sample and its activity expressed in number of decaying atoms per second (becquerels). If E_T is the counting efficiency of the liquid scintillation detector for tritium obtained by calibrating the measuring system with a standard solution, the reading of the water sample R_S will be given by the following equation:

$$R_S = C_T E_T + C_I (E_I)_T \tag{10.3}$$

The values of the efficiencies $(E_I)_T$ and E_T have been determined experimentally, and the ^{131}I concentration has been measured with the NaI(Tl) crystal scintillation detector. Therefore, the tritium concentration of the samples can be computed from the previous equation.

Obviously, the injections should be performed in points of the reservoir where the presence of the infiltration areas is suspected according to the information provided by geological and hydrogeological data, temperature, chemical and isotopic composition, etc. In some cases, it may be interesting to select as injection points depressions in the bottom of the reservoir because these low topography areas can be linked to sinkholes. In this case, it may be advantageous to inject the tracer dissolved in a relatively high volume of water (for example, several cubic metres) with a high concentration solution of common salt (for example, 5 g/l). The higher density of this solution with respect to the reservoir water will facilitate the migration of the tracer solution to the deepest zone of the depression, avoiding its dispersion in the water body.

Makowski (1967, 1970) applied the above-described multitracing technique in the reservoir Pszeczica, Poland, using ^{82}Br, ^{51}Cr and ^{131}I as tracers. Single tracing of reservoir water was performed by Andreu Ibarra et al. (1979) in the Lajas reservoir, Mexico.

In other cases, instead of depressions, other structural accidents are identified in the bottom, such as areas intensively fractured, through which water may infiltrate. In these cases, the tracer to be injected can be confined at the suspected permeable zone by means of a circular barrier that prevents the tracer from migrating far away from this zone. This barrier can be made of a flexible plastic plate of about 1 m width. An iron chain is fitted at the lower border of the plastic plate, and a line of small buoys is fitted to the upper border. Under these conditions, when the plastic plate is submerged in the reservoir, a vertical wall is formed. A closed circular zone can be delimited, inside of which the tracer solution is injected.

10.4 FREE MIGRATION OF THE TRACER CLOUD

The method of free migration of the tracer cloud was first used by Guizerix et al. (1967) at the reservoir Krith-Wildenstein, France. In this method, a relatively high activity of a radioactive tracer is injected at a point of the reservoir where the presence of leakage is suspected. The tracer injection is made at a short distance from the bottom, for example, 5 m. Commonly, a point injection is performed, but depending on the nature and shape of the bottom area, the injection along a line may be more advantageous. If the flow prevailing at the injection point or area is mainly governed by the leak, the tracer cloud will migrate in the direction of the sinkhole or seepage zone, following a pathway more or less parallel to the bottom. The migration of the tracer cloud is controlled from one or more boats by means of submerged NaI(Tl) scintillation detectors. When the tracer arrives at the infiltration zone, the migration stops and the cloud intercepts the bottom of the reservoir (see Fig. 10.2). After a certain period of time, the tracer cloud vanishes due to the infiltration of the radioactive material. The position of the boats during the control of the tracer cloud is carried out by means of GPS or, more precisely, by using a laser-based theodolite according to the technique described in Section 10.2. The information obtained through this method is used to plot on a chart the position of the tracer cloud at different times after the injection, as well as the area where the tracer underwent infiltration.

The most appropriate radioactive tracers for this kind of experiment are ^{82}Br and ^{131}I. The former radionuclide allows the labelling of water volumes much higher

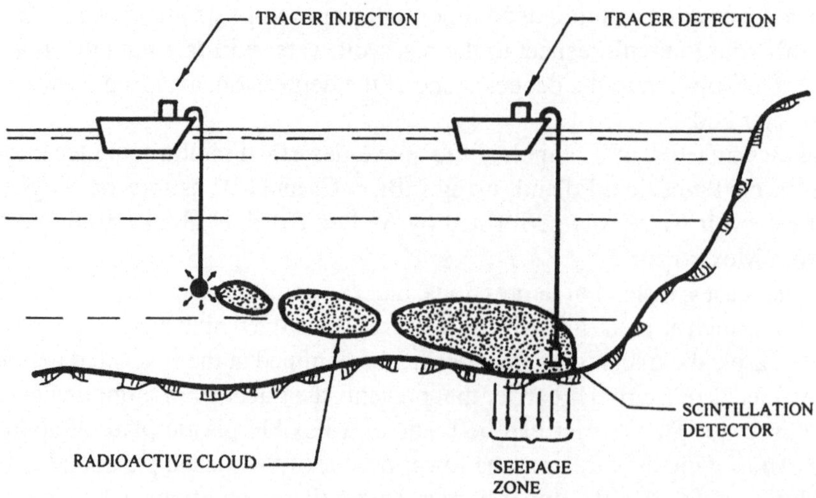

Fig. 10.2. Illustration of the method for the localisation of preferential infiltration areas at the bottom of the reservoir, based on the migration of a radioactive tracer cloud.

than ^{131}I. For example, an activity of ^{82}Br of 1 Ci can be easily detected in the reservoir after its dilution with 25×10^6 m^3 of water (see Section 7.6.2). However, disadvantages of ^{82}Br are the need for heavy lead transportation containers (between 350 and 500 kg) and its short half-life (36 hours). In some cases, the half-life can be too short to finish the experiment. In any case, the radioactive material has to be prepared 1 or 2 days in advance of its use in the field, creating some logistical problems. When these difficulties cannot be easily overcome, the only choice is ^{131}I. Activities of several curies of this radionuclide can be necessary for this kind of experiment.

This experiment should be carried out together with a parallel control of the eventual appearance of the tracer at the springs or boreholes located downstream from the dam. This requires the implementation of a periodic sampling campaign in the springs. The sampling can be done by means of automatic samplers like the one shown in Fig. 9.3.

The described technique was successfully used by Plata et al. (1989) in the Colbún reservoir, Chile. The results obtained are described in Chapter 12. Usually, the zone where the tracer infiltrates remains labelled with the radioactive material for a long period of time after finishing the experiment. This allows the identification "a posteriori" of the infiltration zone, as in the case of the Colbún reservoir. In this case, the injected tracer was also detected in the emerging water downstream of the dam.

The main difficulty for the application of this technique lies in the interference produced by the natural flows prevailing in the lake. However, these flows are usually small at short distances from the bottom, particularly, in cases of deep stratified lakes. Therefore, this type of experiment should be implemented only when sufficient information is available on the zone where the escaping water infiltrates.

On the other hand, it is obvious that the technique can provide positive results only when the infiltration of water takes place in a concentrated form, that is, in a small area. This type of infiltration is more frequent when the bottom of the reservoir consists of consolidated rocks but also can be found in granular materials, as was found in the Colbún reservoir. Concentrated infiltration can also occur through the contacts of granular materials with concrete structures of the dam.

The described method can also be applied using uranine or rhodamine as tracers. The detection can be made, in this case, using a fluorometer fitted with a coupled fibre optic probe, which has been recently developed. If this fluorometer is not available, the tracer detection can be achieved by means of a fluorometer fitted with a continuous flow cell. In this case, the reservoir water is pumped through this cell and the tracer concentration is continuously recorded on paper or on a portable computer. The positioning of the boats is performed in the same way as described in Section 10.2. Calmels and Saleçon (1992) have applied this technique for the investigation of water flows in the reservoir Pareloup. The flows were studied both at the epilimnion and hypolimnion of the reservoir.

10.5 METHOD OF FILTER TUBES

The filter tube method has been suggested and applied by Guizerix et al. (1967) at the reservoir Kruth-Wildenstein in France. The method is illustrated in Fig. 10.3. A filter tube is sunk to the bottom of the reservoir in the zone suspected of being affected by leakage. The lower stretch of the tube sunk in the terrain is perforated with holes or slots. The tracer solution is then injected in this tube and its eventual outlet at the downstream springs or boreholes is controlled in the usual way. The main advantage of this technique compared with the free injection of tracer at the water body is the disappearance of the dispersion effect on the tracer derived from the natural water flows existing in the lake. As a consequence, the tracer amount required for the interconnection experiment is drastically reduced. A disadvantage may be the point character of the test. A negative result of the experiment can be obtained just because the filter tube was placed in the wrong site. In the case of concentrated outflow, it can be difficult to find the right site to sink the tube.

To avoid experimental failures with this technique, before injecting the tracer in the filter tube, it is recommended to prove that a vertical descending flow is present inside the tube. This can be done using the techniques described in Chapter 8 for measuring vertical flows. If no vertical flow exists, no tracer injection should be performed.

For obvious reasons, the installation of filter tubes can be very difficult in the case of deep reservoirs. However, this work can be done at a time when the reservoir is empty or contains a small layer of impounded water. The filter tube can be connected to the surface of the reservoir by means of a flexible plastic tube through which the tracer can be injected. The method can be very interesting in the case of new reservoirs. Filter tubes can be fitted in points of the flooded area where a high risk of leakage has been identified.

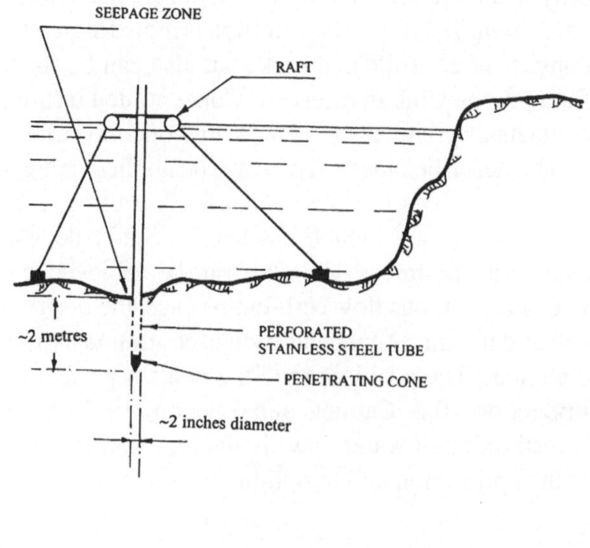

Fig. 10.3. Illustration of the method suggested by Guizerix et al. (1967) for the identification of infiltration areas at the bottom of the reservoir based on tracer interconnection experiments. The tracer is injected in a perforated tube to avoid its dispersion in the water body.

10.6 LABELLING OF THE INFILTRATION ZONE (ADSORBED TRACER)

The method based on the labelling of the infiltration zone is illustrated in Fig. 10.4. Basically, the method is similar to the procedure described in Section 10.4. The only difference is that instead of using a tracer that shows a similar behaviour as water, the tracer used shows a strong tendency to be adsorbed in the terrain. The best tracer with which to implement this technique is the radionuclide ^{198}Au, although other tracers, such as ^{59}Fe, ^{51}Cr, ^{110}Ag, etc., can also be used. The radioactive tracer is injected at a short distance from the bottom of the reservoir in the zone where the presence of leakage or seepage is suspected. The tracer will migrate according to the prevailing flow. Most of the tracer mass will be adsorbed in the infiltration zone, which will become labelled for a certain period of time. The localisation of this labelled zone is carried out 1 or 2 days after the tracer was injected using a submerged scintillation detector. Depending on the conditions at the bottom of the reservoir, the detector can be dragged along the bottom from a boat or point by point measurements can be performed from a boat. The results obtained with any measuring system can be plotted in the form of contour maps corresponding to equal responses of the detector, which will show the location and extent of the infiltration area.

The ^{198}Au is used in the form of chloroauric acid ($HClAuO_4$), which is prepared by dissolving gold, irradiated in a nuclear reactor, with aqua regia (three parts of HCl and one part of HNO_3).

This technique has been successfully applied by some authors, for example, Easey and Hanna (1994) at the Pedu reservoir, Malasia, and by Andreu Ibarra et al. (1979) in the Lajas reservoir, Mexico. In the first case, the injection of

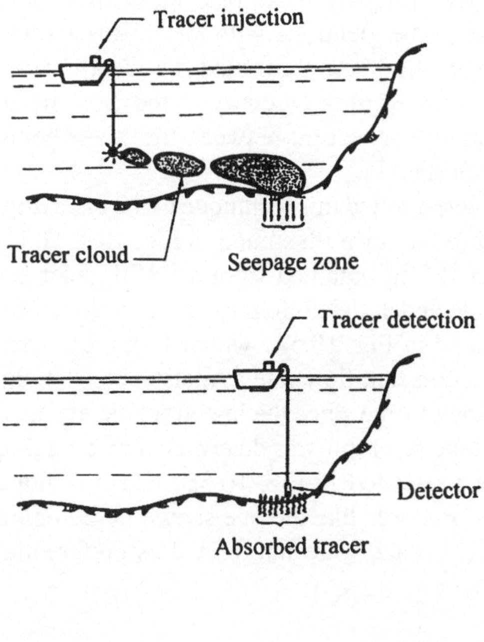

Fig. 10.4. Illustration of the method for the localisation of infiltration areas at the bottom of a reservoir based on the use of absorbable radioactive tracers.

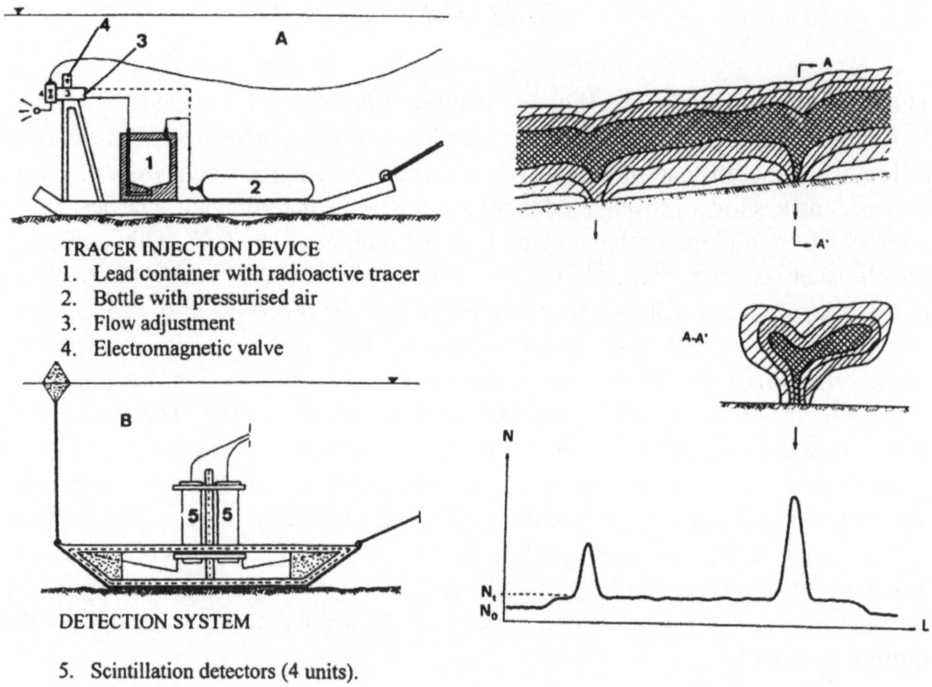

Fig. 10.5. Illustration of the method described by Owczarczyk and Szpilowski (1984) for the localisation of infiltration areas at the bottom of the reservoir using a technique based on the fixation of a radioactive tracer in the bottom sediments.

^{198}Au was accompanied by the simultaneous injection of tritium in the form of tritiated water as a conservative tracer, with the purpose of proving the hydraulic connection between the infiltration zone (to be identified with the ^{198}Au) and the emerging waters. The result was positive: the ^{198}Au permitted the identification of various preferential infiltration areas located in the vicinity of the dam, while tritium provided evidence on the hydraulic connection between the water being infiltrated through these zones and the springs.

Owczarczyk and Szpilowski (1984) have applied this technique to the reservoirs Przeczyce and Wista-Czarne, in Poland in the form illustrated in Fig. 10.5. Gold-198 in colloidal or $HClAuO_4$ form and ^{113m}In obtained from a ^{113}Cd generator were used as tracers. The tracers were injected at the bottom of the reservoir along a line using the injection device illustrated in Fig. 10.5A, which is dragged over the bottom. The tracer is injected with a constant flow rate at a distance from the bottom close to 1 m. After a certain period of time since the injection, the eventual deposition of the tracer on the bottom of the reservoir was determined by dragging on the same line the measuring device illustrated in Fig. 10.5B, which is fitted with four NaI(Tl) scintillation detectors. Records like the one shown in the figure were obtained, in which the radioactivity peaks were interpreted as preferential infiltration points in the bottom of the reservoir. Obviously, this technique can be applied only in reservoirs whose bottom is covered with sediments that allow the dragging of both the injection and the measuring devices. The presence of plant remains or rocks will make the use of this technique impossible.

Other alternative techniques have been proposed for the labelling with radioactive tracers of the infiltration zones at the bottoms of lakes and reservoirs. The most interesting are those based on the injection of a certain product or material labelled with a radioactive tracer whose density is slightly higher than water. In this way, the labelled product will be slowly deposited at the bottom of the reservoir and will concentrate at the preferential infiltration or seepage zones. The following materials have been suggested for this purpose:

1) Fine sediments with the appropriate grain size distribution, chosen so that their deposition at the bottom of the reservoir takes place several hours after the injection. The labelling of these sediments can be easily done by facilitating the adsorption by the sediment particles of a solution containing ^{198}Au. The sediment sample is stirred up for a short period of time with the ^{198}Au solution in the form of $HClAuO_4$ (Azoeuf et al., 1982; Bougault, 1970), and the tracer is firmly adsorbed on the surface of the sediment particles.
2) Anionic exchange resins with absorbed ^{131}I, whose density is very close to that of water.
3) Molinari et al. (1970) suggested the use of an emulsion of a bituminous material mixed with a small amount of triolein labelled with ^{131}I.[1] The emulsion is prepared by stirring the bituminous product with a detergent. The tracing material is injected in the reservoir and transported to the infiltration zone by the flow generated by the seepage or the leak. The emulsion breaks down in the infiltration zone due to the accumulation of the labelled particles and due to the effects of the bottom materials. The tracer is then detected by means of submerged scintillation detectors. The technique was used in the Rhine–Marne channel, France, but never was used in a reservoir. Additional studies are necessary to evaluate the possibilities offered by this technique.

10.7 DIRECT MEASUREMENT OF INFILTRATION

10.7.1 Generalities

Several methods have been suggested and/or used for the direct measurement of the seepage or infiltration rate at the bottom of lakes and reservoirs. Most of these methods are based on the measurement of the vertical descending flow generated

[1] The triolein labelled with ^{131}I is commonly used in nuclear medicine for the study of fat metabolism. The labelling procedure is very simple.

at the infiltration zones. As indicated below, all the measuring devices or methods present important limitations and can be successfully applied only under concrete favourable circumstances. In the opinion of the authors, the adaptation of some of the modern geophysical methods currently used for hydrological investigation should lead in the near future to the development of a reliable and efficient measuring device for the solution of this important problem. The proposed techniques to date are summarised in the next sections.

10.7.2 DWICA meter

The DWICA meter is a measuring system developed by Aebersold (1964). The name of DWICA given by the author corresponds to the initials of the expression "deep water isotopic current analyser." The measuring system consists of eight scintillation detectors mounted on a circular frame, forming an octagon. In the centre of this frame, an injector of radioactive material is adapted. Pulse injections of tracer are performed according to the commands given from the boat, where the measuring system is located. The orientation of the measuring system underwater is achieved by means of a device fitted with a compass. The injected tracer arrives at the position corresponding to one or two detectors, depending on the flow dispersivity. The orientation of one or two detectors provides the flow direction, and the time needed for the tracer to travel from the injection point to the detectors provides the flow velocity. The measuring system allows the measurement of flow velocities above 1 mm/s.

No details were reported about the application of this measuring system to the investigation of real leakage problems. Therefore, this method should be considered just an idea. Obviously, such a measuring system can only provide valuable information in cases of sinkholes with a high leakage rate.

10.7.3 Method proposed by Zuber

The method proposed by Zuber (1970) for the measurement of infiltration can be of interest in cases of reservoirs or lakes of shallow depth with almost flat bottoms that are covered with sediments. The method is illustrated in Fig. 10.6. The measuring technique consists of a bell-shaped container or cloche deposited at the bottom, in the manner shown in the figure. A small diameter metallic tube is fitted at the upper part of this cloche, in such a way that all water infiltrating across the covered zone has to flow through this tube. Due to the high ratio between the diameter of the cloche and that of the metallic tube, the device acts as a flow amplifier. The velocity of the water flowing through this tube is measured by means of a radioactive tracer. A pulse of tracer is injected at a point of the small tube, and its passage through the position of a collimated scintillation detector, fitted at a short distance downstream, is detected and recorded. The transit time of the tracer, together with the distance between the injection and the detection points, provides

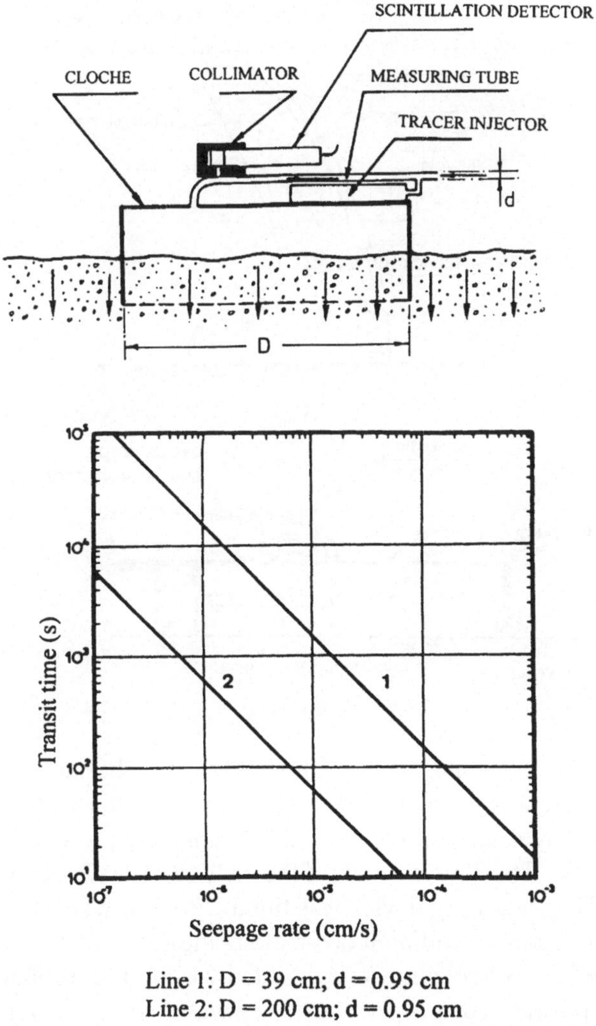

Fig. 10.6. Illustration of the method described by Zuber (1970) for the measurement of the seepage rate at the bottom of a lake using an inverted cloche.

the flow velocity. The value of this parameter, multiplied by the cross-section of the tube, provides the flow rate of the infiltrated water.

For obvious reasons, the method is difficult to apply in cases of deep reservoirs or lakes, as well as in cases of reservoirs whose bottoms are covered with plant debris or rocky blocks. The use of a radioactive tracer to measure the flow velocity through the tube is not justified. This flow velocity can be easily measured in a much simpler way, as shown below. In any case, the same method could have been developed using common salt solution as a tracer and a conductivity meter as detector.

The senior author used a similar method for the detection of fissures at the body of the Atazar concrete dam, Spain, which were produced by a sitting-down process

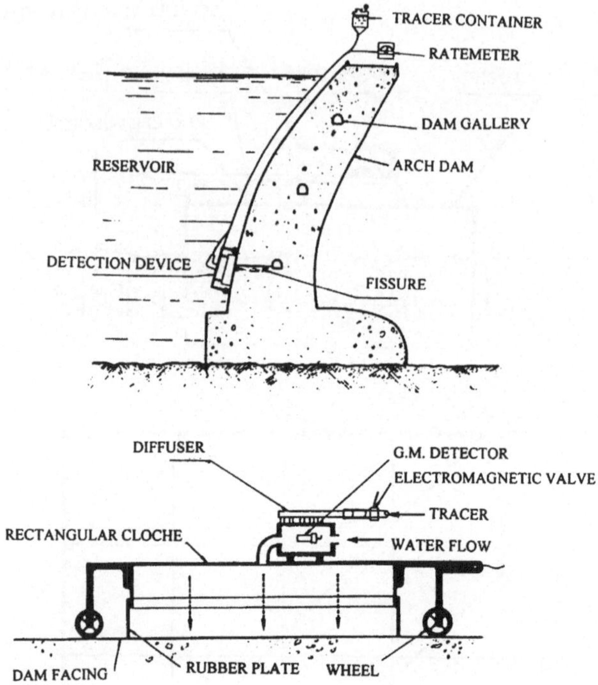

Fig. 10.7. Illustration of the method used by Plata for the localisation of fissures at the body of the concrete arc-dam El Atazar, Spain.

(see Fig. 10.7). The fissures were located at a depth of about 80 m from the surface. In this case, the impounded water was absolutely necessary for the water supply of the city of Madrid, and for this reason, it was not possible to empty the reservoir so as to undertake the repair works. The fundamentals of the method were similar to those illustrated in Fig. 10.6. The measuring device was fitted with four wheels so it could be lowered to different depths by sliding it down along the dam walls. The lateral water ingress in the cloche was kept almost negligible by using four rubber plates. The seepage rate at the position covered by this device was measured using a dilution technique instead of the method proposed by Zuber. A small activity of ^{131}I was injected in a 3-l container, inside of which a Geiger-Müller was fitted. The exponential decrease with time of the tracer concentration provided quantitative information on the seepage rate. This technique could also have been applied using a saline tracer, such as a saturated solution of common salt, and a conductivity meter as detector.

10.7.4 Method proposed by Easey

Easey and Hanna (1984, 1985) proposed an alternative method to measure the seepage rate, as illustrated in Fig. 10.8. In this case, the cloche is connected with the water surface by a flexible plastic tube. The flow along this tube induced by the infiltration in the area covered by the cloche is measured by means of a flow meter

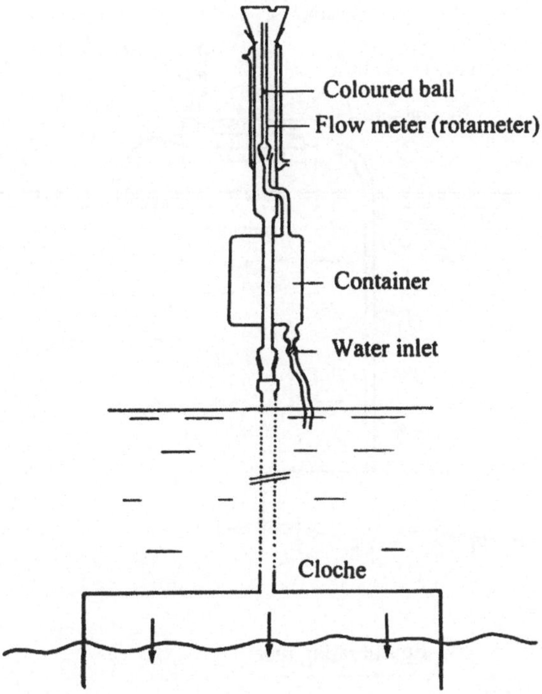

Fig. 10.8. Illustration of the method used by Easey and Hanna (1984, 1985) for the measurement of the seepage rate at the bottom of a reservoir.

specifically designed for liquids, called a rotameter. The inlet of this rotameter is directly connected to the reservoir water by means of an additional plastic tube. The tube connected to the cloche is also connected to the outlet. Under these conditions, any fraction of water infiltrating along the cloche area has to circulate through the rotameter. The position of the coloured ball of this flow meter in a graduated vertical scale provides the flow velocity. The method is much simpler than the procedure suggested by Zuber.

The authors of this book have applied at the Ulldecona reservoir an even simpler and more precise technique then the previous procedure for the measurement of the flow velocity in the tube connected to the cloche. This tube is connected to a measuring device consisting of a silicon tube mounted on a graduated scale of about 60 cm length, as shown in Fig. 10.9. The device is submerged into the reservoir at a shallow depth in such a way that the operator can directly verify whether the water inside the tube is flowing through. For example, the existence of water flow can be easily detected by putting the fleshy tip of the finger at the inlet of the silicon tube. In the affirmative case, flow velocity is measured by injecting with a syringe a pulse of a coloured tracer (for example, fluoresceine) at the zero point of the scale. The time needed for the tracer to travel a certain distance along the silicon tube is determined. Measuring the time elapsed and knowing the cross-section of the silicon tube, the flow rate is computed. If the cloche is well-fitted to the bottom of the lake, this flow rate will correspond to the seepage rate.

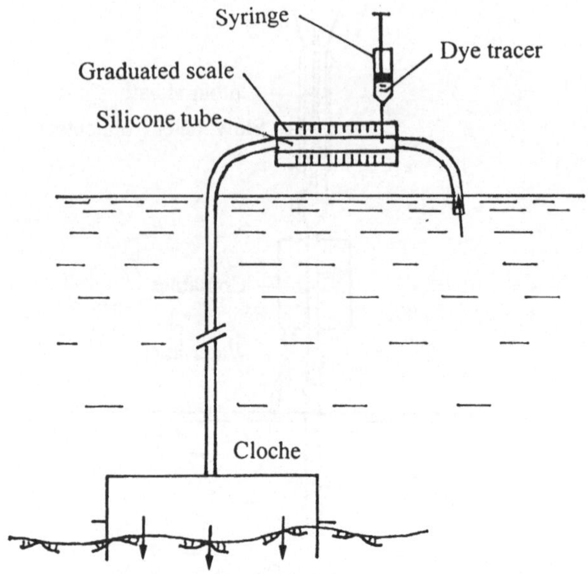

Fig. 10.9. Illustration of the method used by the authors for the measurement of seepage rates at the bottom of the Ulldecona reservoir, Spain. The flow velocity along the silicon tube is measured by means of a point injection of a dye tracer.

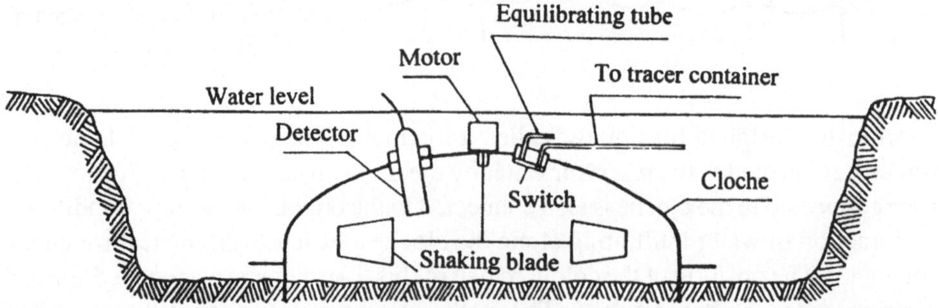

Fig. 10.10. Illustration of the method suggested by Gaillard et al. (1991) for the measurement of the very low seepage rates at the bottom of shallow water bodies.

10.7.5 Method suggested by Gaillard et al.

The method proposed by Gaillard et al. (1991) was designed to measure extremely low flow velocities. The method is appropriate for the case of very shallow lakes or ponds, especially when polluted wastewaters are impounded and slow seepage rates can be of relevance. The principle of the method is illustrated in Fig. 10.10. The cloche is deposited at the bottom of the lake in such a way that no lateral water ingress exists. In the upper part of the cloche, the following devices are fitted: (i) a scintillation detector with NaI(Tl) crystal for the measurement of the concentration of a radioactive tracer (for example, ^{131}I), (ii) a shaker driven by a motor that ensures a permanent constant tracer concentration in the inner volume of the cloche, and (iii) two access tubes. One of these access tubes (the short one in Fig. 10.10) is used to equilibrate pressure between the water in the pond and the volume inside the cloche as the whole system is deposited on the bottom of the reservoir. The second tube (the long one) is connected with a small container filled with the radioactive solution.

The measurement starts with the short tube closed and the long one open. Under these conditions, the infiltration occurring at the area covered by the cloche creates a certain under-pressure, and an equivalent water volume of the radioactive solution will ingress in the cloche. If the seepage rate is very low, this volume of radioactive solution will also be very low, but the increase in the response of the detector will be easily measurable due to the high sensitivity of the technique. This sensitivity will depend on the concentration of the radioactive solution, which can be adjusted to the most appropriate value. It can be stated that this sensitivity can be as high as the operator wishes. For example, the ingress of several drops of the radioactive solution over one day can be sufficient to detect the increase of radioactivity to be measured inside the cloche. The only limitation of the method is the diffusion of the radioactive solution from the tube to the inner volume of the cloche. However, this diffusion is kept very small because the inner cross section of the tube is also very small.

The shaker is used for the homogenisation of the tracer concentration in the water volume inside the cloche, so that the tracer amount injected in the cloche can be precisely determined. A calibration of the measuring system allows the determination of the amount of tracer solution that has entered the cloche during a certain period of time, which provides the seepage rate.

10.7.6 Point dilution method

The point dilution method has been proposed by Plata and Iragüen (1992) and used in Lake Laja, Chile. The measuring technique is illustrated in Figs. 10.11 and 10.12. The device consists of a stainless steel container perforated along its whole surface to make it permeable to the water flow. A Geiger-Müller detector is fitted in the centre of this container. Four perforated injection tubes are adapted close to the inner wall of the container. These tubes are connected to the tracer solution container, which is located in the boat, through an electromagnetic valve that controls the tracer injection in the container.

This container is deposited at the point on the bottom of the reservoir suspected to be an infiltration area. Then, an appropriate amount of a ^{131}I solution is injected into the container by opening the electromagnetic valve. The injection can be done by means of a small peristaltic pump or by keeping the container of the radioactive solution under low pressure. Once the Geiger-Müller detector has reached a counting rate, for example, between 30,000 and 500,000 counts per minute, the injection is stopped. A higher counting rate will produce significant counting losses due to the detector dead time. Of course, better than a Geiger-Müller will be the use of a scintillation counter because it allows the measuring of higher counting

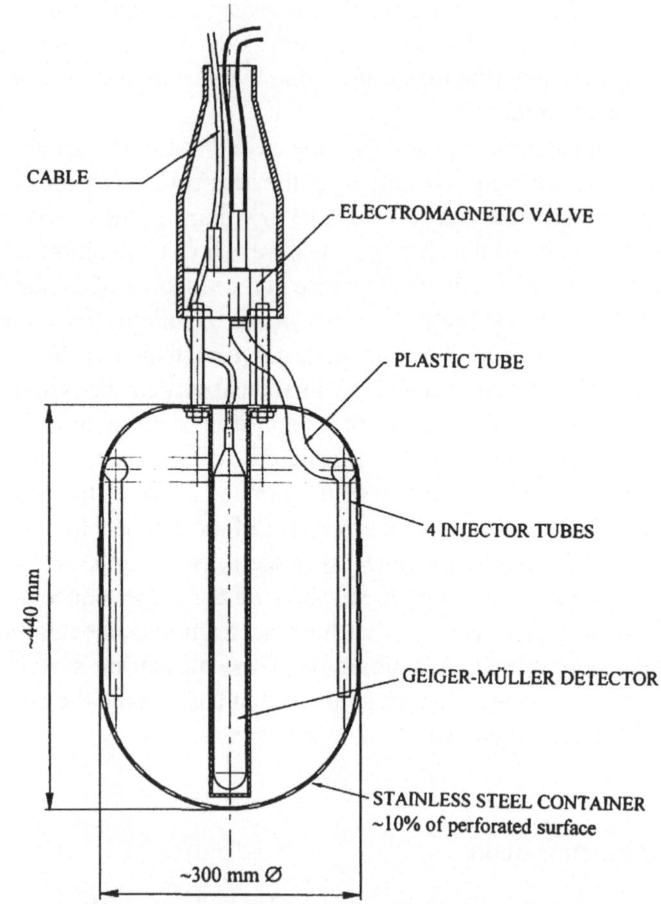

Fig. 10.11. Illustration of the method used by the senior author for the localisation of leakage areas at the bottom of Lake Laja, Chile, based on the point dilution of a radioactive tracer.

rates with no significant losses and the use of lower tracer amounts due to its higher sensitivity.

After the tracer injection, the variation of the counting rate as a function of time is recorded for an appropriate period of time. If the water flow at the bottom of the lake is negligible, the tracer dilution is mainly produced by diffusion, that is to say, very slowly. On the contrary, if the flow velocity is considerable, the tracer concentration inside the perforated container will decrease in an exponential form. Obviously, the slope of the dilution curve increases with the flow velocity. This parameter can be determined by calibrating the measuring system.

This method can provide satisfactory results only in cases of concentrated leakage, when the flow velocity induced by the leaks is much higher than the velocity

Fig. 10.12. Measuring device shown in Fig. 10.11.

derived from natural currents. This condition was fulfilled in the case of Lake Laja, where the total leakage was on the order of 25 m^3/s. The results obtained in this study are shown in Fig. 10.13, where the flow velocity distribution in the studied area is presented. About 500 point measurements of the flow velocity were performed. The leakage area was very well identified in this case. Further information on this case study is presented in Chapter 12.

It is important to point out that this technique can also be applied using common salt as tracer and a conductivity meter as detector. In fact, the authors of this book have applied a similar technique in the Ulldecona reservoir, Spain. The main difficulty appeared at the time of ensuring that the measuring device was placed close to the leakage point. Therefore, this technique should be used only when the

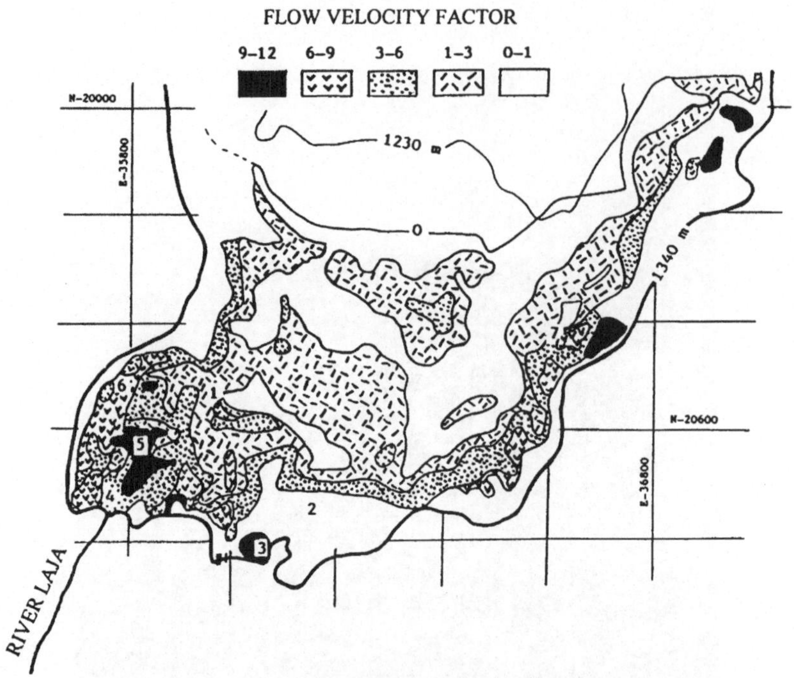

Fig. 10.13. Results obtained by the senior author on the leakage distribution at the bottom of Lake Laja, using the measuring device illustrated in Fig. 10.11. The velocity factors indicated correspond to relative values of the seepage flow rate in a scale from 0 to 12.

location of the leakage zone is suspected, and a confirmation on the location of the infiltration area is needed. For example, the geological nature and texture of the sediments in the bottom of the reservoir can provide a clear indication on the location of the leakage area. This was the case of the reservoir Contreras, Spain, where the infiltration points were clearly associated with some subvertical fractured strata of Liassic limestones. Therefore, the extent of the involved area was relatively small, and the implementation of this technique was possible.

10.7.7 Additional suggestions

In this section, some ideas of the authors, which can help in the development of new measuring systems for the localisation of preferential leakage areas at the bottom of the reservoir, are described. Therefore, measuring techniques already developed or used will not be included. Of course, many other ideas can be added to this list.

10.7.7.1 *Point dilution system for the identification of infiltration through the walls*

The suggested system for measuring point dilution to identify infiltration through the walls is schematically illustrated at Fig. 10.14. The system is similar to the

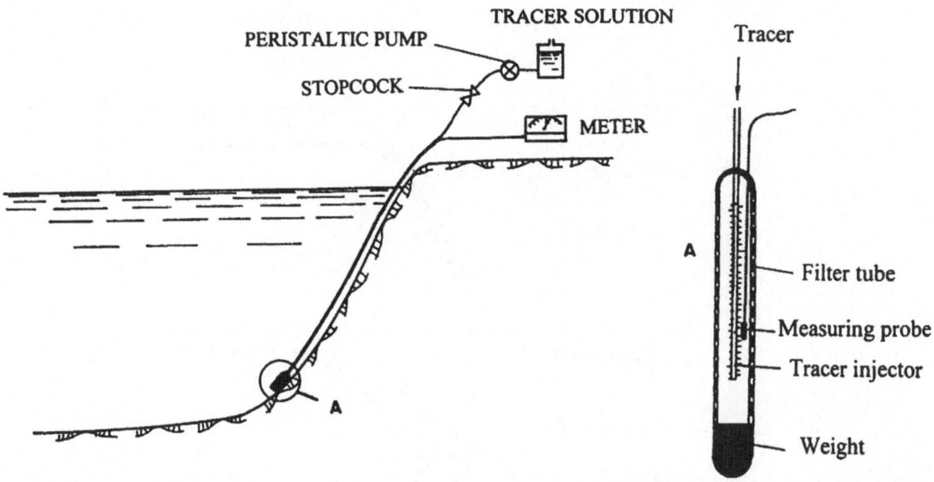

Fig. 10.14. Method suggested by the authors for the identification of seepage zones at the walls of the reservoir using the point dilution technique (see Chapter 8). A common salt solution is used as a tracer.

point dilution technique described in Section 10.7.6. A filter tube is used, which is slipped through the bottom of the lake if the relief allows it. The measuring probe of a conductivity meter and a tracer injector are placed inside this filter (perforated) tube, the size of which is about 5 cm in diameter and 30 cm long. A small weight is fitted at the lower end, just enough to ensure the deposition of the filter tube at a fixed point in the bottom. The injector is connected to the tracer container by means of a flexible plastic tube. A saturated solution of common salt is used as tracer. This is injected with a small peristaltic pump.

The measuring device is situated at a point on the bottom of the reservoir located at the deepest end of the vertical line to be studied. The operator, as well as the injection and measuring systems, preferably remain on the lake or reservoir shore, and if this is not possible, on a well-anchored boat. A tracer amount, sufficient to obtain an electrical conductivity, for example, between 5 and 10 mS/cm, is injected. The pump is then stopped and the stopcock closed. The variation of the conductivity signal with time is determined. The measuring process will be stopped when the conductivity reaches the original value corresponding to the reservoir water or once a reasonable period of time has elapsed, for example 5 min. The slope of the dilution curve will provide an indication on the prevailing flow velocity. When the measuring process is finished, the filter tube is pulled up, for example, 2 m, and a new measuring process starts. The procedure continues until the full measuring profile has been covered. Then, the measuring line is moved to the next position.

This method can also be applied in a continuous way. The filter tube can be slowly dragged along the bottom of the lake from a boat or from a fixed position at the shore of the reservoir. At the same time, a constant flow rate of tracer solution

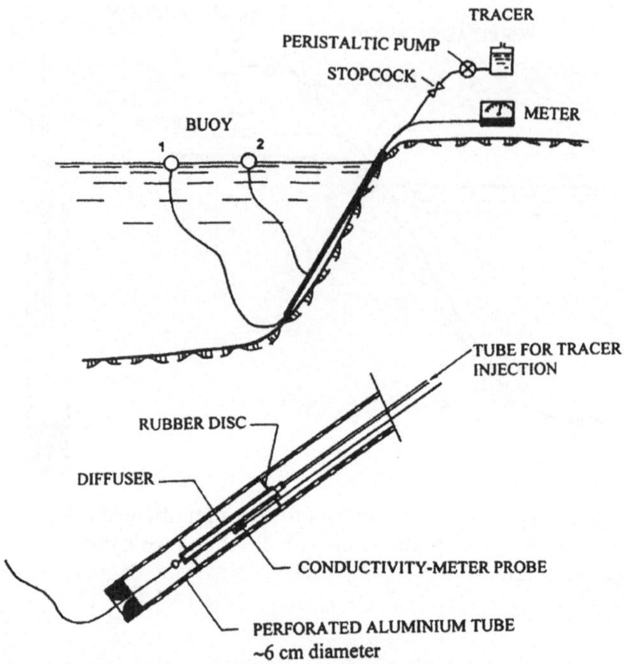

Fig. 10.15. Method suggested by the authors for the identification of outflow zones at the walls of the reservoir using a point dilution technique. The measuring system is pulled up along a rigid aluminium tube deposited in the bottom of the lake. A common salt solution is used as a tracer.

is injected, which will produce a more or less constant signal of the conductivity meter. The response of this meter will drop drastically when the filter tube crosses a sinkhole or a preferential seepage zone.

10.7.7.2 *Point dilution using a long porous tube*

When the bottom relief of the reservoir is very irregular, the use of the technique described in the previous section may create serious logistic problems and the porous tube may get stuck. The method illustrated in Fig. 10.15 allows the measurement of the prevailing flows in the lateral walls of a reservoir using the same point dilution method. A perforated tube is placed on this wall in the form illustrated in the figure. The best choice for this tube is to construct it of aluminium, as is used for inclinometric measurements in landslide studies, with a diameter of about 5 cm. This tube is not very heavy (it has a density higher than that of water), facilitating its deposition on the bottom of the reservoir. It always keeps its straightness, even in the case of lengths of several tens of metres, due to the presence of longitudinal grooves.

The measuring probe of the conductivity meter is attached to the centre of a diffuser tube with a length of about 50 cm, which is connected to the container of the tracer solution (common salt solution with an electrical conductivity of about 10 mS/cm). This measuring assembly is completely introduced inside the perforated tube. A certain volume of the tracer solution (2 or 3 l) is injected using a peristaltic pump, and the tracer concentration is measured for several minutes. A rubber rod is used at each end to confine the tracer solution in the stretch where the measuring probe is located. In the presence of a high flow velocity, the tracer rapidly escapes from the aluminium tube. In any case, the dilution rate gives relative values of this velocity. Once the measurement is completed, the measuring line is pulled to the next point inside the aluminium tube, and the measuring procedure is repeated.

After finishing the measuring profile, the aluminium tube is moved to a different position on the rocky wall. This operation is made by pulling up from the rope number 2 (see Fig. 10.15). In this way, the perforated tube is transported to the new position by pulling from both ends. The rope number 1 is used to bring the injection line to the end of the perforated tube. This work will be better done while the tube is kept in the horizontal position.

The technique described in Section 8.5 for the labelling of the whole water column can also be alternatively applied when the position of the aluminium filter tube offers favourable conditions. In this case, a continuous profile of the flow velocity is obtained, which will allow the identification of the stretch where water is moving faster. In stretches corresponding to areas with considerable infiltration, the injected tracer will disappear almost immediately. The technique illustrated in Fig. 10.11 can be applied afterward to confirm the presence of a rapid flow velocity and to define more precisely the involved stretch.

10.7.7.3 *Measurement of the vertical component of the bottom flow velocity using tracer techniques*

The measurement of the vertical component of the flow velocity at different points on the bottom of the reservoir is, for obvious reasons, very important for the identification of leakage and seepage areas. This measurement can be performed with small influence of the natural currents existing in the lake. The natural flows or currents existing in areas where the bottom of the lake is impermeable tend to be parallel to the bottom surface. Under these conditions, the vertical component of the flow will be very small or negligible. However, a vertical component of the flow may exist if the bottom relief is not horizontal. The direction of this flow component can be both ascending or descending, depending on the sign of the bottom slope in the direction of the flow prevailing in the reservoir (ascending in case of a positive slope and vice versa). In spite of that, the presence of a vertical descending flow with a magnitude higher than the value that can be expected from the bottom relief will indicate the presence of water infiltration.

The device illustrated in Fig. 10.16 allows the measurement of the vertical flow component at the bottom of the lake using a tracing technique. The measuring system can be constructed in such a way that it always keeps a stable vertical position when placed at the bottom of the reservoir. This position is achieved by a

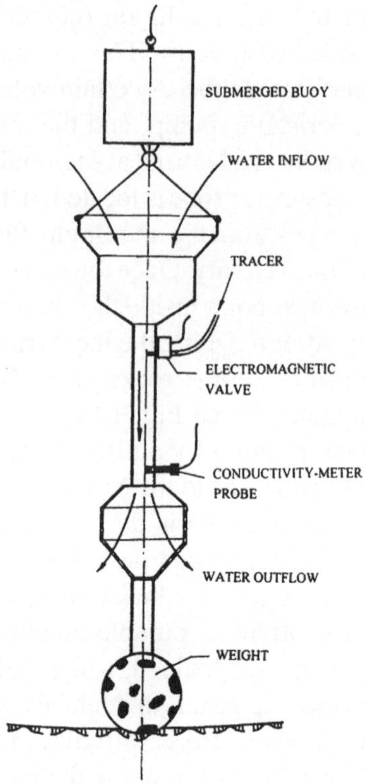

Fig. 10.16. Illustration of the method suggested by the authors for the measurement of the vertical component of the water flow at the bottom of a reservoir using a tracing technique.

weight placed at the lower end and a submerged buoy at the upper end of the device (see figure). The flow measuring system is located at the central part of the device between these two components. The upper component of this central part consists of a kind of funnel with a diameter, for example, of 40 cm. Below this funnel, the measuring tube for the vertical flow is located, the diameter of which can be on the order of 5 cm. This difference in the diameters creates a considerable amplification of the vertical descending component of the flow.

10.7.7.4 *Measurement of the vertical descending flow component using current meters*

The techniques commonly used for the measurement of vertical flow in wells and boreholes described in Section 8.6.2 can also be applied for measuring the vertical descending flow component on the bottom of the reservoir. Concretely, the current meters based on (1) the use of propellers, (2) transmission of a thermal pulse and (3) induction of an electromagnetic field can be applied. These methods are illustrated in Fig. 10.17.

The suggested measuring technique is based on the use of the same combination, consisting of a weight and a submerged float, to keep the measuring device in an upright position (see Fig. 10.17). The central unit of this figure for the injection

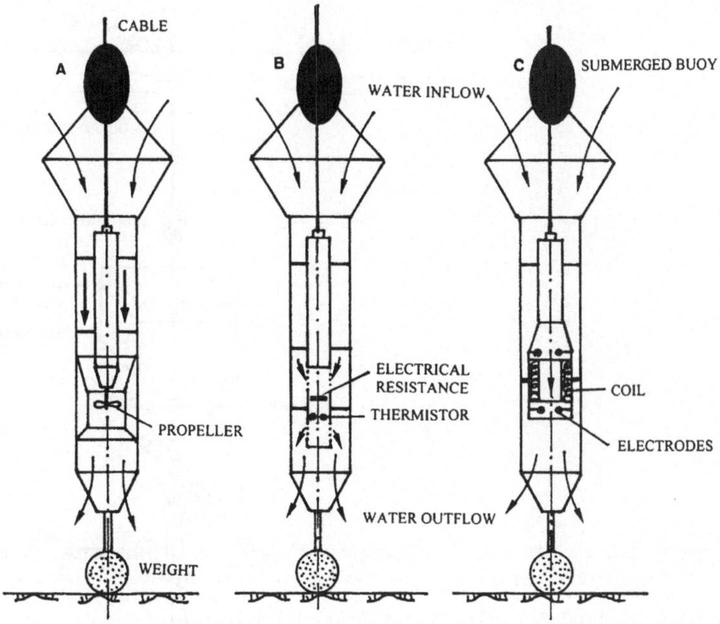

Fig. 10.17. Illustration of different flow meter techniques that can be applied for the measurement of the vertical flow component at the bottom of a lake or a reservoir. (A) Propeller flow meter; (B) thermal pulse flow meter; (C) electromagnetic flow meter.

and measurement of the tracer is replaced by the unit that contains the above-mentioned current meter devices. Usually, water is forced to circulate through the corresponding sensor by means of a collar to increase the sensitivity of the measuring system.

In the case of the propeller current meter, the commercially available measuring probe used for the well-logging systems can be used. However, a measuring system developed ad hoc will provide some advantages, mainly in terms of sensitivity. The same is valid for the thermal probe and for the electromagnetic method. The last method offers, in principle, the following advantages, with respect to the first two methods:

1) The response of the probe is almost instantaneous.
2) The probe response is linear for a wide range of flow velocities.
3) The minimum measurable flow velocity is lower than in the case of a propeller current meter.

10.7.7.5 *Three-dimensional measuring systems*

Some measuring methods for flow at depth in lakes are appropriate for the development of three-dimensional measuring devices. This is the case of the electromagnetic method. A multiprobe device, as illustrated in Fig. 10.18, can be used

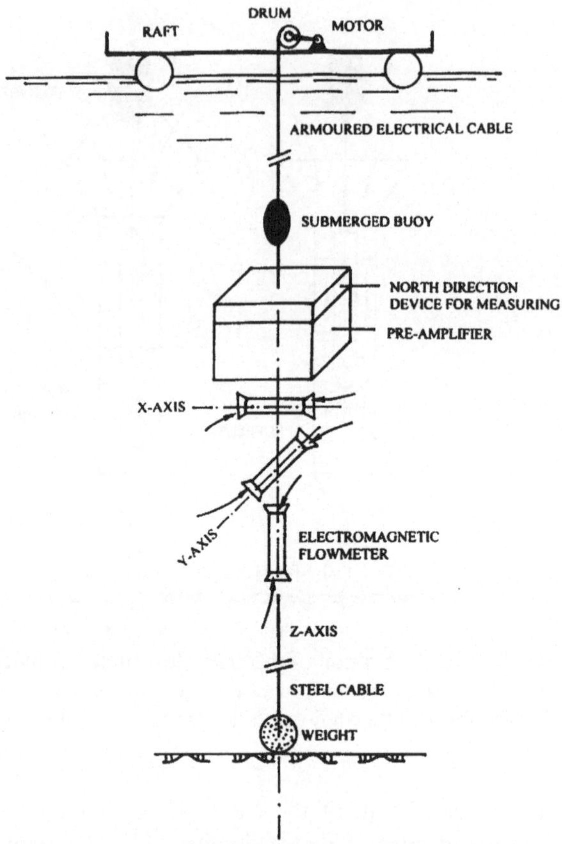

Fig. 10.18. Schematic view of the main components of a proposed device for the three-dimensional measurement of water flows at the bottom of lakes based on the use of three bidirectional electromagnetic flow meters.

for the determination of the flow direction and velocity in the directions X, Y and Z. Three bidirectional electromagnetic flow meters are mounted on a frame oriented along these axes. These probes will provide different velocity values when the measuring system is submerged at a certain depth in the reservoir. Each probe can determine flow velocity in two directions, which correspond, respectively, to positive and negative potential values. Therefore, six velocity values can be obtained. These measured values can be electronically processed to provide the true flow direction and the flow velocity. The combination of weights and floating devices can be used to keep the measuring device in the upright position once it reaches the bottom of the reservoir. Obviously, upward flows will occur only when the measuring system is placed close to the wall of the reservoir and seepage or leakage is taking place at a point located at a lower depth. A motorised drum will facilitate the lowering of the measuring device.

It is obvious that to obtain the flow direction, the orientation of the measuring system inside the lake has to be known. In the case of shallow reservoirs, this can be achieved by means of a mechanical form, by lowering the device suspended from a rigid rod, so that it can be rotated from the surface. In the case of deep lakes, the use of a compass fitted to the measuring device will be necessary. The direction marked by the needle can be determined using the optical technique developed by Drost, which has been described in Section 8.4.3. An alternative technique is based on the labelling of the needle end with a phosphorescent product or with a low activity radioactive source, such as ^{137}Cs. The position of this needle is determined using, respectively, a set of six photocells or six small Geiger-Müller (G.M.) detectors. In both cases, the signals of these six detectors can be processed to obtain the orientation of the measuring device. The last method (use of a radioactive source) has been used by Godinho (1996) in Brazil and Jiansheng (1999) in China. For the identification and processing of the signal generated by different G.M. detectors, Godinho suggests a method based on the sequential measurement of the counting rates provided by each detector using electronic switches. Jiansheng transforms the individual pulses provided by the detectors into square-shaped pulses of characteristic height. The pulses provided by the different detectors are measured separately by means of a multichannel pulse analyser, developed especially for this purpose. A barium-133 radioactive source is used in this case.

The main advantage of such a measuring system is the instantaneous response to changes in the flow direction and velocity. The method will provide different flow directions for different zones of the lake. The confluence of flow directions in a point or zone of the reservoir will indicate the location of the seepage or leakage area. The changes of flow velocity will be aligned in the proximity of this area. The measuring system can also be used suspended from a well-anchored raft or boat, instead of lying down on the bottom.

A second possible method for the determination of deep flow direction and velocity of water is based on the injection of a radioactive tracer (for example, ^{131}I) in a point surrounded by a set of small Geiger-Müller detectors. Fifteen or twenty detectors are mounted on a frame, distributed, for example, with a more or less spherical arrangement. A tracer pulse of some microcuries is injected at time zero. The tracer will be detected by one or more G.M. detectors. The identification and processing of the signals emitted by the detectors can be carried out using the above-mentioned techniques.

The previous technique can also be developed using common salt solution as a tracer and replacing the G.M. detectors with electrical conductivity probes. However, the technique for the determination of the orientation of the measuring system at the bottom of the reservoir based on the above-mentioned techniques has to be kept (no alternative method seems to be technically feasible).

REFERENCES

Aebersold, P.C., 1964. *Deep water isotopic current analyser.* Report JLI 2748-0-9, William H. Johnston Laboratories Inc., Baltimore, MD, USA.

Andreu Ibarra, B., Galvez Cruz, L., Ruiz Pena, O. and Del Arenal, R., 1979. Estudio isotópico de las fugas de la presa Las Lajas, Chihuana, México. In *Isotope hydrology 1978*. International Atomic Energy Agency, Vienna, pp. 125–143 (in Spanish).

Azoeuf, A., Gourlez, P., Hoslin, R. and Tola, F., 1982. *The study of the dynamics of a suspension of silt sediments off Bedok (Singapore) by means of radioactive tracers*. International Atomic Energy Agency, Vienna, Project Report SIN/8/008.

Bougalt, H., 1970. *Etude de la sorption de quelques radioelements artificiels par les sediment pelitiques en vue de son application au marquage radioactif de ces materiaux*. Thèses pour obtenir le titre de Docteur-Ingénieur, Faculté des Sciences, Université de Paris, France (in French).

Calmels, P. and Salençon, M.J., 1992. Technique de marquage por l'étude des circulations des masses d'eau dans les lacs: Application à la retenue de Pareloup. In *Isotope techniques in water resources development*. International Atomic Energy Agency, Vienna, pp. 107–122 (in French).

Easey, J.F. and Hanna, G.L., 1984. The use of radioisotope tracers to identify the location of seepage areas in a dam. Personal communication.

Easey, J.F. and Hanna, G.L., 1985. *A study of leakage from Pedu dam, Kedah, Malaysia*, IAEA Technical Co-operation Project, Report AAEC/C48, Australian Atomic Energy Commission, Melbourne.

Gaillard, B., Calmels, P., Margrita, R. and Andre, P., 1991. Methode et appareillage pou la mesure a l'aide d'un traceur radioactif des tres faibles debits d'infiltration verticale de l'eau au travers des sols en place. In *Isotope techniques in water resources development*. International Atomic Energy Agency, Vienna, pp. 648–650 (in French).

Godinho, R., 1996. Personal communication. Centro de Desenvolvimento da Technología Nuclear, Belo Horizonte, Cidade Universitaria, M.G. Brazil.

Guizerix, J., Molinari, J., Gaillard, B., Santos-Cottin, H., Mornas, A. and Corda, R., 1967. Localisation des fuites sur un grand reservoir à l'aide de traceurs radioactifs. In *Isotopes in hydrology*. International Atomic Energy Agency, Vienna, pp. 587–599 (in French).

Jiansheng, C., 1999. Using 131-I as a tracer to determine the leakage of Nanfu Dam in Haizi reservoir. *Nuclear Techniques*, 12(3), 17–25.

Makowski, J., 1967. Radiometric method of seepage control in dams and their environs. In *Isotopes in hydrology*. International Atomic Energy Agency, Vienna, pp. 601–616.

Makowski, J., 1970. Radiometric method for investigating the permeability of reservoir beds. In *Isotope hydrology*. International Atomic Energy Agency, Vienna, pp. 727–741.

Molinari, J., Guizerix, J. and Chambard, R., 1970. Nouvelle methode de localisation de fuites sur des reservoirs ou canaux au moyen d'emulsions de bitume marqué. In *Isotope hydrology*. International Atomic Energy Agency, Vienna, pp. 743–760 (in French).

Owczarczyk, A. and Szpilowski, S., 1984. Sorption method for leakage localization in water reservoirs of hydrotechnical plants. In *Isotope hydrology*. International Atomic Energy Agency, Vienna, pp. 709–718.

Plata, A. and Iragüen, E., 1992. Leakage study at lake Laja Chile. In Hötzl and Werner (Eds.), *Tracer hydrology*. Balkema, Rotterdam, pp. 378–396.

Plata, A., Iragüen, E. and Heresi, N., 1989. Estudio de las filtraciones de la presa "El Colorado" con técnicas isotópicas. In *Actas de un Seminario Regional para América Latina sobre el empleo de técnicas isotópicas en Hidrología, México*. International Atomic Energy Agency, Vienna, Austria, 21 pp. (in Spanish).

Zuber, A., 1970. Method for determining leakage velocities through the bottom of reservoirs. In *Isotope hydrology*. International Atomic Energy Agency, Vienna, pp. 761–771.

CHAPTER 11

General strategy of studies

11.1 INTRODUCTION

In previous chapters, the most important techniques developed for the investigation of leakage from lakes and reservoirs have been described. Obviously, not all these techniques are applicable in all cases nor do all of them provide valuable results in a general way. Therefore, for each particular case, the most appropriate techniques must be selected for the investigation. For this reason, the correct planning of the strategy to be adopted is very important for the success of the investigation. Many examples are known where a concrete technique leads to the solution of the problem, and sometimes, this key technique is the less expected one.

In the present chapter, the guidelines suggested to define the most appropriate strategy for a particular investigation are given. Some new indications not given in previous chapters are also included. As a general comment, it must be pointed out that the investigation of leakage from reservoirs should be tackled in the four following stages, as mentioned in Chapter 1:

1) Studies performed before the dam construction (preconstruction studies), that is to say, during the feasibility stage of the dam construction.
2) Studies performed during the first filling of the reservoir and during the first few months after this filling.
3) Studies directly related to the characterisation of the flow pattern of the leak, once the problem is clearly manifested.
4) Long-term investigations on the hydraulic connection between the lake and local groundwater.

In some cases, even when the reservoir is in the decommission phase, studies of the local hydrological conditions might be required. This is particularly the case when polluted water and/or sediments have been impounded in the reservoir.

11.2 PRECONSTRUCTION STUDIES

11.2.1 Generalities

The construction of a dam always entails a certain risk in relation to the future appearance of a leakage problem. The nature and the hydraulic behaviour of the involved geological formations are usually so difficult to evaluate that this risk is always present even if exhaustive studies of the dam site have been performed. However, it is also true that most of the leakage problems are linked to reservoirs where insufficient (or sometimes, no) previous hydrogeological studies have been carried out. Practical experience shows that in many cases, leakage problems could have been foreseen and solved during the dam construction, if thorough enough studies on the hydrogeological characteristics of the dam site formations would have been carried out.

According to the previous statement, it must be stressed that the execution of preconstruction studies is almost always worthy from the safety and from the financial points of view. The reason behind this statement is that the impermeabilisation or grouting works are usually less expensive during the dam construction stage than when the reservoir is in operation. The difference in price of such works can be extremely high, especially if the water losses and the delay in the beginning of the reservoir operation are also considered. On the other hand, it must be taken into account that the investment needed to perform a detailed and thorough study of the dam site is always negligible in comparison of the total cost of the dam construction.

It is true that when a dam is constructed, especially a large dam, preliminary studies of geological and hydrogeological reconnaissance of the dam site are always carried out. However, most of the techniques described in this book very rarely are applied. The main reason for that is the incomprehensibly scarce diffusion attained to date by these techniques. Most of the techniques described in this book are usually not considered in the university curricula for geologists or civil engineers.

The preconstruction investigations of the dam site should aim at two well-differentiated objectives. One objective needs to be included in the prefeasibility studies for the dam construction. As shown below, through the presentation of some case studies, the feasibility studies may advise against the construction of a dam when the risk of leakage and the associated costs are very high and can not be assumed. The second objective is related to the definition of the grouting or impermeabilisation works that have to be carried out before or during the dam construction to reduce the risks and impacts of leakage.

11.2.2 General recommendations

The reconnaissance studies performed before or during the dam construction always involve drilling of boreholes. Frequently, these boreholes are used exclusively for the extraction of core samples of the geological formations of the dam site. In other cases, permeability tests such as those described in Chapter 3, which are familiar to most drilling companies, are performed. Many of these boreholes are constructed in areas that will remain inaccessible after the dam construction, but this is not the case for many other boreholes. The first recommendation refers to the conservation and maintenance of these boreholes. Very frequently, when a leakage study is launched, new boreholes have to be drilled because those drilled during the dam construction are abandoned and useless due to the lack of conservation. Civil engineers should be aware of the great usefulness of those boreholes and piezometers as control points for the hydraulic behaviour of the local aquifers and reservoir water during the full operational life of the reservoir.

On the other hand, when new boreholes are drilled, their constructive characteristics should be appropriate to allow flow tests to be conducted as described in Chapter 8, as well as to obtain representative water samples for chemical and isotope analysis. Basically, these main characteristics are the following:

1) Inner diameter should be no less than 5 cm (ideally, 8 cm) so that the injection and measuring lines can be easily introduced into it.
2) The casing has to be perforated in the whole stretch below the piezometric level prevailing at full reservoir level.
3) The number and size of the slots or holes of the casing should be sufficient to have a perforated surface not less than 1% of the total casing surface (see Annex 4).

Piezometers with casing perforated only at the deepest stretch (2–3 m) are useful only for water level measurements, being, in many cases, insufficient to obtain the needed information on the flow patterns. Water samples collected from such boreholes may not even be representative of the water flows located through the upper soil strata. Of course, apart from the conservation and maintenance of the exploratory boreholes, the information about their location, original constructive characteristics, lithological composition and geology of the intercepted rock formations, etc., should also be kept and archived as part of the dam documentation.

On the other hand, when a leakage study is initiated, in practically all cases, it is observed that the reservoir water has drastically changed the chemical, isotopic and hydrological characteristics of the groundwater existing in a wide zone around the artificial lake and the dam site. These changes are difficult to establish in advance because usually the initial hydrological conditions prevailing before the dam construction are not known in sufficient detail. For this reason, it is strongly recommended that the studies performed before or during the dam construction should also include the determination of the mentioned water characteristics, that is, the chemical and isotope baseline should be well defined. However, we are aware that very frequently the analytical facilities or even the financial possibilities for these determinations are not easily available. In such cases, we recommend the collection and proper storage of water samples from all springs, wells and boreholes existing close to the dam site within a distance of at least about 15 km, including those drilled within the framework of the exploratory works. These water samples should preferably be kept in well-closed glass bottles. A volume of 2 l for each sample/point will be sufficient. The basic idea is that these water samples can be analysed at the moment considered necessary so that the initial hydrogeological conditions prevailing in the zone can be at least partially reconstructed.

For obvious reasons, the feasibility studies performed before the dam construction should also include the precise determination of the chemical quality of the water to be stored in the future reservoir. Cases are known, for example, at the Guadalhorce reservoir, Spain, in which a reservoir has been constructed in areas with important springs discharging saline water that affected the quality of the impounded water in such a way that it was practically useless. Of course, this type of situation should never happen, especially taking into account the great simplicity of the studies that are necessary to detect the problem. A case study about this problem will be described in Section 11.2.5.6.

11.2.3 Risk of leakage

Obviously, the risk of generating leaks is directly related to the nature of the involved geological formations (see Chapter 1). Sedimentary and metamorphic rocks with fine grain size, such as clay, marls, silt, schist, etc., generally do not present leakage problems. The same can be said for argillite, siltstone, cemented conglomerates, arkoses, and moraines. Reservoirs located on igneous rocks are rarely affected by serious leaks, unless the dams are located on highly fractured massifs. The presence of geological faults may have a tremendous impact on the existence of leaks because the faults may act as preferential flow paths. The highest probability of leakage occurs in the case of the following rocks:

1) Soluble rocks, such as limestone, dolomite, halite and gypsum.
2) Volcanic rocks, especially if they are of recent origin.
3) Coarse granular media, such as alluvial and colluvial materials. Also, conglomerate formations with a calcareous matrix or cement can present serious problems, as was the case of the Escalona reservoir, Spain.

Alluvial materials are present in most dam sites associated with the deposits formed by the river to be regulated. If the thickness of these alluvial deposits is small, for example, less than 10 m, the dam can be founded on the underlying impermeable rocks. But, if this thickness is very high, the alluvial materials cannot be completely removed due to technical and/or economic reasons. In such cases, the construction of impermeable walls may be necessary. These walls consist of grouting injections, continuous concrete walls, walls on piles, among others. The choice of the most appropriate wall type, as well as its extent, requires a detailed reconnaissance of the materials outcropping at the dam site.

Detailed information on the permeability of the geological materials located below the dam and their thickness is absolutely necessary for the calculation or estimation of the expected water losses through these materials once the reservoir is filled. This calculation is carried out through the construction of theoretical flow-nets. In Annex 1, a simplified method is described for the estimation of the water losses under the dam as a function of the permeability and thickness of the underlying materials. An isotropic condition of these materials is supposed; however, very frequently, this condition is not fulfilled.

High probability of leakage may exist when springs located downstream of the dam emerge at an elevation lower than the bottom of the reservoir. In such a case, a thorough investigation of the origin of the emerging water is absolutely necessary. The main goal of the investigation would be to find out whether the emerging groundwater is fed by the river (or rivers) feeding the future reservoir. If the answer is affirmative, the next question is to find out if the infiltration point is located within the flooded area. In the next section, the recommended techniques for this purpose are described. In Annex 5, the estimation of the total volume of the cavities of the rock formation between the infiltration and the emerging zones based on the content on natural tritium is described. The method is based on the assumption of the exponential or good mixing model for the renewal of groundwater. The method gives satisfactory results when dealing with karstic aquifers.

11.2.4 Recommended techniques

An important limitation of the possibilities offered by some of the techniques described in this chapter for preconstruction studies is derived from the fact that at this stage, the phreatic or piezometric water levels of the local aquifers are much lower than they will be once the reservoir is filled. Therefore, the application of the techniques for the investigation of local groundwater flows described in Chapter 8, as well as those based on the use of natural tracers, must be restricted to the prevailing saturated zone at this early stage.

Within the unsaturated zone, the only techniques that can be applied are the following:

1) Geophysical methods described in Chapter 4, including both surface and well-logging techniques.
2) Permeability tests described in Chapter 3.

These techniques, together with the observations made during the drilling of the boreholes, including the lithological data obtained through the extraction of core samples, should provide valuable information on the localisation and size of existing fractures or permeable stretches of the involved rock formations. This information should be very valuable for interpreting flow tests eventually made during the first filling of the reservoir, when the local water level rises up to the elevation where these fractures and/or permeable strata are located.

In the same way, the identification of perched aquifers during the drilling of boreholes will be very important, especially if they are situated at elevations lower than the maximum water level of the future reservoir. If this is the case, the borehole will show a vertical descending flow inside the saturated zone, which will be easily detectable by means of the tracer techniques described in Chapter 8. Groundwater entering the borehole through this perched aquifer should be sampled for chemical and isotope analyses.

Within the saturated zone, most of the techniques described in this book are applicable. The planning of the tests to be performed will depend on the particular characteristics of each case. However, some studies leading to the definition of the hydrogeological baseline should be performed in a general, systematic way. These are the following:

1) Determination of the chemical and isotope composition of the different groundwater types existing in the study area. This requires the sampling of most water points located in an area between 10 and 20 km around the dam site, obviously depending on the local hydrogeological conditions. The list of parameters to be measured will include major chemical components, as described in Chapter 5, stable isotopes deuterium and oxygen-18, and natural tritium contents. Of course, if two or more water samples with the same chemical composition are found, the isotope analyses can be restricted to only one of them.
2) Measurement of temperature and conductivity profiles, as well as natural gamma logs at all existing wells and boreholes.
3) Groundwater flow tests in all wells and boreholes presenting appropriate constructive conditions. The technique based on the labelling of the entire water column and the techniques for the measurement of vertical flows, in case these are present, should be applied (see Chapter 8).

The chemical and isotopic characterisation of the local groundwaters should provide information on the relationships existing between water bodies from different points of the study area, as well as on the replacement of the aquifer water by reservoir water after its filling. Especially interesting is the study, using these natural tracer techniques, of the origin of springs eventually emerging downstream of the dam site and, for obvious reasons, their possible relationship with the nearby river. Extended monitoring periods (for example, a full year) may be necessary to arrive at a definitive conclusion. The correlation of chemical and isotope changes (seasonal or sporadic changes) between the river and the emerging water can provide evidence on this connection. Similarly, the flow rate variation of the spring as a function of the river flow rate can help in obtaining clear evidence for this investigation.

If the chemical and isotope analyses indicate that groundwater from any of these springs is connected to the river, the next action to be undertaken is to investigate whether the emerging water infiltrates in a zone that will be flooded by the future reservoir. Depending on the relative values of the river and spring flow rates, this investigation can be done on the basis of differential gauging of the river stretch to be flooded by the reservoir. The tracer technique described in Annex 2 can be used to this purpose.

However, in most cases, the flow rate of the river will be much higher than that of the emerging spring and, therefore, the previous technique will not be feasible. In this case, a tracing experiment of the river water using uranine, rhodamine B or WT or tritium as artificial tracers can be performed. The tracer will be injected at a point of the river located 1 or 2 km upstream of the area to be flooded by the future reservoir. If the result of such experiment is positive, that is to say, if the injected tracer is clearly detected in the spring, the study can be considered finished. However, a negative result does not mean necessarily that the area of the river where the emerging water is being infiltrated is located upstream of the future flooded zone. This negative result can be explained in two different ways:

1) The transit time of water between the river and springs is longer than the adopted observation time.
2) The fraction of tracer that has been infiltrated was too small to detect.

Of course, most of the injected tracer will be incorporated into the river flow and only a minor fraction will infiltrate. To increase this fraction, such an experiment should be undertaken under low water conditions at the river. Sometimes, a visual inspection of the river stretch may reveal the presence of water escaping from the riverbed and being infiltrated into the ground. This was observed by the senior author in the Saja River, Spain, in a 2-km stretch, where the river was flowing over a limestone formation. If this is the case, this point is where the artificial tracer should be injected for the interconnection experiment. A much smaller amount of tracer will then be necessary. The surface geology of the river stretch is usually a good starting point for the localisation of the infiltration point or area.

Flow tests performed in boreholes always provide valuable information on the permeability of the geological formations close to the dam site, both in porous media and in consolidated rocks. The information obtained through these techniques will help in the definition of the impermeabilisation works to be undertaken. Due to the commonly high heterogeneity of the involved formations and the lack of continuity of the permeable materials, groundwater flows cannot always be defined from the nature and texture of these materials. Therefore, the performance of direct measurements of the prevailing flows is absolutely necessary. In many cases, high investments are made in the impermeabilisation of materials that do not need this kind of treatment, and inversely, other materials responsible for important groundwater flows are not treated.

Flow tests are particularly necessary when the excavation of the total thickness of the alluvial materials is impracticable due to their high volume. In this case, the flow tests should be complemented with appropriate permeability tests already described, such as Lefranc, Lugeon or Gilg-Gavard. Many years ago, Lugeon recommended the construction of grouting walls in all cases where the dam height is above 30 m and the permeability of the underlying materials is above 1 lugeon, that is to say, higher than about 10^{-3} m/s. Alluvial materials are characterised by high permeability variations with depth. Therefore, the previous criteria should be applied to the most permeable stratum. This implies the need for measuring permeability profiles. The flow test based upon the labelling of the whole water column, described in Chapter 8, is very useful for the identification of the permeable strata with preferential groundwater flow.

If permeability tests provide evidence on the need for the construction of an impermeable wall, the flow tests performed in boreholes can be used for the evaluation of the efficiency of such a wall. A set of boreholes drilled downstream of this wall will allow the measurement of groundwater flows existing before and after the wall construction, as well as those generated once the reservoir is filled.

On the other hand, natural gamma ray logs performed at boreholes of the dam site will provide half-quantitative information on the fraction of clay present in different alluvial strata (see Section 4.3.2). This information will be very important due to the existing correlation between clay content and permeability. In the case of consolidated rocks, gamma ray logs, together with other geophysical methods described in Chapter 4 and tracing tests, will allow the identification of fractures, including their filling (air, water or clayey materials).

In case of boreholes with no casing, the caliper method can be used for the identification of fractures and the measurement of their size.

11.2.5 Case studies

11.2.5.1 *Cuyaguateje reservoir, Cuba*

The construction of a dam was planned at the Cuyaguateje River in Pinar del Río, Cuba. More than one-third of the area to be flooded by the future reservoir was formed by limestone and dolomite rocks, highly karstified as revealed by the presence of numerous dolines, trenches and other karstic manifestations. A number of boreholes were drilled for the investigation of the intensity of the karstification at different zones. Tracer flow tests were performed in several boreholes. The results obtained demonstrated the presence of high groundwater flows in all of them. Figure 4.10 (Chapter 4) shows the results obtained in one of these boreholes. The conclusion obtained from the investigation was that the construction of the dam was not feasible due to the extremely high risk of leakage. The responsible authority accepted this recommendation.

11.2.5.2 *Pangue reservoir, Chile*

The Pangue dam located at the Bio Bio River, Chile, has been constructed in recent years. During the preconstruction studies, a paleo-riverbed was identified at the dam site at a depth of about 400 m below the surface. From the very beginning, it was considered that this paleo-riverbed could create important leaks of the water to be impounded in the reservoir. The paleo-riverbed could act as a bypass, connecting the reservoir with the river downstream of the dam. The senior author recommended a simple investigation based upon the measurement of the stable isotopes deuterium and oxygen-18, as well as natural tritium. Due to the fact that water from the Bio Bio River is mostly formed by snow melting from very high elevations of the Andes Cordillera, its isotopic composition corresponds to δ-values much more negative than groundwater derived from the local precipitation (altitude effect). A δ^{18}O-value between $-12‰$ and $-15‰$ was estimated for the river water and between $-5‰$ and $-6‰$ for locally derived groundwater. Therefore, the stable isotopes would provide unequivocal information on whether the paleo-riverbed was hydraulically connected to the river.

The suggested study included the sampling for stable isotopes and tritium of the paleo-riverbed water, river water at different times of the year and groundwater collected from local wells not related to the river. The last set of samples would provide the average isotopic composition of the local precipitation. Obviously, if the isotopic composition of the water collected from the paleo-riverbed was equal to the samples representing local precipitation, the problem would have been solved in the sense that no hydraulic connection existed between the Bio Bio River and the paleo-riverbed.

Besides the stable isotopes of water, the suggested investigation also included the measurement of natural tritium. A tritium concentration for the river water, ranging from 2 to 4 T.U., was expected. The expected concentration for local groundwater should be zero if this water was recharged from precipitation infiltrated before 1952, and between 4 and 8 T.U. if the water were derived from the precipitation of the last 20 or 30 years. In any case, the presence of tritium would indicate that groundwater is rapidly renewed and therefore of a recent origin. Therefore, if the paleo-riverbed water were related to the Bio Bio River, according to the information provided by the stable isotopes, and its origin linked to recent precipitation, the impermeabilisation of the paleo-riverbed would have to be considered.

Unfortunately, the recommended investigation could not be undertaken for reasons beyond our control. Fortunately, no leakage problem has been reported after filling the reservoir. The fact of having included the description of the problem in this chapter is exclusively for didactic reasons because it is an interesting example of the combined use of tritium and stable isotopes. Of course, water chemistry also would surely have provided valuable information for the proposed investigation.

11.2.5.3 *Santa Liestra dam, Spain*

The construction of the Santa Liestra dam on the River Esera, Northern Spain, was planned, and an investigation of the dam site was carried out to identify the hydrogeological conditions of the involved geological formations. Figure 11.1 shows a geological cross-section of the dam site along the dam axis. As can be seen, the alluvial materials deposited by the river at this site are restricted to a small layer of gravel, easily removable. The dam will be constructed on a siltstone

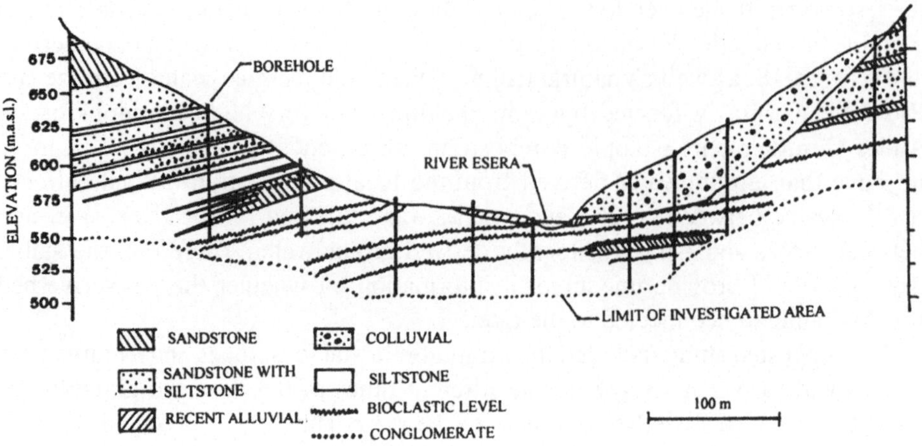

Fig. 11.1. Schematic view of the lithological units in the vicinity of the dam site of the Santa Liestra reservoir, Spain.

and argillite substrate with some embedded layers of sandstone. These materials were deposited during the Upper Tertiary in a marine environment. The substrate is covered at the left side by an important pervious layer of colluvial materials with a thickness of up to 45 m.

The investigation carried out included the determination of major chemical components as well as natural tritium and stable isotopes of water samples collected from 16 boreholes drilled at three different profiles close to the dam site. Lugeon tests were also performed during the drilling of the boreholes. The combined use of these natural tracers allowed the identification at the dam site of different types of waters:

1) Very old (probably, connate) waters with high salinity (more than 5 mS/cm) and no tritium content.
2) Recent waters related to the infiltration of the Esera River water. These waters were recognised by more negative δ-values with respect to groundwater derived from local precipitation, which was derived from the high elevations of the Pyrenees, where the river's main source is located.
3) Water used for drilling was found in some boreholes that had intercepted only very impervious materials. The replacement of this water by local groundwater was in these cases a very slow process.

However, the most interesting information of the investigation came from the flow tests carried out using the tracer techniques described in Chapter 8. A saturated solution of common salt was used as a tracer. These tests allowed the precise identification of the impervious materials at the dam site as shown in Fig. 11.2. The technical staff responsible for the project used this information to define the

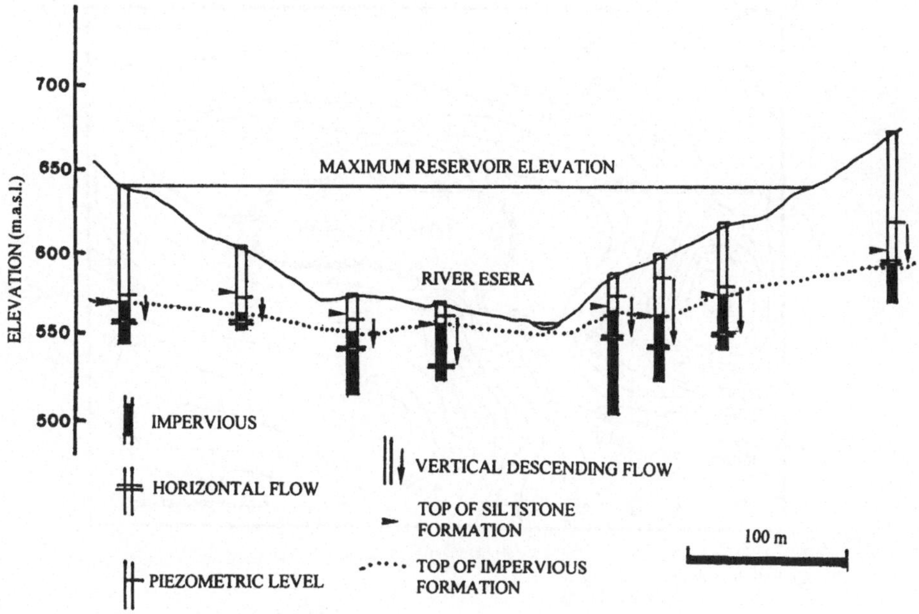

Fig. 11.2. Flows detected in a borehole located near the dam site of Santa Liestra, Spain.

final cross-section of the dam. This was probably the first time this type of study has been conducted on the basis of tracer flow tests. As shown in Fig. 11.2, pervious materials exist at both embankments below the maximum water level of the future reservoir. The impermeabilisation of these materials requires grouting works to be carried out either from the surface or from lateral galleries.

11.2.5.4 *Lake San Marcos, Ecuador*

Lake San Marcos is located at the foot of the Volcano Cayambe at an approximate elevation of 3,000 m.a.s.l. The construction of a dam at the discharge area of the lake was planned with the aim of increasing its storage capacity. The diversion into the future reservoir of the neighbouring River San Pedro was also included in the project (see Fig. 11.3). The stored water in the reservoir was to be used for irrigation purposes at the Tabacundo area located about 150 km downstream.

The lake was formed through the deposition of laharic or glacio-fluvial materials (the true origin of these materials was not clearly identified). As can be seen in the cross-section shown in Fig. 11.3, the lake discharges through a surface channel of the Anzuela river as well as through the groundwater flowing across the laharic or glacio-fluvial materials. The last identified discharge emerges at a deep talus located at a distance of less than 1 km from the lake.

The senior author carried out an investigation to evaluate the feasibility of the dam construction, considering the possible high permeability of the involved rock formation. The study included (1) the measurement of the stable isotopes deuterium

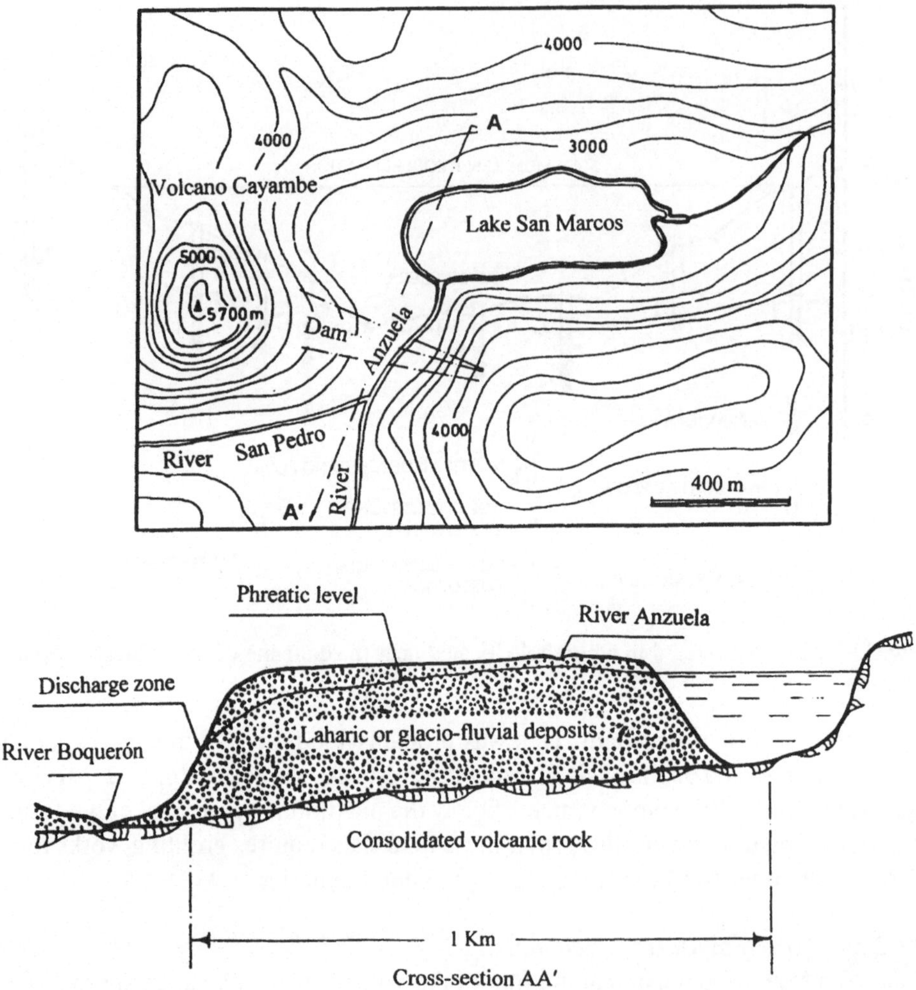

Fig. 11.3. Illustration of the dam site of the San Marcos reservoir, Ecuador.

and oxygen-18, (2) the measurement of thermonuclear tritium, (3) the analysis of the chemical composition of water, (4) the performance of flow tests in the boreholes drilled for this purpose, (5) the measurement of temperature, conductivity and natural gamma radiation profiles in the same boreholes, and (6) the performance of flow rate measurement using tracer methods of the inflow and outflow of the lake. The main goal of these flow rate measurements was the evaluation of eventual water losses in the lake derived from water infiltration (seepage). No measurable losses were found.

The chemical and isotopic composition of water demonstrated that as expected, groundwater from the laharic or glacio-fluvial materials was mostly the result of the

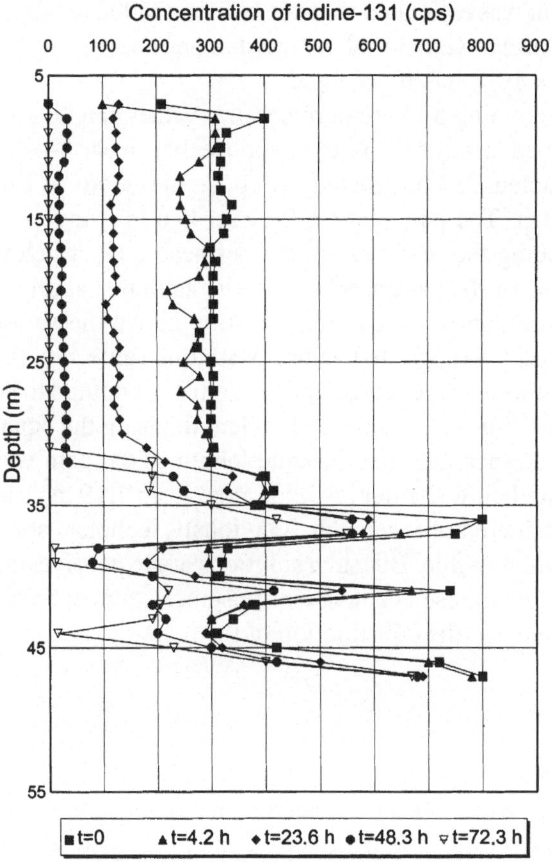

Fig. 11.4. Flow test performed at a borehole of the dam site of the planned San Marcos reservoir (Ecuador). The technique based on the labelling of the entire water column with ^{131}I was used.

infiltration of lake water. The gamma radiation logs showed a great homogeneity of the materials outcropping at the dam site up to a depth of 30 m. Below this depth, the gamma radiation intensity increased considerably due to the parallel increase of the clay content. However, the most important information was provided by the flow tests performed in the boreholes using ^{131}I as a tracer. The results of one of these tests carried out in a borehole located close to the dam axis is shown in Fig. 11.4. The results demonstrated the existence of a homogeneous layer with a thickness of 32 m, in which horizontal flow with a Darcian velocity of about 2.9 cm/d was present. Below this layer, an alternation of impervious and pervious layers was identified. The absence of stratification in the upper layer suggested that we were dealing with laharic instead of glacio-fluvial materials. The opposite was concluded for the lower layers.

These results demonstrated that an impervious substratum located at a reasonable depth appropriate for the dam foundation did not exist. Precise data on the local hydraulic gradient prevailing at the dam site were not available. However, with the available information, this gradient was estimated to range between 0.002 to 0.005. With these values and with the Darcian velocity of 2.9 cm/d, a permeability value between 1.7×10^{-4} m/s and 1.6×10^{-3} m/s is obtained.

The leak under the dam corresponding to these permeability values can be estimated using the procedure described in Annex 1. Let us suppose that an earth-filled dam with an impervious clay nucleus is constructed. Suppose the width of this nucleus at its base is equal to 20 m. The parameter L/b described in Annex 1 is equal to $20/32 = 0.625$ (32 m being the thickness of the permeable layer below the dam). The corresponding value of the parameter $q/(K\Delta h)$ amounts, approximately, to 0.7. Suppose also that a dam with a water elevation above the valley ground level $\Delta h = 50$ m is constructed (projected value). With this value, and the permeability values given before, the seepage rate below the dam is computed to be between 5.95×10^{-3} and 5.6×10^{-2} m/s per metre of dam length, using the equation $[q/(K\Delta h)] = 0.7$. Taking into account that the total length of the dam was about 200 m, the total seepage rate below the dam would range from 1.19 m^3/s to 11.2 m^3/s. These values were considered unacceptable. Obviously, technical solutions to reduce this seepage rate are possible. But such solutions are expensive and their efficiency may be limited. Under these negative circumstances, the technical staff responsible for the project rejected the construction of the dam.

11.2.5.5 *Poçeme reservoir, Albania*

The construction of a dam at the River Vjosa, Albania, was planned with the main purpose of hydropower production. The dam site of the planned reservoir is shown in Fig. 11.5. At the other side of the mountain range, at a straight-line distance of about 2 km, a group of springs, called Poçeme springs, with an average flow rate of 1.7 m^3/s, exists. Groundwater of these springs was emerging through the geological contact between a Cretaceous limestone formation and a clayey Quaternary formation. It was suspected that these springs could be fed from the River Vjosa. If this assumption proved to be true, that is, the river water was really feeding the springs and the infiltration area was located in the zone to be flooded by the reservoir, a new, more appropriate dam site would have to be found.

To investigate the prevailing situation, a long-term sampling program of the river water was initiated, including springs and local groundwater not related to the river. The investigation was conducted by Akiti et al. (1990). Stable isotopes of water, deuterium and oxygen-18, as well as natural tritium, were analysed in these samples. Figure 11.6 shows the variation of the $\delta^{18}O$ values during the sampling period. The average values obtained were −7.85‰ for the Poçeme springs, −8.20‰ for the River Vjosa and −6.55‰ for the local groundwater not related to the river. The more negative value obtained for the river water in comparison to the local groundwater was explained by the fact that most of the river was discharged

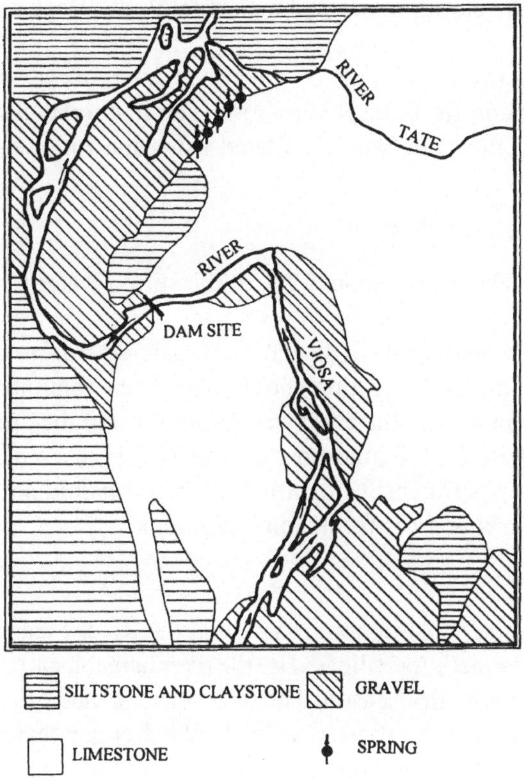

Fig. 11.5. Illustration of the area of the planned reservoir on the River Vjosa, Albania.

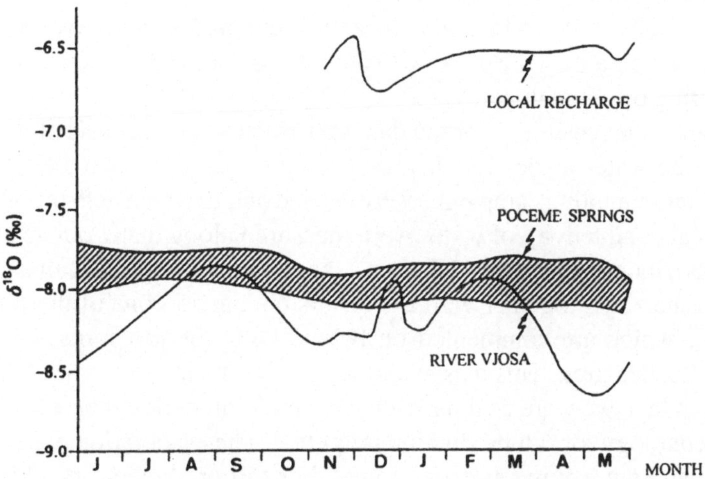

Fig. 11.6. Diagram showing the evolution of the $\delta^{18}O$ contents at the Poçeme spring, the River Vjosa and locally recharged groundwater not related to this river.

from elevations much higher than the elevation of the dam site (altitude effect, see Chapter 6).

The above-mentioned results showed that most of the water emerging at the springs represented, in fact, infiltration from the River Vjosa. The fractional contribution of this river to the spring flow rate was calculated using the following equation:

$$(-8.20)f + (-6.55)(1 - f) = -7.85$$

which provides a value $f = 0.79$. Therefore, about 79% of the spring water was supplied by the River Vjosa.

Due to these negative results, the construction of the dam was disregarded. However, in the opinion of the authors, this investigation should have been continued to investigate whether the infiltration area at the river was located inside the area to be flooded by the planned reservoir. Differential flow rate measurements along the river as well as interconnection experiments using artificial tracers would have been, among others, appropriate working tools for this investigation.

11.2.5.6 *Atance reservoir, Spain*

In the case of the Atance reservoir, the construction of a dam at the River Salado, Guadalajara Province, Spain, was planned. As indicated by the river name, it carries high salinity water, especially during the dry season. This is a consequence of the presence in the river basin of Triassic formations, basically the Keuper facies, which contains important deposits of evaporitic rocks, such as gypsum and halite. Groundwater in the vicinity of the outcrops of these rocks tends to be brackish to saline, reaching salinity values as high as 300 g/l. This type of groundwater is being exploited for table salt production at four different places in the river basin, not far from the reservoir area. One of these exploitations, called the Imon salt mine, even existed during the Roman period.

To evaluate the feasibility for the dam construction on the basis of the chemical quality of the water to be stored, a 2-year investigation was undertaken. Many measuring and sampling campaigns were carried out, in which both the flow rate and the electrical conductivity of water were measured along many points of the main river and its tributaries. Figure 11.7 shows the situation of the four salt exploitations existing in the zone, together with the results obtained in one of these measuring campaigns, which are commented on below. Only the salt mine S-1 is located inside the flooded area. This was abandoned 30 years ago; the other three are still operative. In fact, we were dealing with very small salt mines, where a rudimentary and old technology for salt production is applied. The evaporation ponds are in bad conditions and are leaking at many points. For salt production, brackish water is pumped from the Triassic aquifer using wells of large diameter (2 m) and depth below 10 m. The phreatic or piezometric level at these wells was located at a depth between 2 and 5 m. In the case of the salt mine, the brackish water was emerging at several points of the zone, but with a negligible total flow rate.

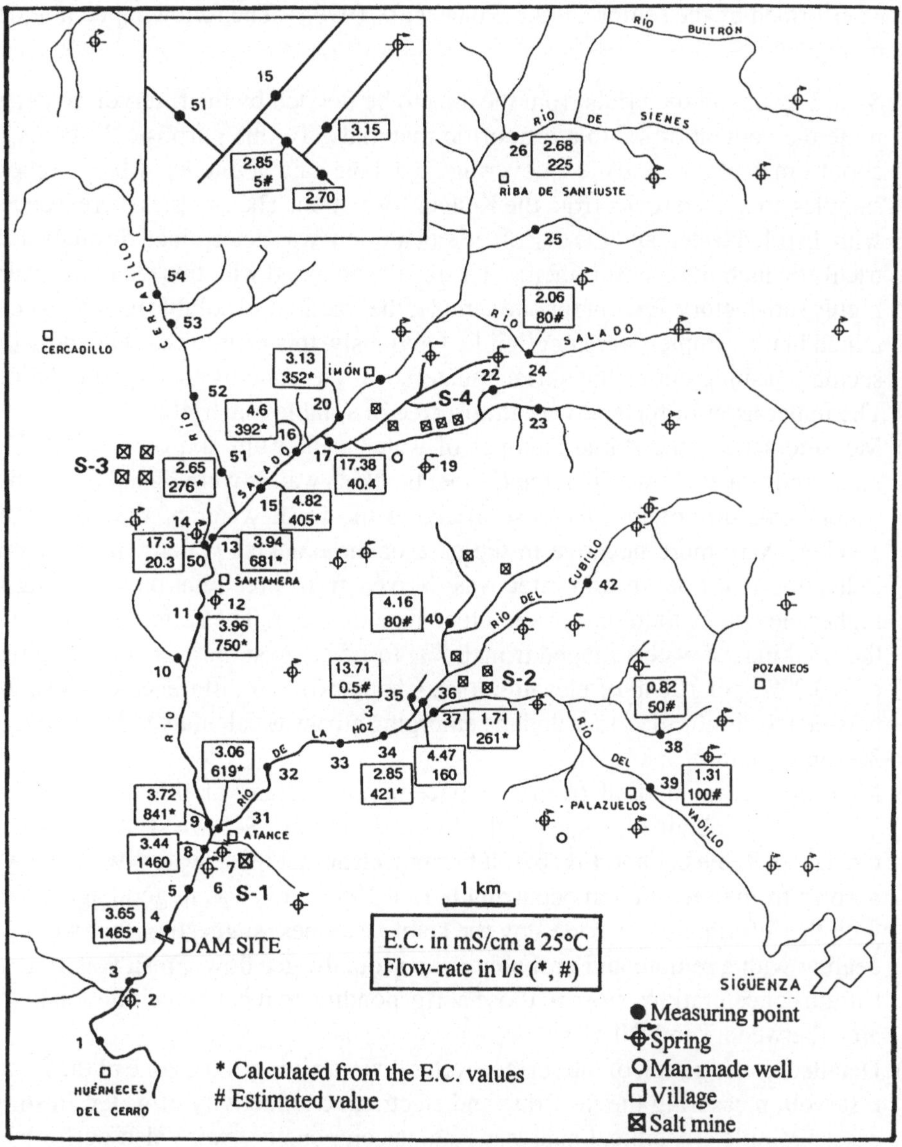

Fig. 11.7. Electrical conductivity (EC) and flow rate values measured in the Salado River and its tributaries during one of the various field campaigns carried out.

Obviously, under the above-described circumstances, the construction of a dam did not seem to be advisable. In fact, the project led to a controversial situation, which seemed to be justified and led to the temporary stop of the field work. A detailed study with the goal of evaluating the feasibility of the reservoir on the basis of the chemical quality of the water to be stored was necessary. The study was performed by the senior author (Plata et al., 1996b). The following techniques were applied:

1) Sampling of soil materials from the area to be flooded by the reservoir to determine the content of soluble evaporitic materials. To this purpose, 27 bores of about 1 m depth were drilled, covering the whole area of interest. Many of these samples contained rocks from the Keuper formation. The samples were leached with distilled water, and the dissolved salt amount was determined through electrical conductivity measurements. The results obtained with this technique were highly satisfactory. Except for one sample, the fractions of soluble materials contained in the samples were negligible. Obviously, this result is explained by the secular washing out of the surface material of the zone by local precipitation. The upper layer of the soil was almost free of soluble materials.
2) Measurement of the stable isotopes of water, deuterium and oxygen-18. The measured values showed that the Keuper brackish waters were linked to groundwater flows different from those affecting the fresh water of the zone. The δ-values were more negative in the case of the brackish water. This fact was indicating that this groundwater was derived from precipitation occurring at higher elevations than that of the other type of water. The difference between the two kinds of waters ranged from 0.7‰ to 1.5‰. Assuming an altitude effect of −0.25‰ per 100 m of elevation (see Chapter 6), the difference of elevation between the infiltration area and the emerging area was calculated to be between 300 and 500 m.
3) Measurement of natural tritium. Concentration values of 0.5, 1.2, 2.3 and 0.7 T.U. were obtained, respectively, for the salt mines S-1 to S-4. According to the data of Table 6.1 and Fig. 6.7, it became clear that these waters were mainly derived from precipitation occurring before 1953. This was in good agreement with the information provided by the stable isotopes, suggesting that we were dealing with a remote and probably deep groundwater flow. Fresh waters gave tritium concentrations close to those corresponding to recent precipitation in the area (between 5 and 8 T.U.).
4) Detailed investigation of the surface water flows arriving at the zone of the future reservoir, measuring the flow rate and electrical conductivity of water. In some cases, complete chemical analyses (major components) were also performed. An example of the results obtained is shown in Fig. 11.7. Surface waters presented large conductivity variations, depending on the origin of the water and on the extent of the mixing processes. These conductivity changes were used in many cases for the determination of the involved flow rates. The method is based on the mixing law. Let us suppose a surface flow of water with conductivity C_1 that flows into another flow with conductivity C_2, forming a new flow with conductivity C_3. The fraction f of the contribution of the first flow to the third one is given by the equation:

$$C_1 f + C_2(1 - f) = C_3 \tag{11.1}$$

For the calculation of the flow rate downstream of the confluence point of the two surface flows, it is only necessary to know the flow rate of one of these. This method allowed the senior author, in the present case, to perform a water balance of the full flow network with a relatively small number of direct flow rate measurements. These were performed using the method described in Section 4.3 of Annex 2, based on the discrete sampling of the tracer cloud at regular time intervals. Rhodamine WT was used as a tracer.

5) A gauging station was constructed at the dam site and fitted with an automatic device for continuous recording of the water level. This station was also used for the installation of an automatic sampler of the flowing water. Periodic samples were collected over about 8 months with a frequency of one sample every 6 or 12 h depending, on the meteorological conditions. Electrical conductivity was measured in all these samples, and from time to time, complete chemical analyses of the major components were also performed.

The variation with time of the electrical conductivity of water samples collected from the dam site is shown in Figs. 11.8 and 11.9. In the first of these figures, it can be observed that, as expected, the EC of water underwent a drastic decrease when the flow rate of the river increased as a consequence of the high precipitation that occurred during the winter of 1995–1996. The integration of the curves of Fig. 11.8 indicated that if all water crossing the gauging station during a 5,000-h period had been stored, the average conductivity would have reached a value of 2.89 mS/cm, equivalent to about 2 g/l of total dissolved salt. The volume of water impounded would have amounted to 18×10^6 m^3.

On the other hand, the flow rate measurements demonstrated that the average flow rate derived from the aquifers with brackish water was lower than 1 l/s. Most of this water was collected from a drainage channel located close to the salt mine S-4.

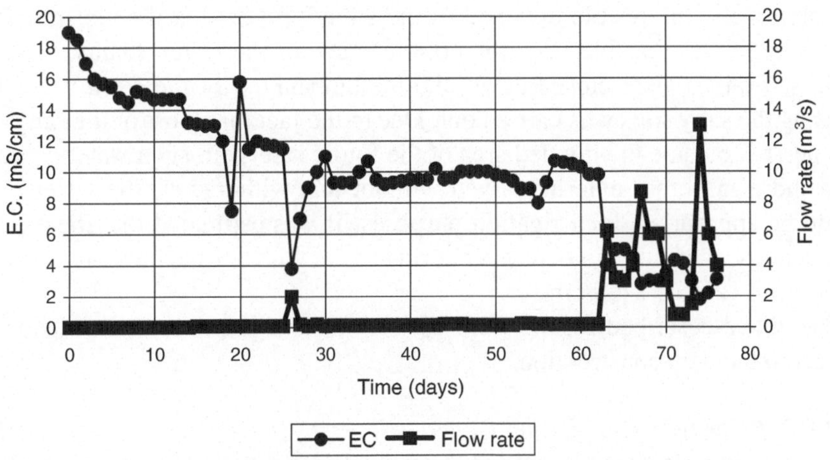

Fig. 11.8. Variation of the EC and the flow rate at the dam site of the El Atance reservoir.

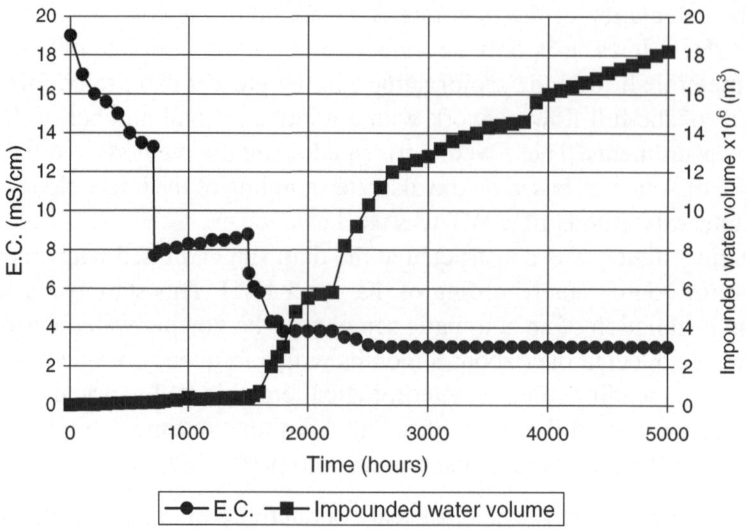

Fig. 11.9. Variation of the EC and the water volume stored at the planned reservoir, assuming all the water crossing the dam site during a 5,000-h period would have been stored.

A significant fraction of this groundwater was derived from the infiltration of salty water at the evaporation ponds of the mining installation. Differential flow rate measurements of the salt water surface flows proved that an important fraction of the saline contribution to the dam site was derived from the washing out by precipitation of the salt deposited on the soil surface in small areas close to the salt mines. The salinisation of these soils mainly took place as a consequence of the dispersion due to wind (aerosol formation) and the salt deposits inside the mine zone.

Definitively, the results obtained in the study proved that the construction of the reservoir was feasible from the point of view of the water quality, especially if the salt mines were closed down and the embanking of the fresh water flows crossing the salty soils was carried out. Due to the fact that most of the salty water is generated outside the flooded area of the future reservoir, such water could also be withdrawn before entering. Even without this withdrawal, the water quality would be appropriate for irrigation purposes. It was estimated that the electrical conductivity of the impounded water during a year with normal precipitation would range from 1.5 to 2.2 mS/cm.

The above-described results were used to make a positive final decision with respect to the dam construction.

11.2.5.7 *Rules reservoir, Spain*

The Rules dam is being constructed at the River Guadalfeo near the town of Motril, southern Spain. The study was conducted by the senior author (Plata

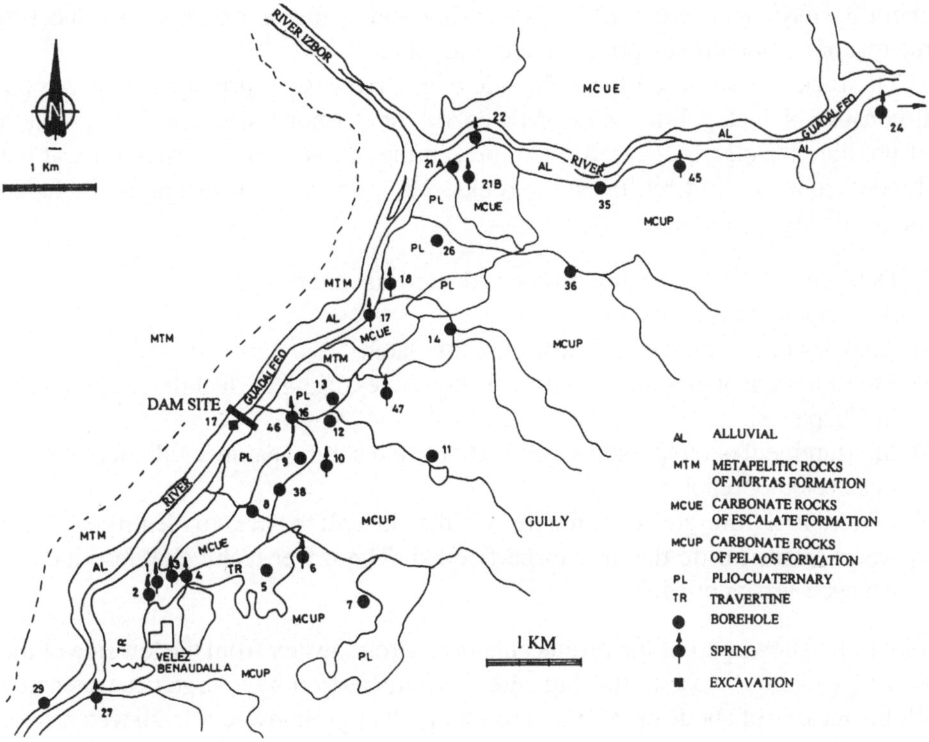

Fig. 11.10. Surface geology of the left side of the Rules reservoir. The numbers correspond to the reference of the sampling points for chemical and isotopic analyses.

et al., 1997) during the construction stage to investigate the permeability of the materials to be flooded and to characterise the chemical and isotopic composition of the local groundwater before the first filling of the reservoir. As shown in Fig. 11.10, the materials to be flooded by the future reservoir consist of impervious metapelitic rocks of Paleozoic or Permotriassic age and pervious Triassic limestone from the formation known as Sierra de Lújar. This limestone formation is located at the left side of the river and shows faint karstification. Groundwater flow through the limestone formation is mainly taking place through the rock fractures and discharges mainly through several springs located downstream of the dam, near the town of Vélez de Benaudalla. The average flow rate of these springs is about 250 l/s.

A second group of springs was located inside the flooded area. They are known as the Rules springs (point 22 of Fig. 11.10), discharging an average flow rate between 400 and 500 l/s. A new spring, called La Colorá, is located at a distance of about 200 m downstream. Groundwater emerging through this spring is red-coloured due to the presence of ferruginous compounds. Its average flow rate is about 10 l/s, the water temperature is about 26°C (thermal water) and the electrical conductivity is about 3.3 mS/cm. Therefore, this is a spring with different origin with respect to the rest of the identified springs in the study area.

The rock formation on the right side of the river is, in principle, impervious. It consists of metapelitic rocks of Paleozoic or Permotriassic age. The problem of possible leakage is focused, therefore, in the left side and, particularly, at the flooded zone covered by limestone rocks. The performed investigation included the following techniques:

1) Determination of the stable isotopes deuterium and oxygen-18.
2) Analysis of natural tritium.
3) Analysis of the chemical composition of water.
4) Measurement of groundwater flows in boreholes, using the techniques described in Chapter 8.
5) Measurement of temperature, conductivity and natural gamma radiation profiles in selected boreholes.
6) Differential flow rate determinations in the Guadalfeo River to identify eventual water losses inside the area to be flooded. The tracer technique described in Annex 2 was applied.

Table 11.1 shows part of the results obtained. Groundwater from the left side of the River Guadalfeo, close to the dam site, is characterised by a large heterogeneity. Within an area of about 2 km^2, modern water, older groundwater, fresh water, salty water, thermal water, springs, and boreholes with emerging water, that is, different types of water, were found, with extremely variable chemical and isotopic compositions. The most outstanding conclusions obtained from the investigation are the following:

1) Contradictorily to the previous statements made by different authors, groundwater from the Rules springs is not recharged in the limestone aquifer of Sierra de Lújar but by the alluvial materials of the River Guadalfeo. This conclusion is based both on the chemical and isotopic composition of the water. As can be seen in Fig. 5.20 (Chapter 5), the points of the Piper diagram corresponding to groundwater from the carbonate aquifer are located close to the left vertex of the diamond-shaped diagram. The points of the alluvial aquifer are close to the upper vertex due to the important contribution of the ions $SO_4^=$, Cl^- and Na^+. On the other hand, the δ-values of water from the Rules springs are more negative than the groundwater tapped in the carbonate aquifer due to the altitude effect (water from the Guadalfeo River is formed at very high elevations in the Cordillera Penibética). The average $\delta^{18}O$-values for these waters were, respectively, −8.88‰ and −7.88‰. It must be pointed out that these springs do not represent any leakage risk for the future reservoir because they are located inside the area to be flooded.
2) The isotopic composition of groundwater discharging at the spring La Colorá corresponds to groundwater from the carbonate aquifer of Sierra de Lújar.

Table 11.1. Results obtained for some parameters of the more representative groundwater types from the dam site of the Rules reservoir. The numbers in parentheses correspond to the codes shown in Fig. 11.10. V.B. = Vélez de Benaudalla. G.S. = Geological Service. T.U. = tritium units.

Sample	Temp. (°C)	Cond. (μS/cm)	$\delta^{18}O$ (‰)	Tritium (T.U.)	$[Cl^-]$ (mg/l)	$[SO_4^=]$ (mg/l)
Spring Nueva (V.B.) (1)	19.7	505	−7.89	6.3		
Spring El Nacimiento (V.B.) (2)	19.1	510	−7.89	5.0	12.4	39.0
Spring El Torchal (V.B.) (3)	19.0	510	−7.84	5.5	17.7	39.6
Spring La Perilla (V.B.) (4)	19.0	510	−7.88	5.3		
Spring El Mesoncillo (6)	18.4	358	−6.30	9.7		
Spring Rafael Pérez (10)	20.1	430	−6.10	10.1		
Spring Don Baldomero (16)	27.5	509	−6.24	5.0		
Spring La Colorá (18)	26.0	3300	−7.86	0.2	355	1620
Springs of Rules (22)	19.3	1600	−8.88	7.8	284	300
Spring Estévez (27)	18.1	520	−7.47	6.4	24.8	51.6
Spring El Algarrobo (47)	18.5	343	−7.78	5.2	14.2	31.9
Spring El Castillejo (24)	18.7	515	−8.19	5.2	60.0	25.1
Borehole P-5 (5)	21.0	640	−6.10	0.1	140	17.0
Borehole A of G.S. (36)	18.0	660	−6.26	0.4	106	85.2
Borehole B of G.S. (11)	19.0	266	−7.51	4.2	14.2	3.6
Borehole C of G.S. (7)	19.2	398	−7.70	4.5	10.6	27.2
Borehole E of G.S. (35)	20.0	1560	−8.90	8.4	213	344
Borehole H of G.S. (45)	18.0	1338	−9.01	7.5		
Borehole P-24 of G.S. (8)	17.9	415	−7.44	3.8	17.7	22.6
Borehole P-22 of G.S. (9)	18.0	480	−7.82	0.0	15.0	22.6
Borehole P-21 of G.S. (12)	19.1	610	−7.66	3.7	17.7	99.2
Borehole P-13 of G.S. (13)	18.7	242	−6.27	7.8	23.1	3.2
Borehole P-23 of G.S. (14)	18.7	542	−6.47	0.4	49.7	102
Dam excavation (17)	18.5	2100	−8.69	4.6		
Well Llano Moreno (21A)	24.4	5200	−8.65	0.4	1300	2600
Well Llano Moreno (21B)	21.8	1400	−6.35	1.3	1450	1250
Well in the alluvial (river) (29)	19.2	943	−8.74	8.1	135	122
Borehole Oloroso (38)	23.0	3640	−7.30	4.1	852	504
Well in dam gallery (46)	17.5	1373		4.7	256	235

This groundwater is probably circulating at great depths, explaining its high temperature. The contact of groundwater with the underlying saline substrate can explain its different chemical composition. The absence of tritium shows that we are dealing in this case with groundwater recharged from precipitation that occurred before 1953.

3) Old waters with tritium concentrations lower than the average values for the last 45 years' precipitation (see Fig. 6.8 in Chapter 6) are also found in points close to the dam site. This fact, together with the different chemical and isotopic compositions found in local groundwaters from this area, demonstrates the

absence of significant groundwater flows in the left embankment of the dam and therefore a low permeability of the involved materials. This result presages a favourable behaviour of these materials once the reservoir is filled.

4) The Vélez de Benaudalla springs constitute the main discharge of the carbonate aquifer of Sierra de Lújar. Using the exponential or well-mixed model described in Chapter 5 for the interpretation of tritium data, a turnover time above 100 years is obtained for the groundwater stored in this aquifer. As mentioned in Chapter 5, this turnover time represents the ratio between the stored volume in the aquifer and its recharge rate. Therefore, assuming a recharge equal to the discharge of the Vélez de Benaudalla springs (250 l/s), a value of 790×10^6 m^3 is obtained. On the other hand, it has been observed that these springs reduce their flow rate considerably, and some of them even become dry, during long-term dry spells. This observation is in contradiction with the high storage volume of the aquifer and suggests that it is a closed aquifer, bordered by materials of very low permeability. The Vélez de Benaudalla springs would represent the overflow of this closed system, and therefore, they become dry when the water level decreases several metres.
5) The risk of leakage in the future reservoir seems to be, in principle, minor. The maximum risk is concentrated at the flooded area where the limestone formation outcrops. This area is located immediately upstream of the Rules springs. Obviously, if water from the future reservoir infiltrates at this zone, it would emerge through the Vélez de Benaudalla springs. The probability of this leakage is incremented by the observed fact that two boreholes located at this zone contain water related to the alluvial aquifer and not to the carbonate units, as indicated by the stable isotope composition.

However, apart from the previous statements, the main outcome of the investigation is related to the characterisation of the groundwater existing at the vicinity of the future reservoir. This characterisation will provide a better understanding of the hydraulic processes occurring in the affected zone once the reservoir is filled. The modifications of groundwater flows and the replacement of the original aquifer water by reservoir water will be easily identified on the basis of the results gained.

11.2.5.8 *Bellús reservoir, Spain*

The Bellús reservoir is located on the River Albaida, near the village of the same name (Bellús) in the province of Valencia, eastern Spain. The dam has been constructed on a Cretaceous limestone outcrop where groundwater of the carbonate aquifers Sierra Gorda and Sierra de la Creu (see Fig. 11.11) is discharging. It is foreseen that this reservoir will impound water of poor chemical quality due to the discharge of industrial and domestic wastewaters upstream. Therefore, certain risk exists for groundwater from the carbonate aquifer to become polluted with the reservoir water. The evaluation of this risk was the main objective of the investigation conducted by the authors (Plata et al., 1996a). This investigation was

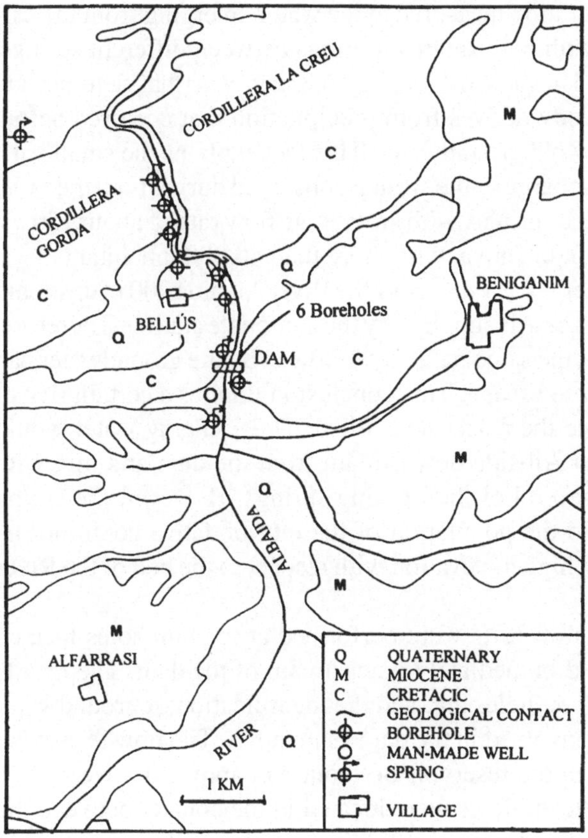

Fig. 11.11. Geology and location of the sampling points in the area of the Bellús reservoir, Spain.

performed at the end of the construction of the dam but before the filling of the reservoir.

The following techniques were used: stable isotopes of water, that is, deuterium and oxygen-18, thermonuclear tritium, and flow tests in boreholes and flow rate determination of the river to identify points where groundwater from the carbonate aquifer was emerging.

The River Albaida cuts the carbonate formation, forming the two above-mentioned sierras in an almost perpendicular direction, delimiting a deep and narrow pass. The studies conducted in several campaigns allowed the localisation and quantification of the discharge of important springs in this narrow pass, with a total flow rate close to 1 m^3/s. Many of these springs were not known because groundwater was emerging in the riverbed and was mixed with the polluted river water. The emerging water is of good quality for human consumption but changes to unacceptable after mixing with the polluted river water. The construction of the dam makes feasible the separation of the polluted river water from the good quality groundwater of the carbonate aquifer, allowing the use of the latter type of water for human consumption.

The average tritium concentration measured at the water emerging from the carbonate aquifer was 2.6 T.U., with only small variations between different springs. According to the data presented in Table 6.1 and Fig. 6.8, it is clear that an important fraction of this groundwater body derives from precipitation that occurred before 1953. Therefore, it is relatively old groundwater. This fact explains the small variations in the flow rate of the discharge of the springs, observed during both the rainy and the dry periods. One of these springs, with an average flow rate of about 150 l/s, is used to supply the population of Játiva, with more than 40,000 inhabitants.

A flow rate above 100 l/s emerges at a stretch of the River Albaida 200 m upstream of the dam. Groundwater in this case is supplied by the carbonate aquifer. Therefore, a relatively high risk exists that the same pathway followed by the groundwater can be used for the reservoir water to escape. This conclusion implies a certain risk of the appearance of leakage once the reservoir is filled. The escaping water would eventually emerge in the River Albaida downstream from the dam and mix with groundwater emerging from several of the existing springs. However, the spring being used for water supply to the population of the city of Játiva could not be affected because it is located at higher elevation with respect to the bed of the River Albaida.

On the other hand, several flow tests were performed at the boreholes located both inside the dam gallery and immediately downstream of the dam. The results obtained for the first group of boreholes allowed the identification of groundwater flows existing in the materials involved in the dam foundation. The flow tests were performed with a water level in the reservoir of about 5 m above the riverbed. It was found that the most significant flows were located in the contact between the concrete structures of the dam and the terrain. The information collected helped to better plan the grouting works carried out later on.

An old thermal water spring discharges at the left side of River Albaida, about 300 m upstream of the dam site. The same water quality was also found in some boreholes of the dam gallery. The absence of tritium and the different chemical composition of water demonstrated that, at this stage, no connection of these waters with the river was taking place.

The flow tests performed in the external boreholes located at the foot of the dam showed that only one of these boreholes located at the right side was clearly connected to the river. The results obtained for this borehole are shown in Fig. 8.11. Water recharged from the river was entering the borehole through some strata located above the piezometric level. A groundwater flow with a Darcian velocity of about 4 cm/d was detected at the boreholes on the right side. Therefore, a significant risk of leakage through the involved materials (mostly, calcarenite) exists once the reservoir becomes filled with water. Most probably new grouting works will have to be implemented.

As in the case of previous examples, the information obtained during this investigation permitted an improved understanding of the local hydraulic problems that eventually will be generated after the filling of the reservoir.

11.2.5.9 *Algar reservoir, Spain*

The Algar dam is being constructed at the River Palancia near Segorbe, Valencia. The dam and most of the flooded area of the future reservoir are located over carbonate rock formations (limestone and dolomite) of Muschelkalk and Jurassic age. Civil engineers responsible for the dam construction were aware from the very beginning of the possible appearance of significant water leaks once the reservoir is filled. However, in spite of the unfavourable conditions, the construction of the dam was accomplished because it was considered that, even in the most unfavourable case, the reservoir could have a positive impact by increasing the recharge of the limestone and/or alluvial aquifers located downstream. On the other hand, the dam is expected to play a key role in flood control.

The Algar dam is located at a distance of about 25 km from the coast. The area under the influence of the reservoir is extremely important from the agricultural point of view, with citrus products being the main crop. During the last 25 years, most of this area has been irrigated with groundwater extracted from the aquifers. The overexploitation of the local aquifers was considered the decisive factor for recommending the construction of the dam. However, in the opinion of the authors, the studies carried out during the feasibility stage of the project were not sufficient, and the applied tools were not fully appropriate. The present investigation was performed at a stage when the concrete dam was almost finished (Araguás and Plata, 2002). Only the installation of the gates and some other small complementary works were still missing. It must be mentioned here that the dam itself was constructed on a karstic sinkhole whose hydraulic connection with the downstream aquifer is not known.

The hydrogeological investigation included the following tools: (1) chemical analyses of water samples, (2) stable isotopes of water (deuterium and oxygen-18), (3) measurement of natural tritium in groundwater samples, and (4) flow tests in the boreholes located inside the perimetral gallery of the dam, to evaluate the permeability of the rock formations at the dam site. The investigation extended for approximately one year. During this time, a heavy flash flood affected the catchment area of the reservoir. The flow rate determined in the River Palancia at the time of the flood was more than a hundred times higher than the long-term average values. The reservoir became almost full in a matter of hours, in spite of the absence of gates. The same happened with the dam galleries and with the boreholes that had been drilled inside. This circumstance was used to obtain valuable information on the response of the involved geological formations to an extreme precipitation event.

Water chemistry About 220 chemical analyses (major components) of water samples collected over the last 30 years were available, besides 40 samples collected and analysed during the present study and an exhaustive field determination of parameters such as electrical conductivity, temperature and alkalinity. The most important conclusions obtained from these data are summarised in the

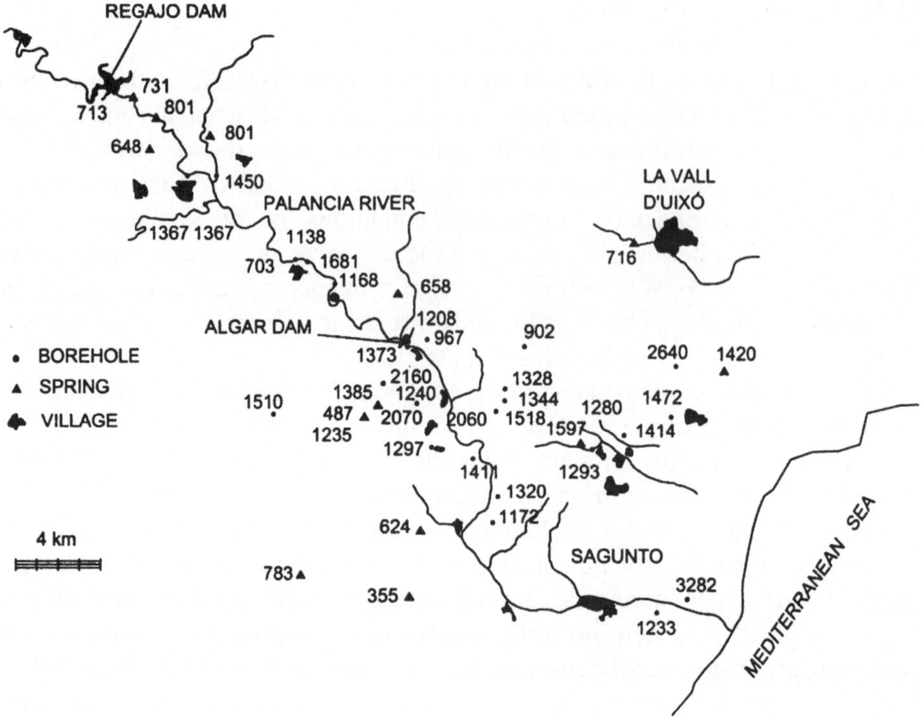

Fig. 11.12. Distribution of the electrical conductivity of the groundwater downstream of the Algar dam, Spain.

following points:

1) Most of the wells existing in the study zone show conductivity values between 1,000 and 1,500 μS/cm, while the values for the springs located at high elevations present values around 500 μS/cm. The distribution of these values in the zone downstream of the dam is shown in Fig. 11.12. All water samples are saturated in calcite. Nevertheless, the high conductivity of water is also partially derived from the recycling of irrigation water.
2) As expected, the major components of the groundwater samples from the karstic springs are Ca^{++}, Mg^{++} and HCO_3^-. In samples collected from wells, the higher conductivity values are mainly governed by the dissolution of gypsum from the impervious substratum formed by Triassic materials (mainly Keuper and Muschelkalk) with a high content of evaporitic rocks. This is clearly shown in Fig. 11.13, where the electrical conductivity is plotted against the concentration of calcium and sulphate ions. The same observation can be applied for the springs located at low elevations, through which the carbonate rock aquifer has been secularly discharging. These results show that most of these waters are flowing at the deep levels of the aquifer. The difference in chemical composition

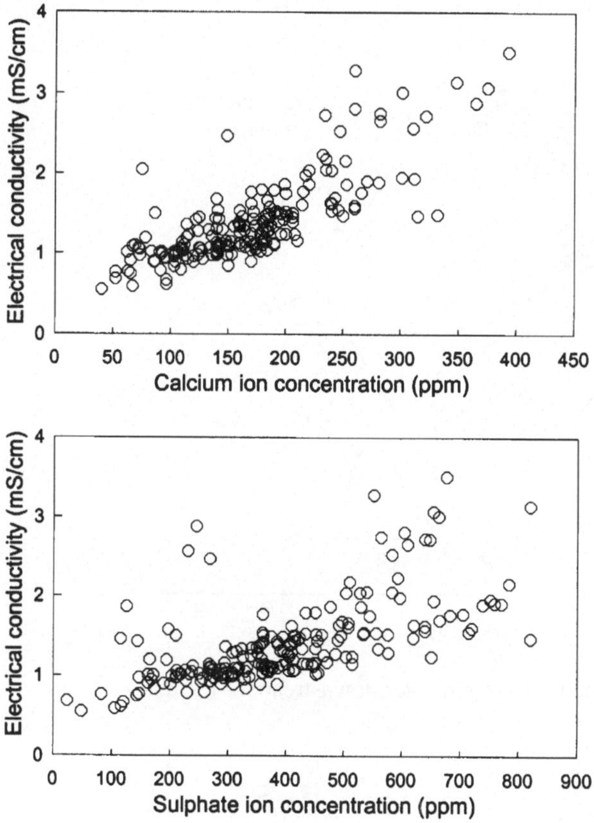

Fig. 11.13. Relationship between the electrical conductivity and the concentration of ions Ca^{++} and $SO_4^{=}$ for groundwater of the Algar reservoir area.

between karstic springs and groundwater from wells is clearly shown in the Piper diagram of Fig. 11.14.

3) An important result of the investigation is related to the high contents of the nitrate ion observed in most of the wells and low elevation springs. Obviously, this is a result of the agricultural activity of the zone. This is also demonstrated by the important increase in nitrate content observed during the recent years, as shown in Fig. 11.15, where the evolution with time of the average electrical conductivity and ion nitrate concentration for the two most important low elevation springs is shown. This result suggests that probably most of the recharge of the carbonate aquifer derives from the local precipitation and from the recycling of irrigation water. The contribution of the River Palancia seems to be minor and probably restricted to small areas close to the riverbed.

Stable isotopes, deuterium and oxygen-18 More than 60 water samples were analysed for stable isotopes. The most important factors determining the isotopic composition of groundwater of the study area are the evaporation of water before its

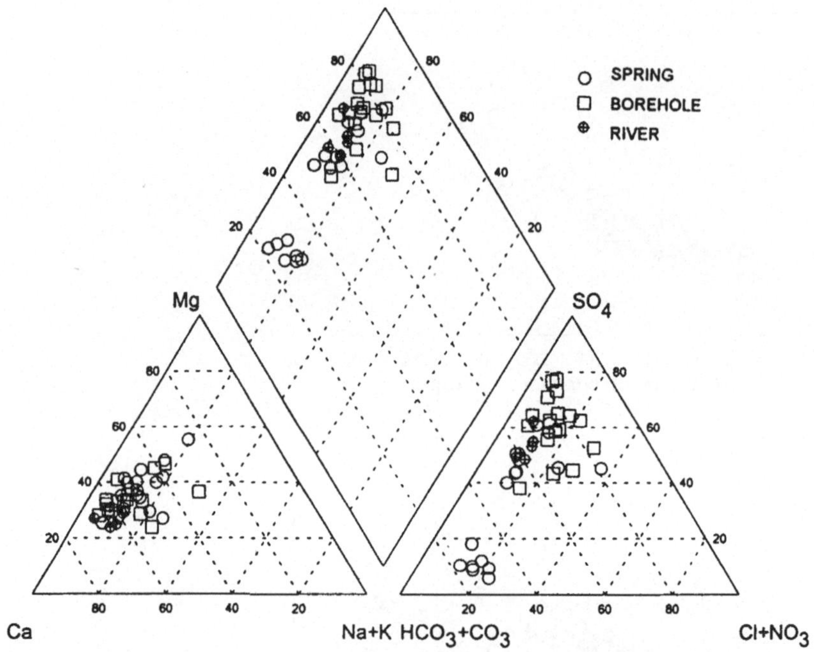

Fig. 11.14. Piper diagram obtained for groundwater downstream of the dam.

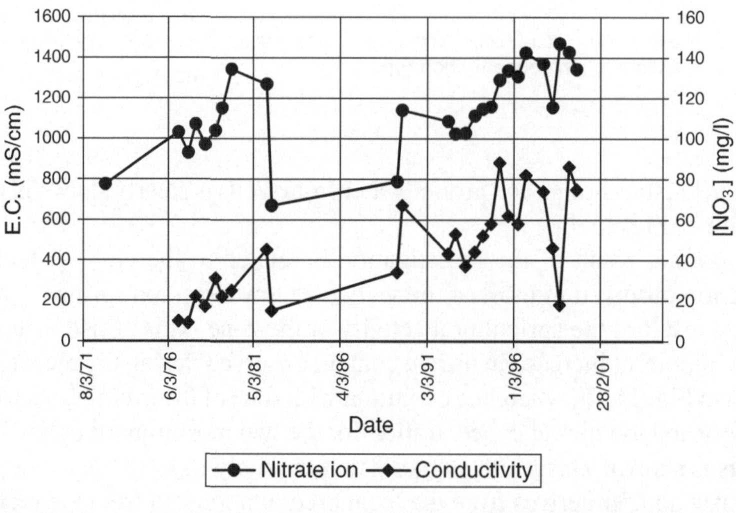

Fig. 11.15. Evolution of the electrical conductivity and NO_3^- ion concentration of groundwater downstream of the Algar dam over the last 30 years.

infiltration into the soil and the recharge elevation, which, is related to the distance of the sampling point to the sea.

Figure 11.16 shows a diagram relating the isotopic deviation δD and $\delta^{18}O$. The points define two different correlation lines: the Global Meteoric Water Line and an

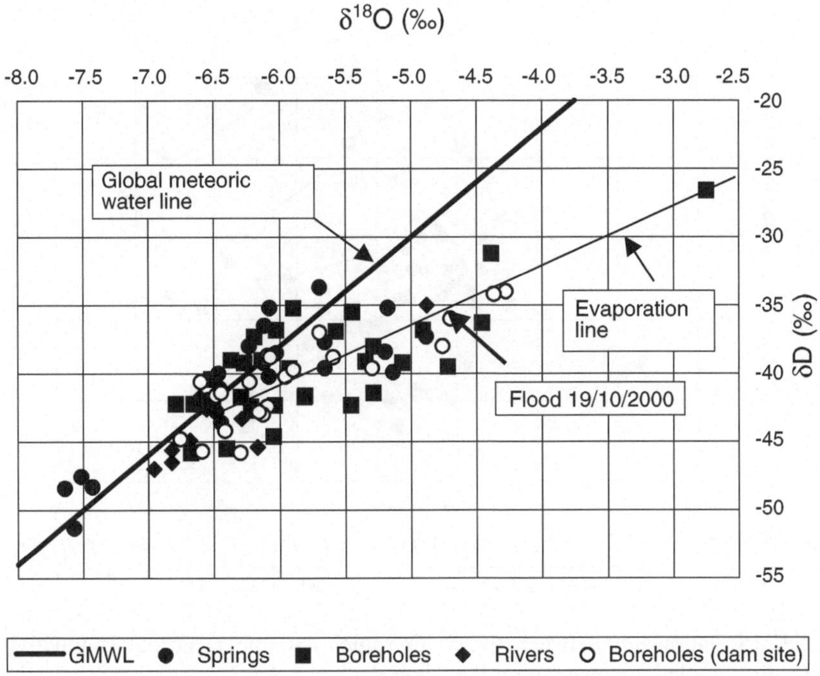

Fig. 11.16. Diagram relating the isotope deviations δD and δ^{18}O for groundwater sampled downstream of the Algar dam.

evaporation line. The first line corresponds to water derived from local precipitation that rapidly infiltrates into the aquifers. The evaporation line corresponds to waters that have undergone a significant rate of evaporation before infiltration. Obviously, these points include river water and recycled irrigation water sampled in wells.

The average isotopic composition of the meteoric water (nonevaporated) of the study area corresponds to the approximate values $\delta D = -38‰$ and $\delta^{18}O = -6.2‰$. This is also the approximate average isotopic composition of the local precipitation. On the other hand, the intercept of the two lines of Fig. 11.16 corresponds to the values $\delta D = -46‰$ and $\delta^{18}O = -7.0‰$. Due to the altitude effect, water from the River Palancia is characterised by more negative δ-values than the local precipitation close to the dam site. Therefore, the difference between the isotopic composition of the local precipitation and the values corresponding to the intercept of the two lines of Fig. 11.16 suggests that water from the Palancia River upstream of the dam will contribute in some extent to the recharge of the aquifer.

The heavy storm that occurred in October 2000 produced a significant change in the chemical and isotopic composition of groundwater in many wells and springs. This change is illustrated in Fig. 11.17, where the electrical conductivity of water is plotted versus the δ^{18}O-values. Obviously, the infiltration of low conductivity

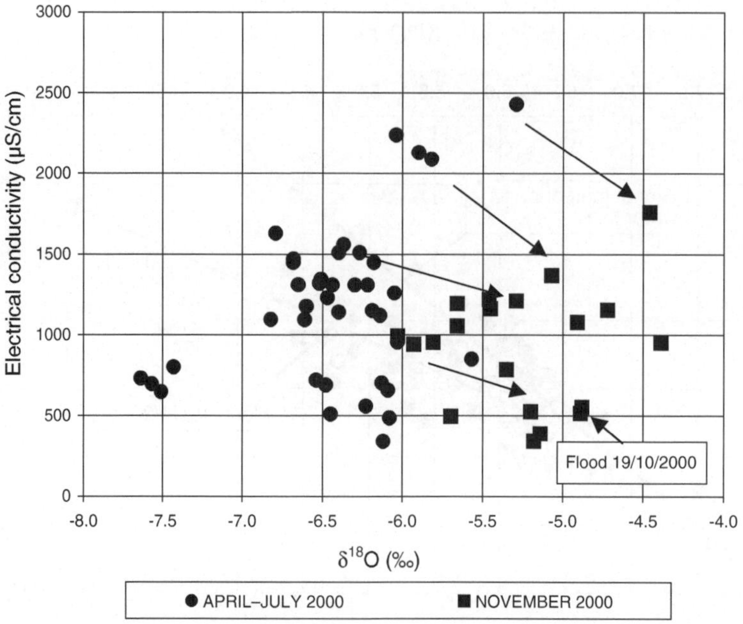

Fig. 11.17. Relationship between the $\delta^{18}O$ values and the electrical conductivity for groundwater sampled downstream of the Algar dam.

(precipitation) water was responsible for the decrease of conductivity. The same can be said for the $\delta^{18}O$-values because this value for the storm water was about −4.8‰. These results indicate that most of the groundwater of the study zone is of a recent origin, which is also supported by the measured tritium contents.

Tritium concentration The air masses and associated cloudy fronts that are responsible for most of the precipitation occurring on the Iberian Peninsula originate in the Atlantic Ocean and reach the study area highly weakened. For this reason, most of the precipitation in the study area is related to air masses generated in the Mediterranean Sea. Such humid masses travel at low elevation and do not undergo a complete exchange with the upper atmosphere, where most of the tritium is stored in the form of gaseous hydrogen, H_2. This is the reason why local precipitation in eastern Spain shows average tritium contents lower than those observed in the precipitation when compared to other areas of the Iberian Peninsula (between 6 and 7 T.U., according to Table 6.1).

Figure 11.18 shows a histogram of the tritium values obtained in the study zone. Disregarding the few values lower than 3 T.U. and higher then 5 T.U., an average value of 3.8 T.U. is obtained. This will be the approximate value corresponding to the local precipitation for recent years. The real value is probably a little bit lower due to the presence, in some wells and springs, of small fractions of water related

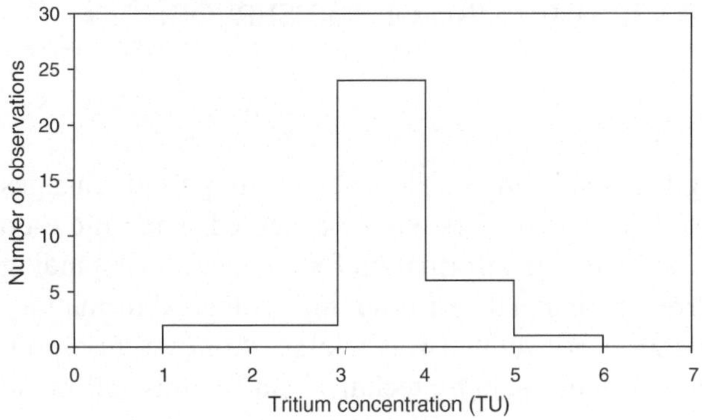

Fig. 11.18. Histogram of the tritium concentration values measured in groundwater sampled from wells located downstream of the Algar dam.

to the precipitation of the 1960s and 1970s, when the concentration of tritium in precipitation were much higher, even after the decay correction. Therefore, a value of 3.5 T.U. will be valid, most probably, for recent local precipitation.

According to these considerations, it was concluded that most of the groundwater in the study area represents precipitation from recent years. Only two wells, with tritium concentrations of 1.4 and 1.9 T.U., respectively, contain significant fractions of water from precipitation that occurred before 1953. These results are in good agreement with the observations made on the variation of the chemical composition during the last years and the changes observed after the flood of October 2000.

Conclusions The results obtained in the investigation of the Algar reservoir suggest that the influence of the future reservoir in the recharge of the downstream aquifers will be restricted to concrete areas probably located close to the River Palancia. Many of the existing wells will not be affected at all by the assumed recharge. Therefore, the drilling of new wells located in the influence area of the reservoir will be necessary for an appropriate exploitation of the reservoir water.

The changes produced in the chemical and isotopic composition of groundwater downstream of the dam once the reservoir is filled will surely provide a precise definition of the extent of the influence area. However, better information on this point would be obtained by means of a tracing experiment using artificial tritium as a tracer. The water to be impounded in the reservoir can be labelled with tritium, so that an average concentration between 100 and 200 T.U. is obtained. Such low concentrations will not represent any risk at all for human consumption. The eventual arrival of the injected tracer in the downstream wells and springs would be determined during a long enough period of time. The drilling of some new boreholes at specific points would be necessary. The described problem is a typical example of a case study where the use of artificial tritium as a tracer is the only available choice. Fluorescent tracers would not allow the labelling of such a high water volume (about 150×10^6 m^3).

11.3 STUDIES DURING THE FIRST FILLING OF A RESERVOIR

11.3.1 Introduction

It is pretty obvious that during the first filling of the reservoir, important changes in the piezometric levels of the adjacent aquifers will be observed. The follow-up on these changes always provides valuable information for a better understanding of the relationship between the reservoir and the surrounding geological formations and, of course, for the investigation of eventual water losses from the reservoir. However, insufficient attention is usually paid by technical staff to this follow-up investigation, and it is almost never carried out under appropriate conditions. Apart from this follow-up study, some other important tests and measurements should be performed during the first filling of the reservoir. They are described in the following sections.

11.3.2 Water balance

Obviously, a significant fraction of the water entering the reservoir during its first filling is used in filling up cavities of the surrounding rock formations. The piezometric or phreatic levels of these formations will rise as a function of the water level in the reservoir with a rate that will depend on their permeability and their hydraulic connection with the reservoir. The volume of the rock formation affected by this piezometric level rise and, consequently, the volume of the cavities to be filled with water, are usually very high.

A water balance performed during the first filling of a reservoir can provide quantitative information on the total volume of the cavities being filled with reservoir water. The method requires the determination, with sufficient precision, of the water volume V_{IN} that has entered the reservoir during a specific period of time, as well as the same volume that eventually has been discharged, V_{OUT}. The difference between these two water volumes has to be compared with the water volume ΔV that has been stored during the same period of time in the reservoir. The value of the last volume can be obtained from the curve reflecting the storage capacity of the reservoir as a function of the water level (rating curve), which usually is determined using topographic data. Obviously, its value can be positive (increase of water level) or negative (decrease of water level), depending on the variation with time of the inflow and outflow water amounts. The water loss V_L due to infiltration through the geological formations is then obtained using the following equation:

$$V_L = V_{IN} - V_{OUT} - V_E - \Delta V \tag{11.2}$$

where V_E represents the water losses due to evaporation; this value can be estimated using the theoretical equations developed for these purposes. Due to the high value of the parameter V_L in comparison with the water losses derived from evaporation, the error made in its estimation will be negligible in most cases.

Obviously, the influence of time in this case is very important due to the inertia in the filling process of the cavities of the involved rock formations. For example, the filling process will continue even if the water level at the reservoir is kept constant once the maximum level has been reached.

The whole process can be illustrated by plotting diagrams that reflect the changes as a function of time, both of the difference between the inflow and discharged water amounts ($V_{IN} - V_{OUT}$) and the total water volume stored in the reservoir (V_E). In the same way, diagrams relating the difference between these two parameters as a function of time and of the water level at the reservoir can be very illustrative.

11.3.3 Follow-up of piezometric levels

As mentioned in the previous section, the phreatic or piezometric water levels at the surrounding geological formations, usually undergo an important increase during the first filling of the reservoir. Therefore, the follow-up on these changes of levels in the existing boreholes or piezometers as a function of the water level in the reservoir will always provide very valuable information on the permeability of the rock materials intercepted by them. In principle, three cases are possible:

1) The borehole might not have any hydraulic connection with the reservoir, and as a consequence, its level does not vary at all with the reservoir level. However, in this case, long-term level measurements should be performed because the hydraulic response of the borehole can occur with a long delay.
2) The permeability of the materials connecting the borehole with the reservoir is very low. In this case, a high hydraulic gradient between the reservoir and the borehole exists at the beginning of the filling process. The magnitude of this gradient will decrease with time with a rate that depends on the permeability of the involved formation. During the periods of time when the water level at the reservoir is kept constant, the water level at the boreholes will tend to an equilibrium value. The rate of variation of this level will provide valuable information on the relative permeability of the formation. In certain cases, this information can be quantitative or almost quantitative.
3) If the hydraulic connection between the reservoir and the borehole takes place through a high-permeability geological formation, both water levels will vary in a parallel way with a short time delay.

Obviously, in the last two cases, if a water flow is continuously escaping from the reservoir toward the aquifer, the water level in the borehole will always be lower than the level at the reservoir due to the loss of hydraulic head associated with the groundwater flow through the geological formation. However, it should be pointed out that a parallel variation of water level in the reservoir and the borehole does not necessarily mean that a significant water flow infiltrating in the reservoir is flowing through this borehole. The hydraulic head of the reservoir can be transmitted through the geological formation to the borehole without significant water flow. This will only occur if the borehole is located in an area of the aquifer where groundwater can freely flow, but not if it is located in a stagnant area of the aquifer (for example, end of a permeable stratum).

Very frequently, the permeability of the geological formations intercepted by the borehole presents important vertical changes. If this is the case, the response of the borehole to water level variations at the reservoir can drastically vary with this level. For example, a borehole can be practically disconnected hydraulically from the reservoir during a wide interval and show a rapid response once the water level reaches a certain elevation. This occurs very frequently in formations made of consolidated rocks and particularly in karstified limestones. Obviously, the information obtained through the identification of these permeable strata or horizons connecting the borehole to the reservoir is absolutely valuable for the leakage study. The results obtained in this way will provide the elevation at which the reservoir water is escaping.

11.3.4 Follow-up on water flow inside the boreholes

The importance of follow-up on groundwater flow through the boreholes located at the surrounding formations of the reservoir is pretty obvious, particularly taking into account the statements made in the previous section. The measurement of this flow can be carried out using the techniques described in Chapter 8, both for the horizontal and the vertical components of the flow.

In most cases, the only study carried out by the technical staff responsible for the reservoir management consists of the water level control in the existing boreholes. However, it is clear that this information is not sufficient to understand the dynamics of the involved geological formations around the reservoir. Groundwater may flow either very rapidly or very slowly, either with a water level in the boreholes very close to the reservoir water level or with a much deeper level. Groundwater flow velocity through the borehole will depend on its situation with respect to the reservoir and the discharge zone (for example, springs). A borehole located close to the discharge zone will have a piezometric level much deeper than the reservoir level and very close to it, if it is located close to the reservoir. But also a borehole only tapping impervious materials and therefore with no flow, can show a water level much deeper than the reservoir level.

In conclusion, only with the measurement of the groundwater flow circulating through the boreholes existing at the dam site as a function of the water level at reservoir is it possible to know the flow pattern prevailing in the geological formation adjacent to the reservoir.

11.3.5 Follow-up on the flow rate of emerging waters

The importance of the determination of the flow rate variation in selected springs as a function of the water level at the reservoir has been described in detail in Chapter 2. Also very important is the control of the flow rate of the springs that eventually were identified downstream of the dam site before its construction. Very frequently, the flow rate of such springs increases with the presence of the reservoir. However, that does not always mean that the reservoir is directly contributing to the flow discharging through the springs. The increase of the phreatic or piezometric level in the surrounding aquifers induced by the reservoir can produce an increase of the emerging flow rate at the spring, even with no direct contribution from the reservoir. Only the local piezometric gradient responsible for this flow rate has increased due to the transmission of the hydraulic head through the geological formations.

The information on whether a real contribution of the reservoir to the emerging water exists can be obtained only by determining the natural chemical and isotopic tracers described in Chapters 5 and 6.

11.3.6 Study of the water infiltration through the fractures or cavities of the rock formation

Due to practical reasons, it is not possible to determine the distribution of the permeability of all materials forming the flooded area of a reservoir. A large number of boreholes and flow tests would be necessary. However, the qualitative identification of the most permeable strata or fractured areas of the rock formation can be done in a relatively simple way before the first filling of the reservoir. Fractures, cavities, diaclases, karstified areas, etc. can be identified at the reservoir walls and bottom. The same can be said for the eventual presence of alluvial strata that will be covered by water. Also, the lithological columns of the formations tapped by the existing boreholes will provide valuable information on the relative permeability of the intercepted materials. For this preliminary study, it is important to take into account the large heterogeneity and anisotropy of the geological formations, especially when dealing with consolidated rocks.

The infiltration of water through the lateral walls and the bottom of the reservoir during the first filling can be investigated with very simple tracer tests using fluorescent tracers. The recommended technique is based on the implementation of multiple tracer injections in suspected areas of infiltration. Several grams or tens of grams of uranine or rhodamine B or WT will, in most cases, be sufficient for each individual injection. In the case of infiltration through the walls, the tracer should be injected at some distance from the bank. The information is visually obtained. If infiltration through the formation is taking place, the tracer will move in the flow direction and will disappear after several minutes. Obviously, measuring campaigns have to be periodically carried out accompanying the rise of water level in the reservoir. The points of tracer injection can be chosen according to the observations previously made for the identification of the places where the infiltration risk is higher.

When dealing with consolidated rocks, the orientation of the fractures can be important. Negative infiltration results can be obtained when the fracture shows a positive slope and the water level reaches the visible opening at the wall. It can happen that, due to a siphoning process, the water admission of the fractures starts only when the water level has reached a certain value above this visible opening.

To overcome the previous difficulty, another method can be applied, which is based on the measuring of water flows in front of the points showing a high risk of infiltration once the reservoir is full or has reached a high filling level. For this method, an access line is firmly fixed to the rock at the point where it is suspected the reservoir water could enter the formation. Of course, this is done before the water level reaches this point. The access line may simply consist of a nylon rope or a flexible iron armoured cable. Depending on the type of rock, the fixation of the line can be done by inserting a small concrete block or using a nail of the type used by climbers. At the other end of this access line, a buoy is tied. The length of the line is great enough to cover the depth difference between the fixation point and the water surface for the maximum level expected. The buoy will permit the upper end of the access line to be kept at the water surface at any time.

Once the water level has reached its maximum value or an appropriate level above the point to be tested, a device similar to the one described in Section 10.7.6 and shown in Figs. 10.11 and 10.12 for the measurement of deep water flow is lowered to the point where infiltration is suspected. The access line allows the proper positioning of this device in front of the point where water might infiltrate in the rock formation. A tracer injection is then performed from a boat and the dilution curve (concentration versus time) is measured. The tracer may consist of a saturated solution of table salt, and a conductivity meter may be used as a detector. Usually, a straight line will be obtained, the slope of which will depend on the flow velocity existing at the selected points. Of course, a certain background dilution may be derived from natural flows. But the magnitude of the velocity investigated is much higher than the natural currents so that this will produce only a small interference. The measuring device can be positioned at lower depths along the access line for a better scanning of the prevailing water flows. This measuring system can be calibrated to obtain half-quantitative values of the flow velocity. The information obtained can be improved if the flow measurement is repeated for different water levels of the reservoir.

If the previous experiment gives a positive result, that is, a high flow velocity is detected, the same access line can be used to place in front of the infiltration point the lower end of a tracer injection line (plastic tube), which will allow the performance of interconnection experiments between this point of the reservoir and the springs and/or boreholes eventually existing downstream.

In the case of a positive result, the described technique can lead to important economic benefits because the costs of the repairing works will be drastically reduced. It always will be much cheaper and more efficient to impermeabilise a small area responsible for the water ingress into the rock formation than to create a grouting wall in a zone located downstream. The risk of having concentrated water infiltration inside the reservoir is usually high when dealing with consolidated rocks, such as limestone, dolomite, sandstone or gypsum. Frequently, the reservoir floods old karstified strata located at elevations higher than the present elevation of the riverbed. The old karstic pathway may then be reactivated and create an important leakage problem. In such cases, the point of water inflow can be easily identified using the above-described method.

This technique was used by the senior author in the Aguacate reservoir, Dominican Republic, where an almost vertical wall of limestone more than 20 m high existed. Several systems of fractures and caves were identified in this wall. Fortunately, the results obtained in this study were negative. No water inflow was detected at the six points where access lines were fixed. The same technique has been used in the Tous reservoir, Spain, also with negative results.

11.3.7 Under-dam seepage

Suppose that significant amounts of water start emerging at a short distance from the dam during the first filling of reservoir, and the techniques described in Chapters 5 and 6 have proven that the water is coming from the reservoir. The immediate action to be undertaken should be to carry out a tracing experiment to see if the emerging water derives from infiltration occurring at a short distance from the dam. This information is necessary to evaluate the risk derived from the under-dam water flow. Obviously, the risk derived from such a leak will depend on the water velocity under the dam and, therefore, on the flow pattern and distance of this flow to the dam body.

In such a case, the injection of an appropriate tracer (for example, a fluorescent tracer like uranine or rhodamine) is performed at a short distance from the dam, and the eventual breakthrough curve in the existing springs is measured. The tracer should be injected at a point close to the bottom of the reservoir (for example, 5 m above the bottom). Depending of the length of the dam, several point injections may be necessary to cover the whole area. In case of very long dams, most probably the infiltration area at the reservoir will be located in front of the downstream springs. Therefore, it is recommended to start the tracer injections at this area. In some cases, a longitudinal injection could be advantageous. The tracer will then be injected along a line parallel to the dam axis and always at a distance of a few metres from the bottom to eliminate unnecessary tracer dilution.

If the leakage amount is very small, the previous experiment can provide negative results because most of the tracer injected in the reservoir is swept out by the natural current to areas different from the infiltration area. In such a case, the technique described in Section 10.3 (Chapter 10) can be used instead of the previous one. Porous tubes for tracer injection are installed at the bottom of the reservoir at a short distance upstream of the dam. The penetration of these tubes in the terrain can be only 4 or 5 m, and their external diameter can be 50 mm or even less. These tubes can be connected to the water surface through small diameter plastic tubes (for example, 10 mm diameter). Obviously, the installation of the tubes will be easier with an empty or low-level reservoir. If the porous tube is located at a place where seepage is taking place, a vertical descending flow will be present inside the tube. This can be easily detected by introducing the upper end of the plastic tube into water. This test should be made before the tracer is injected.

The above-mentioned technique can be directly applied when the risk of important under-dam seepage is very high due to the nature of the rock formation existing under the dam foundation. This can be the case when the dam has to be founded on a thick layer of alluvial material. Then, the porous tube can be installed before starting the first filling of the reservoir. The detection of the injected tracer will be made, as in previous cases, in the springs located downstream of the dam. The experiment can be repeated for different water levels in the reservoir. The results will allow the localisation of possible preferential pathways of the escaping water flow.

If an important under-dam flow is detected, the depth of the permeable layers that are responsible for the seepage can be determined by drilling boreholes penetrating from the top of the dam to the impermeable rock under the dam. The technique based on the labelling of the entire water column described in Chapter 8 can be applied to localise the permeable layer through which water is escaping from the reservoir. This technique has been successfully applied by the senior author in the reservoirs Maguaca, Chacuey and Sabaneta, all of them in Dominican Republic. The results obtained in the latter reservoir will be described in full in Chapter 12.

11.3.8 Dam sitting-down studies

When dealing with earth-filled dams, differential sitting-down can occur, particularly during the first filling of the reservoir. As mentioned in Sections 4.3.5.2 and 4.3.6.2, the geophysical methods (γ, γ) and (n, n) can provide valuable information to detect and quantify this sitting-down. It is recommended to incorporate these techniques among the standard methods for dam auscultation.

11.4 LEAKAGE INVESTIGATION AFTER ITS APPEARANCE

In this section, some indications are made on the recommended strategy to be adopted once the leakage problem has become evident. Frankly speaking, a unique general strategy to be followed does not exist. But some general rules can be given that will help in the initial planning of the investigation. The magnitude and location of the leakage, as well as the eventual risk derived from the leak for the stability of the dam structures and the geological formations that are supporting them, have to be taken into account in this planning.

An immediate action to be undertaken after the appearance of a leak is to investigate the possible sweeping out of solid particles by the emerging flow. This can be easily done by filtering a relatively high water volume from the emerging flow (or flows), for example, 50 litres. The presence and the mass of solid particles are obtained by gravimetry or turbidity measurements. Even better is the use of a continuous filtration system. A small flow rate from the emerging water is diverted to this filtration system by gravity or using a small peristaltic pump. Periodically, the filter is removed and the eventual content of solid particles is determined by weighing.

For obvious reasons, as is done for many other studies, before delineating a work program, all the available information related to the stated problem has to be compiled. For example, this information should include: (a) geological and hydrogeological information on the affected area; (b) recent history of the springs and their relationship with the water level variation at the reservoir; and (c) hydrogeological conditions prevailing in the affected area before the dam construction. Sometimes, the information obtained from persons who were living in the region before and during the dam construction can help a lot for the identification of the origin of the leakage. The analysis of all available information should assist in planning a preliminary work program. Of course, this program has to be periodically reviewed, according to the results obtained during its development.

An important action to be undertaken at the previous stage of the investigation is to check whether the emerging groundwater downstream of the dam really represents reservoir water. Hydraulic engineers tend usually to admit that it is the case without any technical data supporting this assumption. However, experience demonstrates that very frequently groundwater from the springs does not come from the reservoir, or it represents a mixing of reservoir water with aquifer water not related to the reservoir. Only in isolated cases, for example, for reservoirs located in very arid zones, can one be absolutely sure that the emerging water can only be related to the reservoir. Cases are known in which very high investments have been made trying to eliminate or reduce water flows that were not coming from the reservoir, of course, unsuccessfully. Therefore, even if it seems evident that the emerging water is exclusively fed by the reservoir, the authors recommend verifying this, using the methodology described in this book (Chapters 5 and 6).

Sometimes, surprising results are obtained. For example, in the Cautillo reservoir, Cuba, there is a spring of about 25 l/s average flow rate that is located at the dam toe and at the central part of the dam. Both the chemical and isotopic compositions of the spring water are completely different from the values observed in the reservoir water. Therefore, there was no doubt that the spring water was not derived from the reservoir, at least not totally (Plata et al., 1999).

In the Giribaile reservoir, located near Jaén, Spain, there are many springs close to the dam site whose water is a mixture of reservoir water and groundwater. Even most of the drainage of the dam gallery (about 35 drains) contains water formed by this mixing. The conductivity of the aquifer water is about 2,500 μS/cm, while that of the reservoir is only 650 μS/cm. The temperature of local groundwater is about 24°C (thermal water), while the temperature of the reservoir water ranges from 14°C to 11°C. The $\delta^{18}O$ value of local groundwater is about −6.5‰, while the one for the reservoir water is −4.5‰. All these parameters allowed the determination of both components in the different water emerging points (more than 40 points). None of these points contained water exclusively fed by the reservoir. Many other examples can be shown of springs located close to the dam site, whose discharge, at least partially, corresponds to the surrounding aquifers.

In the La Honda reservoir, Venezuela, two springs were identified in the left dam abutment and very close to the dam body, with a flow rate of several litres per second each. The distance between the two springs was about 20 m. Groundwater emerging through these springs had similar chemical and isotopic compositions, but only one of them was responding to the water level variations at the reservoir. Obviously, this phenomenon was surely derived from the different hydraulic conditions affecting the flows feeding these springs.

Figure 11.19 shows the changes in flow patterns at the dam site, which can explain the appearance downstream of the dam of emerging waters not related to the reservoir. Figure 11.19A shows the flow patterns prevailing before the dam

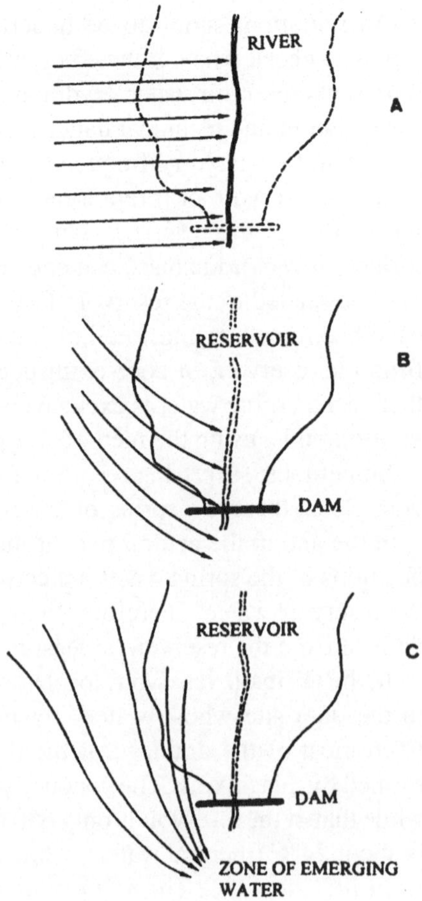

Fig. 11.19. Schematic view illustrating the hydraulic barrier effect produced by the water stored at a reservoir.

construction, when groundwater was discharging in the river. Once the dam is constructed and the reservoir filled with water, the flow patterns may change in the way shown in Figs. 11.19B and 11.19C. In both cases, springs are generated downstream from the dam. However, in the case of Fig. 11.19B, the emerging water is completely controlled by the aquifer, while in the case of Fig. 11.19C, it is a mixture of reservoir and groundwaters. In both cases, the flow rate of the emerging water responds to the water level variation in the reservoir because this level is governed by the local hydraulic gradient that is responsible for the flow rate of the springs. In the case of Fig. 11.19B, a certain fraction of the aquifer water continues discharging into the reservoir. This fraction will also vary with the water level variation of the reservoir.

The identification of the origin of the water emerging downstream of the dam and/or flowing through a borehole drilled in this zone can be made by means of the techniques described in Chapters 5 and 6, based on the analysis of natural tracers (chemical and isotopic components and sometimes, temperature). In some cases, the application of these and other techniques requires long observation periods. This occurs, for example, in the following cases:

1) When the correlation between the flow rate of the existing springs and the reservoir water level is needed. In these cases, the duration of the period of observation will depend on the variations of the water level which, in turn will depend on the meteorological conditions and on the exploitation regime of the impounded water.
2) When the interest is the correlation of seasonal or sporadic variations of chemical and/or isotopic parameters (and sometimes, also temperature) identified in the reservoir with the same changes taking place in the springs and boreholes. Such correlation will inform the dynamic characteristics of the water flow escaping from the reservoir.
3) When trying to obtain information on the depth (or elevation) at which the emerging water is being infiltrated in the reservoir on the basis of the thermal and/or saline stratification. This study requires the measurement of many temperature and conductivity profiles in the reservoir to correlate the changes observed in the reservoir with those observed in the springs or boreholes.

For obvious reasons, these long-term studies should be started as soon as possible, that is, during the first stages of the leakage investigation.

The chemical composition of water or simply its electrical conductivity can provide valuable information on the dynamic characteristics of the water flow escaping from the reservoir. Preferential pathways of this flow can be identified. The measurements should be made both in the springs and in the boreholes. In the last case, vertical profiles of these parameters should be measured. Usually, the conductivity or salinity of the escaping water will be very similar to that of the reservoir water when the residence time of such water in the rock formation is very short. The contrary is also valid, depending on the nature of the rock formations. More precise information on the dynamic parameters of the water flow escaping from the reservoir can be obtained by using the borehole flow tests described in Chapter 8. It is recommended that all these techniques be applied in a parallel way.

Interconnection experiments between boreholes and springs can be initiated once the flow existing in these boreholes has been identified, and it has been verified that the flowing water has similar chemical and isotopic characteristics as those found in the springs. The injection of a tracer in a borehole, when no information is available on the existing water flow or on the water chemical quality, is in no case recommended.

Interconnection experiments between reservoir water and springs should be only performed once sufficient information has been obtained on the elevation at which the water is being infiltrated in the reservoir, for example, through the thermal stratification of reservoir water. Similar information can also be obtained through the identification of structural failures of the rock formation of the reservoir, for example, dolines, sinkholes or highly fractured strata. The only exception is the one described above related to the identification of under-dam seepage.

In many instances, the most important goal of the leakage investigation is the localisation of the zone of the rock formation where the escaping water is flowing. Particularly interesting can be the localisation of the place where water infiltrates in the dam abutment because frequently the most appropriate repair work is the construction of a grouting wall supported in the impervious structures of the dam. It may happen that all possible investigations that can be carried out with existing boreholes are completed without obtaining positive results. In such a case, the drilling of new boreholes might be advisable. To avoid the drilling of unnecessary boreholes, once the drilling of one borehole is finished, it should be exhaustively studied with all relevant techniques described in this book. The results obtained will be used to define the need for drilling a new borehole, and in an affirmative case, these results will help in the definition of the most appropriate location. It is very important to control the variations of the water level in the borehole during its drilling. Usually, a level similar to the one of the reservoir will exist when the borehole intercepts a permeable layer. In the same way, the identification of rock cavities during the drilling can be very valuable. In some cases, it may be interesting to interrupt the drilling of the borehole and to carry out a flow test and water sampling to evaluate the permeability of the intercepted rock.

11.5 RELATION OF RESERVOIR WITH LOCAL AQUIFERS

In some cases, information about the long-term relationship between the reservoir and local aquifers may be needed. This can be the case, for example, when dealing with a reservoir storing polluted water, as well as when the contribution of the reservoir to the recharge of the aquifer has to be evaluated.

Usually, these kinds of studies do not involve serious difficulties, and they are carried out through the analysis of stable isotopes, tritium and/or a specific chemical component of a conservative behaviour contained in the reservoir and absent in the aquifer. In cases of reservoirs containing polluted water, it is recommended to use any component characteristic of the contaminated waters, such as chloride, nitrate or total nitrogen, organic compounds, etc. Obviously, the most appropriate tool will be the one that is conservative and presents the highest contrast between the reservoir water and groundwater.

The method requires the analysis of the selected component(s) in the reservoir water and in the aquifer water in as many points as possible, including areas where no reservoir water exists. The fraction of reservoir water in the aquifer is usually determined using the mixing equation for two components.

An interesting study of this type has been carried out by Aly et al. (1991) in the reservoir High Dam or Aswan dam on the River Nile in Egypt. The study was based on the measurement of the stable isotopes of water, deuterium and oxygen-18. It was found that the influence zone of the reservoir in the aquifer covered a strip of about 8 km. The aquifer is formed by a sandstone formation (the Nubian sandstone formation), whose groundwater is very old. The average isotopic composition of this groundwater body at a distance higher than 8 km corresponds to the values $\delta D = -70 \pm 2‰$ and $\delta^{18}O = -8.9 \pm 0.3‰$. On the contrary, the reservoir water showed values ranging from $\delta D = 21‰$ and $\delta^{18}O = -2.2‰$ at the zone close to the dam to $\delta D = -0.2‰$ and $\delta^{18}O = -1.2‰$ at the zone of Abu Simbel located 280 km upstream of the dam. Evaporation is responsible for the highly positive values of these isotopic deviations. In the zone of the aquifer influenced by the reservoir, intermediate values existed for the isotopic composition of water. Reservoir contributions to local groundwater ranging from 18% to 65% were found.

REFERENCES

Akiti, T., Eftimi, R., Dhame, L., Zojer, H. and Zoetl, J., 1990. Environmental isotope study of the interconnection between the Vjosa river and the Poçeme springs in Albania. Personal communication.

Aly, A.I.M., Froehlich, K., Nada, A. and Awad, M., 1991. Study of environmental isotope distribution in the Aswan High Dam Lake (Egypt) for estimation of evaporation of lake water and its recharge to adjacent groundwater. *Environ. Geochem. Hlth*, 15(1), 37–49.

Araguás, L. and Plata, A., 2002. *Estudio de las aguas subterráneas en el entorno de la presa de Algar sobre el río Palancia.* VII Jornadas Españolas de Presas, Zaragoza, Mayo, 2002. Vol. IV. Comité Español de Grandes Presas, pp. 183–193 (in Spanish).

Plata, A., Greciano, R., Pérez, E. and Díaz, M.F., 1997. *Estudio de la Hidrología isotópica de la zona del futuro embalse de Rules, río Guadalfeo, Granada.* Internal report, CEDEX, 153 pp. (in Spanish).

Plata, A., Kile, S., Linares, N. and Rodriguez, B., 1999. Estudio de filtraciones en el embalse de Cautillo, Cuba. In *Estudios de Fugas en Embalses*, International Atomic Energy Agency, Vienna, pp. 1–52 (in Spanish).

Plata, A., Pérez, E. and Díaz, M.F., 1996a. *Estudio de la influencia del embalse de Bellús en la dinámica y calidad de las aguas subterráneas de la zona.* Internal report, CEDEX, 143 pp. (in Spanish).

Plata, A., Pérez, E. and Díaz, M.F., 1996b. *Estudio de la salinidad del agua del río Salado en relación con la construcción del embalse El Atance, Guadalajara.* Internal report, CEDEX, 151 pp. (in Spanish).

Plata, A., Pérez, E. and Díaz, M.F., 1998. *Estudio de filtraciones en el embalse de Arquillo de San Blas (Teruel).* Internal report, CEDEX, 106 pp. (in Spanish).

CHAPTER 12

Case studies

12.1 INTRODUCTION

In the present chapter, several case studies conducted by the authors are described. The main aim of this presentation is to illustrate the possibilities offered by different techniques when used in a combined or integrated way. Obviously, the final reports describing the studies carried out are too extensive to include here with all the details. Therefore, only short versions with the main conclusions are presented, in which the most interesting aspects from a didactic point of view have been stressed.

12.2 ARQUILLO DE SAN BLAS DAM, SPAIN

The Arquillo de San Blas reservoir is located near Teruel, Spain, on the River Guadalaviar. The gravity dam is located on a massif of a dolomitic limestone formation of Jurassic age at a point where the river has excavated a deep gorge of about 500 m (see Fig. 12.1). The limestone formation covers an area of about 400 m upstream of the dam. Several springs exist downstream of the dam. Groundwater from these springs mainly emerges in the riverbed and is mixed with the river flow, rendering their sampling difficult. The study was initiated in 1997 (Plata et al., 1997).

The Guadalaviar River runs downstream of the dam site on the same dolomitic limestone formation, which is outcropping along an stretch of about 20 km. Therefore, a clear possibility exists that groundwater emerging in many of these springs represents the discharge of the karstic aquifer instead of being a leak from the reservoir.

Figure 12.2 shows the distribution of the springs along the riverbed. The identification of these springs was carried out by differential flow rate measurements, using the technique described in Annex 2. Rhodamine WT was used as a tracer. The measurements were performed in two different campaigns. The results obtained in one of these campaigns are presented in Fig. 12.3 as a function of the distance

Fig. 12.1. General view of the Arquillo de San Blas dam, Spain.

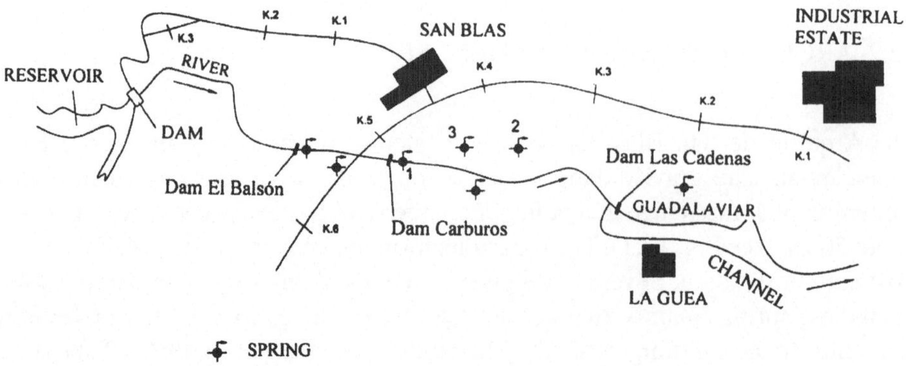

Fig. 12.2. Location of the springs downstream of the Arquillo de San Blas dam along the River Guadalaviar.

to the dam. A constant flow rate of 200 l/s was being discharged at that time from the reservoir (the so-called ecological flow rate). It can be observed that the river in the initial stretches was losing water between the dam and a distance of about 3 km. Downstream of this point, the river flow rate increased up to a value of about 1000 l/s. Most water gain occurred along a stretch of about 1500 m, located between 3 and 4.5 km from the dam site. It must be mentioned that most of these springs were present prior to the dam construction in 1957, indicating that groundwater

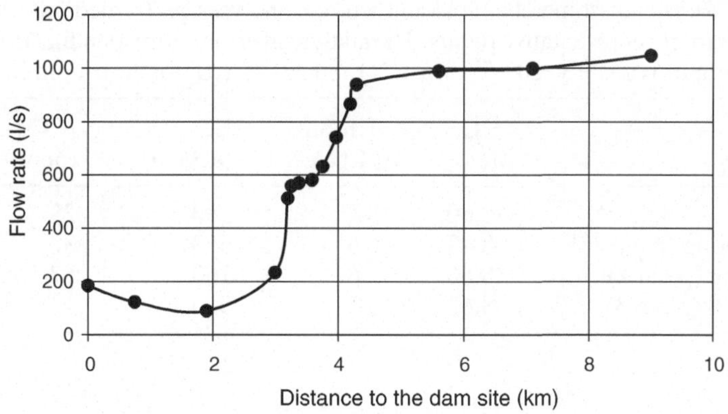

Fig. 12.3. Variation of the flow rate of the River Guadalaviar downstream of the dam as a result of the sinkholes and springs existing along it.

from the karstic aquifer normally discharges along this stretch of the river. However, the possibility that a marked increase in the flow rate discharges was due to the eventual contribution of the reservoir needed to be evaluated. A sufficiently detailed study of the distribution and flow rate variations of these springs never was undertaken prior to the dam construction. Therefore, there was no possibility to compare the results obtained within the framework of the present study with flow rate data obtained prior to the dam construction.

Apart from the springs located along the riverbed, other small springs were identified at both sides of the river, mainly on the left side. A deep well (250 m) was drilled close to the zone where most of the springs are located (point 3 of Fig. 12.2). An almost constant flow rate of 21 l/s was continuously emerging from this deep well. The Ambel spring (point 2 of this figure) also had a constant flow rate of about 30 l/s and existed before the dam was constructed. Five other small springs are also located on the left side of the river, indicating that this zone represents a preferential discharge zone of the adjacent karstic aquifer.

The previous information provides a clue about the main objective of the study. The investigation should provide conclusive information on whether part of the groundwater discharged to the River Guadalaviar, as well as to the other springs, was fed to some extent by the leaks from the reservoir. In an affirmative case, the next goal should be the localisation of the leakage area in the reservoir. To reach these objectives, the following techniques were used, apart from those applied for the flow rate measurement previously described:

1) Chemical analysis of the reservoir and springs water.
2) Isotope analyses, including D, ^{18}O, ^{13}C and natural tritium.
3) Measurement of vertical profiles of conductivity and temperature at the reservoir.
4) Interconnection experiments between the reservoir and the mentioned springs using artificial tritium and rhodamine B as tracers.

Table 12.1. Results of some chemical and isotope analyses performed on water samples from the four more representative points. The analytical errors corresponding to the standard deviation are approximately ±0.05‰ for $\delta^{18}O$ and ±0.25 T.U. for the tritium concentration.

Point	$\delta^{18}O$ (‰)	Tritium (T.U.)	EC (μS/cm)	[Cl^-] (ppm)	[$SO_4^=$] (ppm)
Reservoir	−8.80	8.7	580	23	150
Carburos spring (point 1)	−8.43	6.5	650	32	142
Ambel spring (point 2)	−8.66	1.6	760	57	133
San Blas well (point 3)	−8.62	3.7	820	72	163

The results obtained with these techniques are summarised below.

Chemical analyses The chemical composition of the reservoir water showed similar concentrations to those found in most springs. Calcium bicarbonate was the dominant hydrochemical facies. All water samples were saturated in calcite, and their electrical conductivity was close to 600 μS/cm (see Table 12.1). This range of values is considered typical for most karstic springs but not in the case of the reservoir water, for which a lower conductivity was expected. After a long period of heavy rain, the electrical conductivity of the reservoir water only decreased from 580 to 525 μS/cm. This is explained because most of the water entering the reservoir was the discharge from karstic springs situated along the river basin.

The results obtained by the chemical analysis indicated that the chemical composition of water does not exclude the possibility that a significant part of the waters emerging along the riverbed downstream of the dam, represented by the Carburos spring (Table 12.1), could be infiltrated in the reservoir. On the contrary, groundwater from the Ambel spring (point 2) and from the San Blas well (point 3) showed a chemical composition very different from the typical contents found in the reservoir water and in the springs sampled along the riverbed. Both water points contained water saturated in calcite, but the higher concentrations of ions Cl^- and $SO_4^=$ suggested a possible contact of this water with the impervious substrate. This affirmation is supported by the longer residence time of water, as shown by the tritium contents.

Stables isotopes of water, deuterium and oxygen-18 More that 50 water samples were analysed. The average $\delta^{18}O$ values obtained in the four more representative water types involved in the investigation are given in Table 12.1. The differences in the isotope composition of these waters, that is, the isotope contrasts, are very small. For this reason, these analyses do not provide any additional valuable information on the eventual hydraulic connection between the reservoir and the downstream springs. The effect of the evaporation on the isotopic composition of the reservoir water did not produce a large enough isotope shift to distinguish clearly the isotope signals of the reservoir water from the local groundwater.

Tritium The tritium concentrations were very different from one point to another. The low tritium content measured for a water sample collected from the Ambel spring clearly proved that a major fraction of groundwater was coming from precipitation that occurred before 1953 (see Table 6.1 and Fig. 6.7). The same conclusion is valid for the water emerging at the well San Blas, but in this case, the higher tritium concentration indicates a larger contribution of recent waters. These results suggest that both points are mainly fed by deep groundwater from the aquifer, probably without tritium. This deep component mixes with recent water, characterised by a higher tritium content, before the two components emerge.

The reservoir contained, as expected, water with a tritium concentration more or less similar to the typical values found for recent local precipitation. This is not contradictory with the previous statement indicating that an important fraction of the reservoir water is formed by the discharge of karstic springs. Effectively, the tritium concentration in precipitation during the previous 17 years shows, after correction for radioactive decay, an average annual value of 8.7 T.U. (see last column values of Table 6.1). This value is, by coincidence, equal to the measurement obtained for the reservoir water.

The tritium concentration obtained for the Carburos spring, which is representative of most of the water emerging along the riverbed, was 6.5 T.U., that is to say, a bit lower than the concentration of the reservoir water. This concentration can be explained in three different ways:

1) Taking into account the low tritium concentrations found at the Ambel spring and the well San Blas, one could suppose that the lower tritium concentration of the Carburos spring is derived from the mixing of this water with deep groundwater with a lower tritium content. Assuming for the last component a value between 1.5 and 4 T.U., the contribution of the reservoir to the Carburos spring would range from 29% to 45%.
2) The tritium concentration of this spring could correspond to the characteristic value for the shallow groundwater of the karstic aquifer, suggesting that no contribution of the reservoir was taking place.
3) Groundwater sampled in the springs could be a mixture of deep and shallow groundwater.

Therefore, the tritium content alone did not provide a definitive answer to the question about the origin of the water emerging along the riverbed.

Carbon-13 The results obtained for the carbon-13 are shown in Section 6.4.2 (Chapter 6). As mentioned there, the results obtained suggested that the spring water does not entirely derive from the reservoir, but on the other hand, do not exclude the possibility that a significant part of this water can have this origin.

Temperature of water The temperature of water did not provide valuable information about the origin of the emerging water. Toward the end of March 1998, the temperature of the Carburos spring water was 15°C, while the reservoir showed a temperature ranging from 9.3°C at the surface and 6.6°C at the bottom. Taking into account the season of the year, it was clear that the temperature of the reservoir water had been lower than at least 15°C during the previous four months. This result excluded the possibility of a rapid underground flow connecting the reservoir to the spring.

The above-described information shows definitively that this example constitutes an infrequent case study in which the combination of natural tracers, including both chemical and isotopic tracers, does not provide an unequivocal answer to the question about the origin of the waters emerging along the riverbed. It became clear that at least most of the water from the Ambel spring and from the San Blas well is not infiltrated at the bottom of the reservoir. Nevertheless, these two points represent a negligible water flux compared with the discharge of the springs located along the riverbed. Under these circumstances, it was decided to carry out a tracing experiment using artificial tritium and rhodamine B as tracers.

Tracing experiment On May 22, 1998, about 7.3 Ci of tritium in the form of tritiated water and 5 kg of rhodamine B were injected in the reservoir. The tracers were distributed over an area located between the dam and about 800 m upstream. This area approximately coincided with the outcropping of the limestone formation in the reservoir. The injection of both tracers was made using a peristaltic pump. To obtain a homogeneous labelling of the entire water volume of the selected area, the tracers were injected in the form of vertical columns. About 50 tracer columns distributed all over the area were injected.

The gates of the dam were kept closed for 4 days after the injection in order to facilitate the mixing of the tracers as well as their infiltration through the sediments at the bottom of the reservoir. Then, the reservoir was operated according to the regular needs. Flow rates from 200 l/s to 5 m^3/s were discharged during the following months (see Fig. 12.4). An intensive sampling program was adopted for 3 months after the injection, which included sampling all of the most representative springs and the river water. An automatic sampler, shown in Fig. 9.3, was installed at the Carburos spring. The sampling was extended for 8 months after the injection (until the end of November 1998).

As expected, most of the tritium activity and mass of rhodamine B left the reservoir through the dam gates. The breakthrough curves shown in Figs. 12.5 and 12.6 were obtained. The amount of tracer discharged through the gates was calculated by adding the tracer masses corresponding to periods of time with constant flow rate according to the following expression:

$$M = \sum (C_i \cdot Q_i \cdot t_i) \tag{12.1}$$

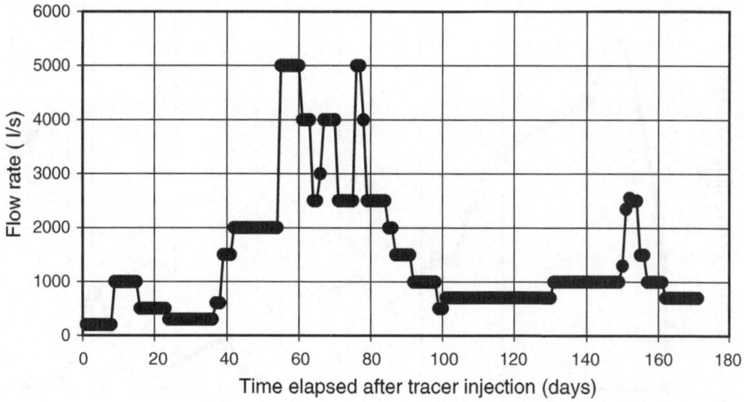

Fig. 12.4. Flow rates discharged from the Arquillo de San Blas reservoir after the tracer injections.

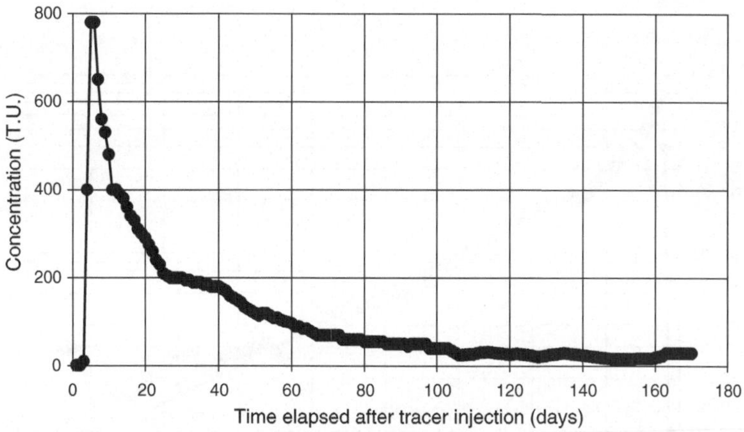

Fig. 12.5. Tritium discharged from the reservoir through the dam gate after its injection. The y-axis represents net concentration values obtained after background subtraction.

where C_i and Q_i are, respectively, the tracer concentration and flow rate observed during the time t_i.

The fraction of tritium discharged through the dam gates over 170 days after the injection was 76.7%. For rhodamine B, the fraction recovered over 40 days after the injection was about 60%. After this time, rhodamine B could no longer be measured because the fluorometer readings were similar to the natural background of natural water.

Figure 12.7 shows the same values of Fig. 12.5 plotted in half-log scale. The points define, more or less, a straight line, which is typical of a well-mixed lake with constant level, that is to say, with the same inflow and outflow. The slope of this straight line corresponds to the ratio Q/V, with Q being the flow rate crossing

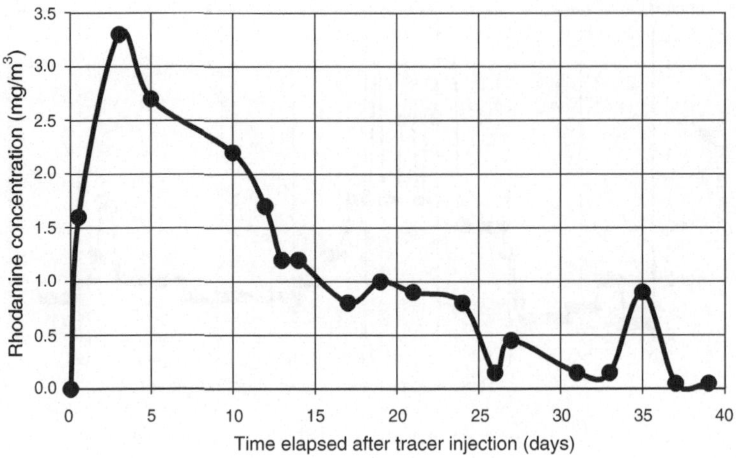

Fig. 12.6. Rhodamine B discharged from the reservoir through the dam gate after its injection.

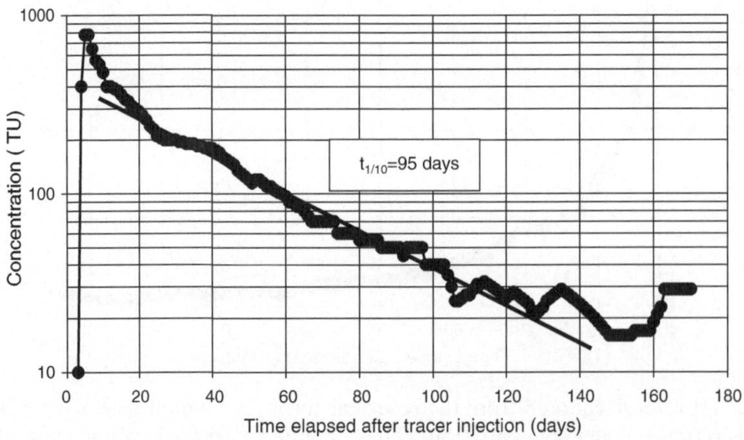

Fig. 12.7. Tritium discharged from the reservoir through the dam gate after its injection. The values are the same as in Fig. 12.5 but plotted in half-log scale.

the lake and V the water volume. This behaviour of the tracer was a consequence of the injection method applied. The tracer was well distributed in the water body and rapid mixing took place.

As indicated above, 23% of the injected tritium was not recovered at the dam outlet. This does not necessarily mean that this tracer fraction was infiltrated through the bottom of reservoir. Tritium losses are usually produced by water exchange between the surface of the reservoir and the atmosphere (Slattery and Ingraham, 1994; Florkowski and Wotjas, 1979), as well as by evaporation. Apart from that, the migration of the injected tracer upstream of the labelled area, for example,

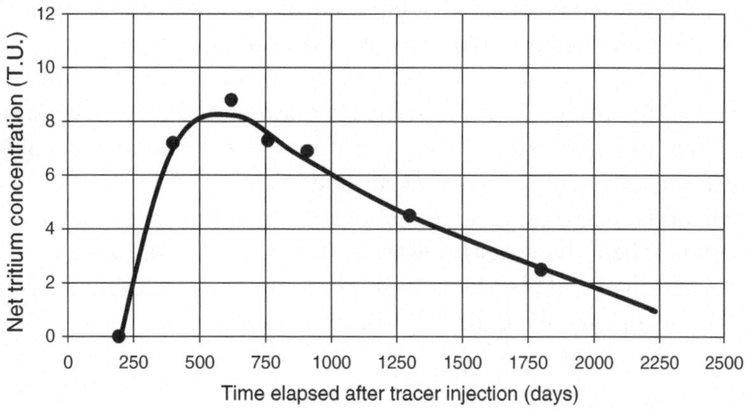

Fig. 12.8. Tritium breakthrough curve measured at the main group of springs downstream of the dam along the River Guadalaviar.

by the effect of the wind, would have generated a very high tracer dilution. This process is surely responsible for the long tail observed in the curve presented in Fig. 12.5. In the case of rhodamine B, significant losses could have taken place by photodecomposition and by adsorption into the rock formation.

The injected tracers were detected about 2 weeks after the injection in a small spring located at the dam site several tens of metres downstream of the dam. Tritium was detected at the Carburos springs only about 1 year after the injection, when the intensive sampling program had been discontinued. After December 1998, only sporadic grab sampling was carried out. The breakthrough curve shown in Fig. 12.8 was obtained. An average transit time of water between the reservoir and the springs of about 2 years was observed. Taking into account that the flow rate discharging through these springs was about 800 l/s, the water volume discharged during this 2-year period amounted to 50.5×10^6 m^3. Considering that the distance from the reservoir to the zone of emerging water was about 4 km, the average cross-section of the flow connecting these two points was around 13,000 m^2. Assuming a porosity of the rock formation equal to 0.1, the cross-section of the underground pathway would be 130,000 m^2, which corresponds to a "flow tube" of about 64 m radius. Of course, this data do not correspond to reality because the water infiltrating in the reservoir is mixed with groundwater flowing through the karstic aquifer before emerging at the spring.

The breakthrough curve presented in Fig. 12.8 allows a rough estimation of the fraction of tritium injected in the reservoir water that is incorporated into the limestone aquifer and discharged in the Carburos and other springs discharging close to the riverbed. The integration of the breakthrough curve gives an area of 6,000 T.U. · day (6 T.U. over 1,000 days), which is equivalent to 19.2 μCi · d · m^{-3}. For a flow rate of 800 l/s, the water volume discharged per day is 69,120 m^3/d. The multiplication of these two values give a value for the recovered activity equal to 1.3 Ci. This represents a fraction equal to 17.8% of the injected tracer (7.3 Ci). Adding this value to the percent of tritium discharged through the dam gates (76.7%), a value of 94.5% is obtained for the total tritium recovery.

The contribution of the reservoir to the Carburos springs can now be calculated as the flow rate that generates a tritium recovery at these springs equal to 1.3 Ci, with the tritium concentration given by the curve shown in Fig. 12.5. To complete these calculations, it should be assumed that the tritium concentration at the area of the reservoir where the water infiltrates was the same as the values measured at the reservoir discharge. This seems to be, in principle, a reasonable assumption. The calculation is made using the following equation:

$$q = \frac{A}{\int_0^{170} C\,dt} = \frac{A}{\sum_0^{170} C_i \cdot t_i} \tag{12.2}$$

where q is the reservoir contribution (l/s), A the recovered activity (1.3 Ci), C is the tritium concentration expressed in μCi/l and t the time in seconds.

A reservoir contribution of 261 l/s was obtained, which represents about 33% of the total flow rate. This value is in good agreement with the contribution estimated from the natural tritium content of the spring, as mentioned above (29% to 45%).

The low tritium concentrations found in the samples collected from the Carburos spring, as well as the large spread of tracer breakthrough curve and its long tail (Fig. 12.8), are explained by the fact that the reservoir water is mixed with the water from the aquifer before emerging.

Neither the tritium nor the rhodamine B were detected at the San Blas well or the Ambel spring, supporting the hypothesis that these points are exclusively fed by very deep groundwater flow paths, apparently crossing detritic materials with minor interaction with the calcareous aquifer.

12.3 LAKE MANAGUA, NICARAGUA

The study performed by the authors at Lake Managua was carried out within the framework of a technical cooperation project convened by the International Atomic Energy Agency, Vienna, Austria, and the Nicaraguan government in 1983 (Plata et al., 1994). In a second phase, the study was continued in the year 2000 within the framework of the hydrologic studies sponsored by the Spanish government after the disaster provoked by Hurricane Mitch (Plata et al., 2001).

Lake Managua is located on volcanic rocks in the vicinity of the city of the same name (Fig. 12.9). The surface of the lake amounts to 1,050 km^2, and the average water volume is 9×10^9 m^3. The lake occupies the lowest part of the Nicaraguan graben, and until 1910 was discharging through the River Tipitapa into the much larger Lake Nicaragua, which has a lower mean elevation (about 9 m). The hydraulic connection between these two lakes was reestablished toward the end of 1998

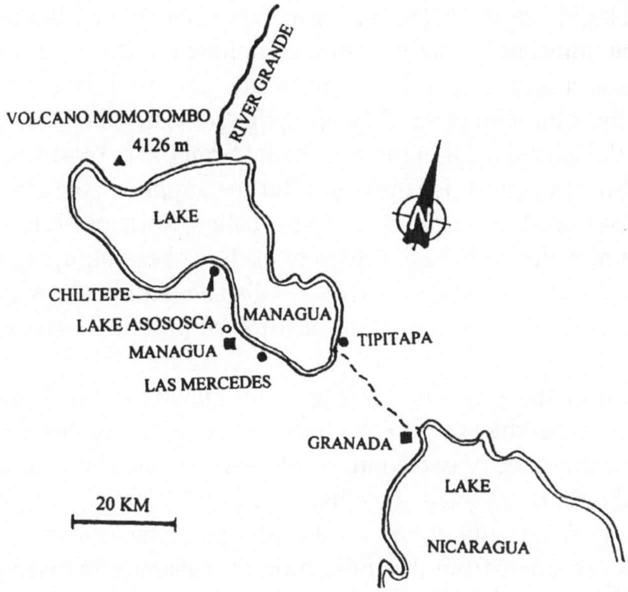

Fig. 12.9. Schematic map showing the area around Lake Managua.

due to the water level increase (about 4 m) produced at the Lake Managua by the floods following hurricane Mitch.

The lake is fed by several rivers, the River Grande being the most important one, with an average flow rate of about 30 m^3/s. For more than 80 years, the only discharge of the lake was evaporation. Obviously, the lake receives the discharge of the surrounding aquifers due to its low elevation.

Lake Managua presents high levels of pollution due to the discharge of domestic and industrial wastes. The pollution problem has become more serious due to the presence of several industrial factories, including a small oil refinery; chlorine, pesticides, and cement production; and a slaughterhouse. Additionally, the acidic geothermal fluids from the geothermal power plant Momotombo were discharged into the lake. Among other pollutants, it has been estimated that more than 25 tons of mercury have been discharged in the lake during the last decades.

A small lake, called Lake Asososca, with a volume of about 40×10^6 m^3 (with a diameter of about 1 km and a maximum depth of 94 m) is located at a distance from the Lake Managua of 2.2 km. This lake occupies the crater of an ancient volcano and does not receive any surface contribution apart from the precipitation over its small catchment area. The lake has been used since 1933 as a source of water supply to the population of Managua. At that time, the water level of Lake Asososca was about 4 m higher than that of Lake Managua. Due to increasing pumping rates, the situation progressively changed, and in 1967, the water level at both lakes was the same. Before hurricane Mitch, the level at Lake Asososca was about 4 m lower than that of Lake Managua. After hurricane Mitch, the level difference increased. Under these circumstances, the intrusion of polluted water from Lake Managua into Lake Asososca could occur. The evaluation of the risk related to such possible intrusion was the main objective of the investigation carried out.

Apart from the intensive pumping carried out from Lake Asososca, groundwater from the Las Sierras aquifer is being used for the supply of the city of Managua. Pumping stations are located in three areas, called Chiltepe, Las Mercedes and Tipitapa. Several wells have been drilled at each of these pumping areas. All the recent wells are located at a surprisingly short distance from Lake Managua. Therefore, a similar and even higher risk of intrusion of polluted water exists to these new wells.

The thickness of the volcanic material is not known, but it is assumed that in some areas, it is higher than several hundred metres. Some studies performed in the past indicate the presence of two aquifers; the shallow one is phreatic and the deep one is confined or half-confined. The shallow aquifer was strongly polluted by the industrial sector, mainly due to the release of significant amounts of chlorinated gases and vapours, apart from the infiltration of wastewaters discharged in Lake Managua through earth channels.

The industrial plants are pumping water from the shallow aquifer to cover their needs. This pumping has created a significant piezometric depression in the zone that governs the groundwater flow patterns of this aquifer.

The main objective of the first study (Plata et al., 1994) was the evaluation of the risk of contamination of Lake Asososca by water from Lake Managua. The following techniques were used:

1) Analysis of the chemical composition of water, including field parameters (electrical conductivity (EC), pH, temperature and alkalinity) and laboratory analyses of major and some minor components (boron and pesticides).
2) Measurement of the stable isotopes deuterium and oxygen-18.
3) Determination of the tritium contents.
4) Measurement of groundwater flow at existing boreholes by means of tracer techniques.

Figure 12.10 shows the distribution of the electrical conductivity at different water points. The low EC value found for the water sample collected in Lake Asososca is remarkable, especially when compared with the wells located in the industrial sector and with Lake Managua.

Similar EC values were found in two deep wells located in the vicinity. This result is a clear demonstration of the absence of contamination at Lake Asososca. It must be pointed out that the EC of groundwater did not change substantially during the period 1985–1999. Only the EC of the water sampled in Lake Managua dropped from 1,870 to 1,293 μS/cm due to the freshwater ingress produced during hurricane Mitch toward the end of 1998.

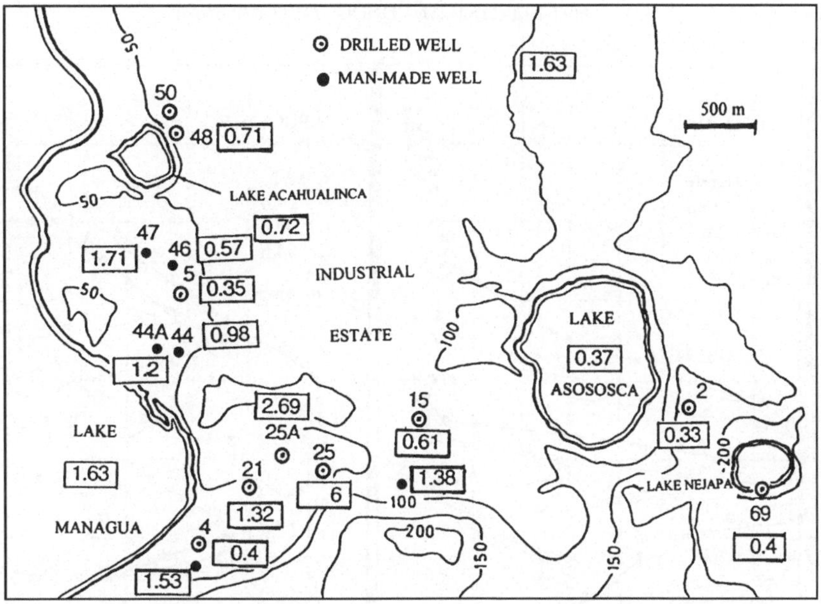

Fig. 12.10. Location of water points between Lakes Asososca and Managua and EC values expressed in mS/cm at 25°C (numbers in boxes).

Apart from the two wells mentioned above, low EC values were also found in two deep wells, which had been drilled about 15 years ago at very short distances from Lake Managua (points 4 and 5 of Fig. 12.10). Flow tests were performed using the technique based on the labelling of the entire water column described in Chapter 8. Table salt solution was used as a tracer. The results obtained, presented in Fig. 12.11, indicate the presence of an important horizontal flow below a depth of 40 m, where the well screens were located. The tracer dilution measured in these permeable stretches corresponded to a relatively high Darcy velocity (between 4 and 5 m/d). The low EC values of groundwater from these two boreholes clearly show that these groundwater flows were not influenced by the polluted water from the shallow aquifer, nor by Lake Managua. The shallow aquifer had been isolated in most wells by placing at the corresponding stretch a nonperforated casing. The high values of the groundwater flow are justified by the high hydraulic gradient derived from the water extractions made at Lake Asososca. Therefore, the flow detected at wells 4 and 5 circulates below Lake Managua, and in spite of that, no influence of the lake is observed.

Apart from the EC values, other chemical and isotopic parameters support the statement that no hydraulic connection exists between the two lakes. These are the following:

1) Boron content Water from Lake Managua contains a boron concentration equal to 2 mg/l, which, obviously, is derived from the residual volcanic activity

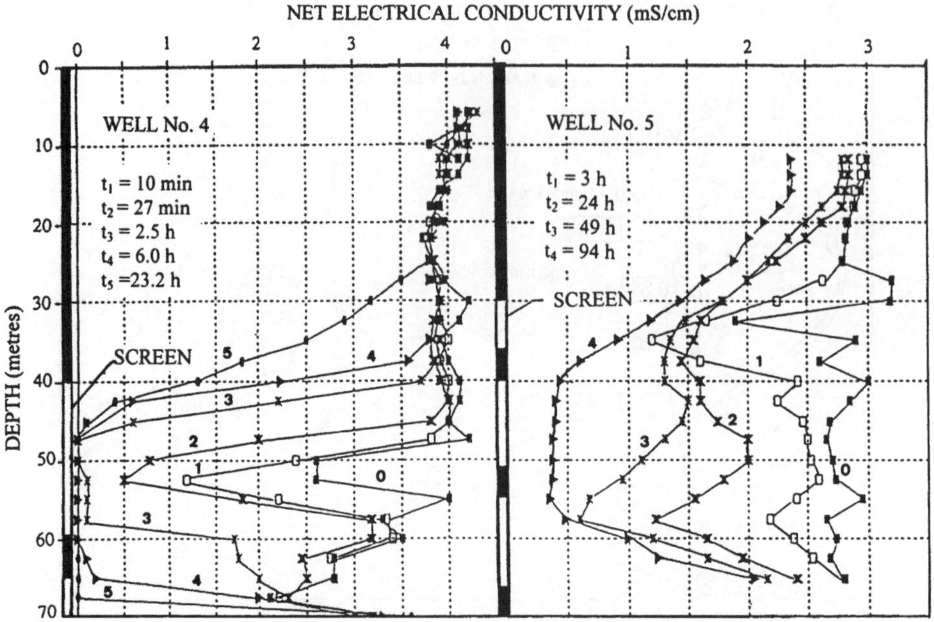

Fig. 12.11. Flow tests performed at boreholes 4 and 5 using the method based on the labelling of the entire water column with a saturated solution of common salt (see Chapter 8). The casing profiles indicate the location of the screens (white stretches).

of the zone. The volcano Momotombo, which continues to be active, is located at a short distance from the lake. Also, the geothermal field that is exploited for electricity production is located at a short distance. On the contrary, water from Lake Asososca does not contain measurable amounts of boron.

2) Tritium content The tritium content of several water samples was measured in 1985 and 1999. The results are shown in Table 12.2. It can be seen that in 1985, the average tritium concentration at Lake Managua was 2.7 T.U. This concentration is justified by the contribution of the River Grande, which flows from the reservoir Apanás, where the tritium concentration was at that time 4.4 T.U. On the contrary, no detectable tritium concentration was found at Lake Asososca and in the majority of the deep wells. This was a clear indication that water from this lake was related to a deep regional groundwater flow, which corresponds to the confined aquifer called Sierras de Managua. However, the high value of the statistical uncertainty of the analyses made in 1985 does not exclude the possibility of small tritium contents in these points. The values obtained in 1999 also show a much higher tritium concentration in Lake Managua than in Lake Asososca. The values obtained for the pumping wells are discussed later.

The low tritium concentrations obtained for Lake Asososca and the deep wells suggested the possibility of finding very old groundwater. To verify this possibility,

Table 12.2. Tritium contents measured at different water points in the vicinity of Lake Managua in 1985 and 1999 (values expressed in T.U.).

Sampling made in 1985		Sampling made in 1999	
Water point	Concentration (T.U.)	Water point	Concentration (T.U.)
Lake Managua	2.7 ± 0.8	Lake Managua	2.3 ± 0.18
Lake Asososca	0 ± 0.8	Lake Asososca	0.6 ± 0.15
Well 2	0 ± 0.8	Well in Las Mercedes	0.0 ± 0.14
Well 4	0 ± 0.8	Well in Tipitapa	0.8 ± 0.16
Well 5	0 ± 0.8	Well in Chiltepe	0.9 ± 0.16
Well 15	0.3 ± 0.8		
Well 21	0.7 ± 0.8	Well 21	0.3 ± 0.14
Well 46	0.7 ± 0.8		
Well 69	0.5 ± 0.8		

analyses of carbon-14 and carbon-13 were performed in TDIC samples (see Chapter 6) collected from this lake and from well number 2. The following results were obtained:

Lake Asososca: 80.1 ± 1.6 percent of modern carbon; $-14.1‰$ for $\delta^{13}C$.

Well 2: 38.3 ± 0.8 percent of modern carbon; $-6.79‰$ for $\delta^{13}C$.

The carbon-14 "ages" obtained, after correction using the available correction methods (IAEA, 1983), gave, in both cases, a negative value. Therefore, such an old component could not be identified with these two analyses. The less negative $\delta^{13}C$ value of well 2 can be explained by the presence of a local, nonbiogenic carbon source.

3) Stable isotopes The stable isotope composition of the water samples collected in the study area was determined several times from 1985 to 1999. The results obtained are shown in Table 12.3. The first set of data corresponds to the deviation $\delta^{18}O$ and the second to δD. Figures in brackets give the number of averaged water samples. As can be seen, the isotopic composition of Lake Managua water drastically changed as a result of hurricane Mitch. Before this extreme climatic event, the lake was highly concentrated in stable isotopes due to the effect of evaporation and the long turnover time of water. Both δD and $\delta^{18}O$ showed positive values, concretely, $\delta D = 11.0‰$ and $\delta^{18}O = 2.43‰$. On the contrary, the average δ-values of groundwater representing the deep pumping wells were approximately $\delta D = -46.8‰$ and $\delta^{18}O = -7.1‰$. For Lake Asososca, intermediate values were obtained, concretely, $\delta D = -37.3‰$ and $\delta^{18}O = -5.72‰$. In principle, these intermediate values could be explained as a mixture of deep aquifer water with water from Lake Managua. However, the calculation of the isotopic enrichment derived from evaporation of the water from Lake Asososca showed very clearly that this

Table 12.3. Isotopic composition of most representative water sampling points from the study area. The first figure corresponds to $\delta^{18}O$ and the second to δD, both expressed in ‰. Figures in brackets indicate the number of averaged water samples.

Water point	May 1985	February 1987	February 1991	December 1999
Lake Managua	2.43/11.0 (8)			−2.09/−16.2
Lake Asososca (surface)	−5.65/−40.3		−6.19/−41.6	−5.23/−35.1
Lake Asososca (bottom)	−6.13/−43.3			−5.40/−36.2
Deep well	−6.90/−48.3		−7.33/−49.1	−6.37/−40.4
(industrial sector)	−7.31/−50.6		−6.60/−47.3	−6.73/−46.8
	−7.33/−48.6		−7.41/−48.6	−6.63/−45.6
Wells in Las Mercedes		−7.18/– (10)	−7.31/−46.7 (2)	−7.22/−46.6
Wells in Chiltepe		−6.75/– (6)	−7.06/−46.7 (2)	−6.92/−43.5
Wells in Tipitapa		−6.28/– (5)		−6.25/−44.5

process can perfectly account for the δ-values found in this water. The more negative δ-values of the bottom water samples with respect to the surface water samples also is explained by the evaporation effect. The nominal turnover time of the lake, given by the quotient between the volume of the lake and the pumping flow rate, was on the order of 400 days in 1985. The evaporation of water from the lake during this period of time is responsible for the observed isotopic enrichment. The calculation procedure of this evaporation effect is described in Plata et al. (1994).

Apart from Lakes Managua and Asososca, there are in the zone a number of small lakes also tapping the same aquifer but with different values of turnover time of water. The isotopic composition of these lakes was also measured in 1985. Figure 12.12 shows a diagram relating the isotopic deviations δD and $\delta^{18}O$. The points define an evaporation straight line that corresponds to the equation $\delta D = 5.6\delta^{18}O - 8.0$. This suggests that all waters of the area have as a common origin the groundwater of the Las Sierras aquifer. The undisturbed isotopic composition of this groundwater body corresponds to the intersect of the evaporation line with the meteoric water line, that is to say, to the values $\delta D = -50‰$ and $\delta^{18}O = -7.3‰$. This is approximately the average isotopic composition of groundwater from the aquifer Las Sierras. The points corresponding to Lake Managua did not match the evaporation line well because most of the water from this lake derived from precipitation at areas located far away and transported mainly by the River Grande.

The results obtained in the study definitively demonstrated that water sampled at Lake Asososca was not related at all to Lake Managua but to the Las Sierras aquifer. Groundwater emerges at the bottom of the lake at the same rate as the

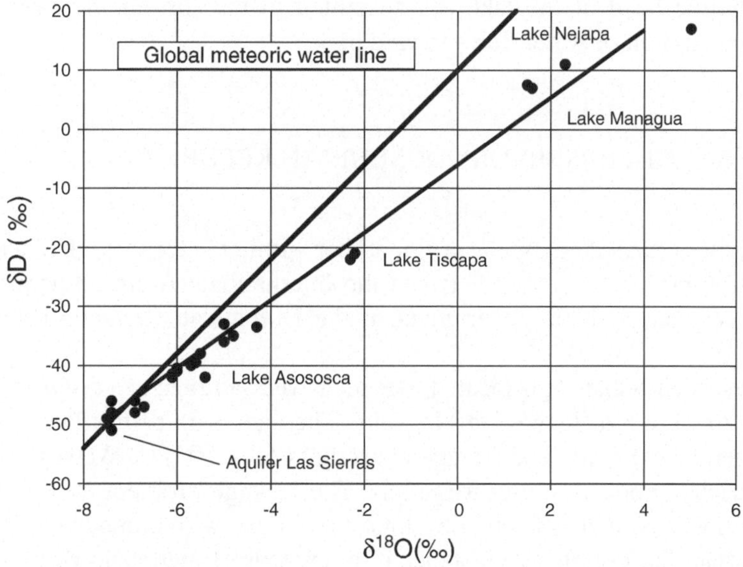

Fig. 12.12. Diagram relating the isotopic deviation δD and $\delta^{18}O$ of different water samples.

pumping rate. Due to the great contrast, the stable isotope of water is an excellent tool to detect any eventual water intrusion from Lake Managua. Such an intrusion would be detected in wells 4 and 5 many years before it arrives at Lake Asososca. The chemical and isotope data obtained over a period of 5 years clearly shows that no intrusion process is at present in progress.

Groundwater from Las Mercedes, Chiltepe and Tipitapa pumping sites The hydraulic situation of the aquifer close to the well fields is similar to the described settings for Lake Asososca. In spite of the proximity to Lake Managua, no water intrusion process is in progress in any of the three areas. The less negative δ-values of the groundwater from the sites Chiltepe and Tipitapa with respect to Las Mercedes and to other deep wells of the zone could be interpreted, in principle, as a mixing of water from Lake Managua with groundwater from the Las Sierras aquifer. However, this interpretation seems to be wrong because the same δ-values were obtained in 1987 as in 1999, in spite of the water level rise at Lake Managua of about 4 m.

The absence of water from Lake Managua at Lake Asososca and other pumping sites can be explained by the siltation of the bottom of the first lake. The rivers that are discharging into the lake have transported huge amounts of suspended sediments. These sediments are distributed by natural currents all over the bottom of the lake. In fact, water from this lake permanently shows the typical brown colour due to the presence of the suspended sediments. A thick layer of sediments is surely deposited at the bottom, which renders difficult the infiltration of water.

The hydraulic head of the lake is transmitted to the surrounding wells but no significant amount of water flows toward them.

12.4 SABANETA RESERVOIR, DOMINICAN REPUBLIC

The study was carried out by the senior author within the framework of a technical cooperation project convened between the International Atomic Energy Agency (IAEA), Vienna, and the government of the Dominican Republic (Plata et al., 1989b).

The reservoir Sabaneta is located over the River San Juan at a distance of 20 km from the town of San Juan de la Maguana. The dam is an earth-filled dam with a clay impermeable core. The storage capacity is 68×10^6 m^3. Water is used both for electricity production and irrigation. The leakage problem is located at the right abutment, which is formed by limestone rocks with embedded marl strata. An important fault exists at a distance of a few tens of metres downstream of the dam. The calcareous strata are folded, presenting in a short distance a transition from almost horizontal to almost vertical disposition. The vertical marl strata act as a hydraulic barrier that was responsible for the existence of a spring, called Seboruco spring, at the fault site. Unfortunately, the dam was constructed just upstream of this spring. No leakage problem would have originated if the dam had been constructed some 50 m further downstream. The spring would have discharged inside the reservoir.

Groundwater from the spring emerged just in the zone where the bottom outlet tunnel had to be constructed. According to the information obtained from workers who were present during the dam construction, serious difficulties arose in the evacuation of the spring water, and the concrete tunnel had to be constructed with a permanent water flow. Most probably, this flow washed away the fine particles of the concrete, leading to its poor quality at the bottom of the tunnel. During the first filling of the reservoir, the bottom of the tunnel collapsed and groundwater started to emerge, with flow rates between 100 and 200 l/s, depending on the water level at the reservoir. The collapse of the tunnel created serious alarm because at the end of the valley, the town of San Juan de la Maguana is located, with a population close to 150,000 inhabitants. During the early stages, the whole leakage was concentrated in a single point at the bottom of the tunnel, but some unsuccessful repair works dispersed the emerging water in several springs along a stretch of several tens of metres.

The filtration law (variation of leakage amount as a function of water level) has been presented in Fig. 2.2 (Chapter 2). Only a fraction of the total emerging flow rate is considered in this figure. It corresponds to an individual water point emerging inside the tunnel. The correlation between the flow rate and the water level shows that water emerges when the water level reaches an elevation of about 605 m.a.s.l. The minimum elevation of the riverbed upstream of the dam is 590 m.a.s.l.

On the other hand, the temperature profiles shown in Fig. 5.5 (Chapter 5) indicate that the leakage at the reservoir is taking place at an elevation between 605 and 609 m.a.s.l. (Fig. 5.6). This result is coherent with the hint derived from the filtration law presented in Fig. 2.2. The zone most likely responsible for the leakage is located at the upstream end of the contour line drawn in Fig. 5.6. The limestone formation of the right side has at this section a chaotic aspect, and according to the information provided by local people, a small waterfall existed at this point where people used to have baths. A tracing experiment was performed in which about 200 mCi of iodine-131 were injected at this point. Unfortunately, however, no positive result was obtained. Probably the period of time during which the detection of the tracer was attempted at the emerging water was too short or maybe the tracer was not injected at the right point.

To identify the pathway of the leakage, seven boreholes were drilled at the right abutment, the location of which is shown in Fig. 12.13. The following tests were performed in these boreholes:

1) Flow tests using tracers using the techniques described in Chapter 8.
2) Interconnection experiments using radioactive (^{131}I) and fluorescent dyes (uranine and rhodamine WT) as tracers.
3) Measurement of vertical profiles of temperature, conductivity and natural gamma radiation (see Fig. 4.5).

The chemical and isotopic analyses showed, as expected, that all groundwater samples from the boreholes and springs were fed by the reservoir. The only

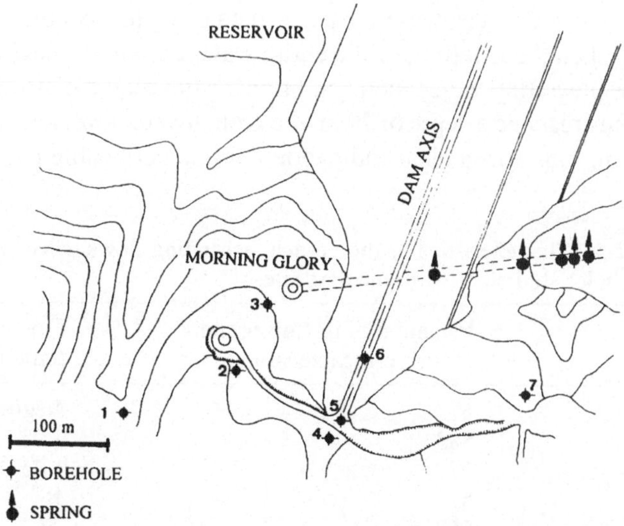

Fig. 12.13. Location of springs and boreholes drilled at the dam site of Sabaneta reservoir in the Dominican Republic.

exception was borehole 1, which did not intercept the groundwater flow from the reservoir.

Very valuable results were obtained from the flow tests performed in boreholes with artificial tracers. All boreholes showed the presence of a vertical ascending flow with the characteristics given in Table 12.4. The analysis of these data was used to make the following observations:

1) The second column of Table 12.4 gives the elevation of the lower end of the fractured stretch where the vertical flow initiates. Due to the high velocity of the vertical flow, it was not possible to define with precision the thickness of this permeable stretch. However, in any case, it was less than 5 m.
2) The elevation of the lower end of this permeable stretch increased with the elevation of the ground surface, that is, from the reservoir to the adjacent formation. This elevation coincided with the dip of the strata of the carbonate rock formation. For example, from boreholes 3 and 2, the lower end of the permeable stretch rose from the elevation 577 to the elevation 582 m.a.s.l. For boreholes 6, 5 and 4, which are located along the axis dam, the elevations were, respectively, 559, 568 and 579. These values permitted the definition of the profile of the permeable strata responsible for the vertical flow and, therefore, for the leakage.
3) The velocity of the vertical flow was very high in all boreholes except for borehole 1, which, as mentioned before, does not intercept the groundwater flow fed by the reservoir. The highest flow rates were obtained in boreholes 4, 5 and 6. The lower value of the vertical flow detected at borehole 7 was due to the proximity to the springs.

In the case of borehole 6, the drilling was stopped at the moment when a depth of 60 m was reached. A flow test was then performed using the technique described in Chapter 8, based on the labelling of the entire water column. A negligible groundwater flow was prevailing under such conditions. After that, the drilling continued. When the drilling reached a depth of 70 m, the groundwater level inside the borehole rose about 10 m. This was a clear indication that the permeable layer responsible

Table 12.4. Characteristics of the vertical ascending flows detected inside the boreholes studied close to the dam site.

Borehole number	Elevation of the lower end of the permeable strata	Flow velocity (cm/s)
1	577	Negligible
2	583	1.25
3	572	1.25
4	579	18
5	568	27
6	559	25
7	561	3.3

for the leakage was then intercepted. This water level elevation explains the high value of the vertical ascending flow detected in certain boreholes. The raising of the water level inside the borehole creates locally an extremely high hydraulic gradient. Under such conditions, the upper layers, of relatively low permeability, can transmit a significant water flow, and a rapid vertical ascending flow is generated.

Only the upper part of the permeable stretch detected in borehole 6 was really participating in the vertical ascending flow. Vertical flow tests were performed using the method described in Section 8.6.2.4 (Fig. 8.27). In these tests, a point injection of the tracer solution is carried out with the detector located, for example, 2.5 m above the injection point. When the tracer injection (^{131}I) was made at a depth corresponding to the elevation of 580 m.a.s.l., the flow test gave negative results. The tracer never arrived at the point where the detector was placed. In a new test, the detector was lowered to the injection point a few seconds after the injection, but the tracer was not there. This was a clear demonstration that in the lower part of the permeable stretch, a pure horizontal flow existed. This behaviour was not observed in other boreholes because of the lower flow velocity. Most of the water connecting the reservoir with the springs was flowing through borehole 6.

The previous statements were confirmed through interconnection experiments between boreholes and springs. Five such experiments were carried out using ^{131}I and uranine as tracers. All these experiments gave positive results. The tracer injected in every borehole always arrived at the springs. One of the breakthrough curves obtained in a experiment, which corresponds to the injection of 3 kg of uranine in borehole 3, is shown in Fig. 12.14. The shape of this curve suggests the existence of two or three different pathways of the groundwater flow. The passage of the injected tracer through boreholes 4, 5 and 6 was detected by collecting water samples from these boreholes at different times. The results obtained are shown

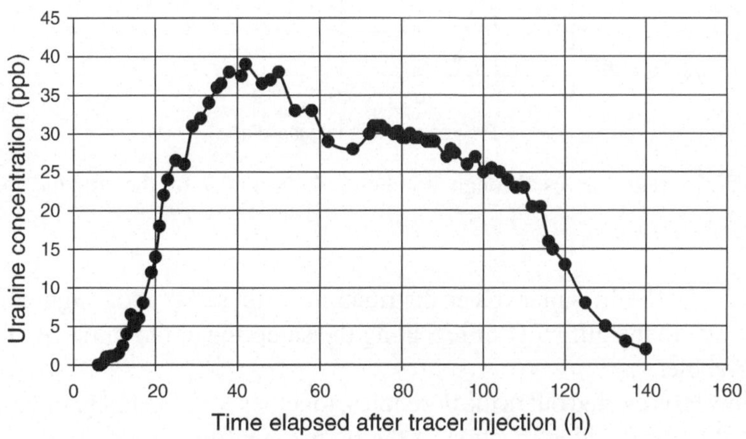

Fig. 12.14. Breakthrough curve measured at the emerging water after the injection of 3 kg of uranine inside borehole 3 (Fig. 12.13).

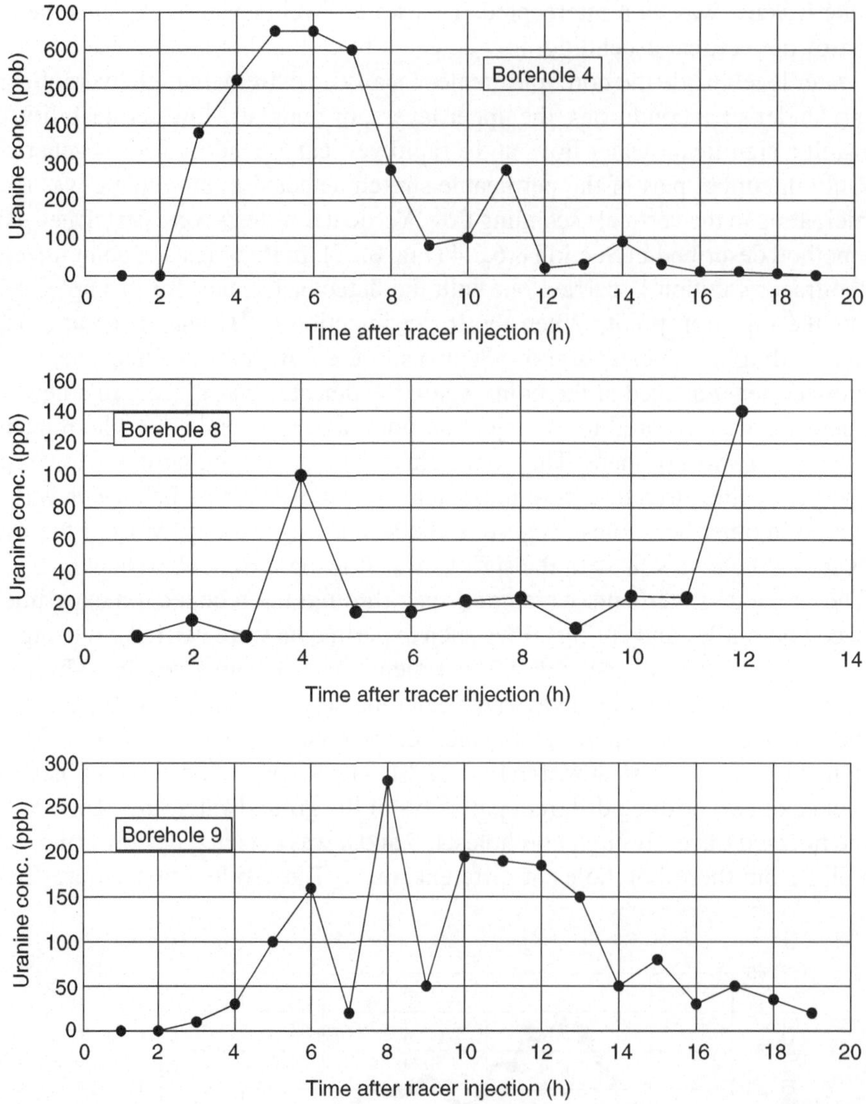

Fig. 12.15. Passage curves through boreholes 4, 8 and 9 of the uranine injected in borehole 3.

in Fig. 12.15. The irregular tracer distribution of these breakthrough curves was probably due to the difficulty of repeating the sampling at the same depth interval in the borehole.

The flow tests carried out in the boreholes, together with natural gamma radiation profiles, allowed the identification of the permeable (due to the presence of fractures) and impermeable strata intercepted by the different boreholes (see Fig. 4.5). The changes with depth of the vertical flow velocity revealed the position of the

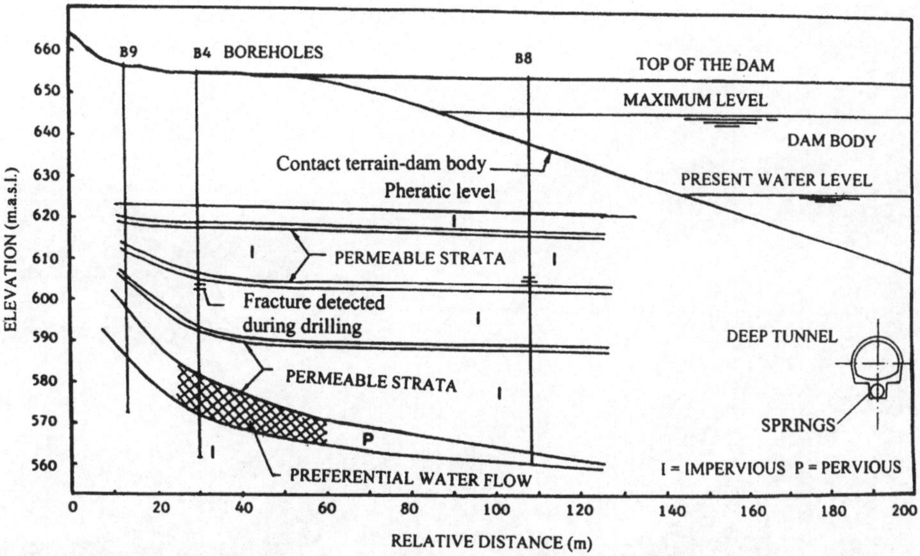

Fig. 12.16. Localisation of pervious and impervious strata on the right abutment of the Sabaneta dam as identified by means of tracer techniques. The preferential flow path of the escaping water is located at an elevation between 570 and 580 m.a.s.l.

strata through which groundwater was escaping from the borehole. Some of these fractures were also identified during the drilling of the boreholes. Figure 12.16 shows the distribution of permeable and impermeable strata on the right abutment of the dam. The stretch in which most of the water emerging in the springs is flowing through the rock formation is also shown. It is important to observe that the rapid underground flow was circulating at a relatively large distance from the dam body. These results helped in the evaluation of the safety risks related to the eventual collapse of the dam.

In the opinion of the authors, the results obtained by the study provided sufficient technical information for the planning of the repair works with reasonable expectation of success. The identification of the groundwater pattern on the right abutment produced significant savings for the grouting works.

12.5 LAKE LAJA, CHILE

Description of the problem As in the previous case, the study of the leakage problem affecting Lake Laja was carried out by the senior author within the framework of a technical cooperation project convened between the International Atomic Energy Agency (IAEA), Vienna, and the government of Chile (Plata and Iragüen, 1992).

Fig. 12.17. General view of Lake Laja from the site of the natural dam.

Lake Laja is located in the Andes Cordillera, about 600 km south of Santiago and at about 1,340 m.a.s.l. (Fig. 12.17). Its maximum storage capacity is 8×10^9 m^3. The lake is used for hydropower generation with two plants: one of 400 MW using water falling through a waterfall (El Toro) about 520 m high, and the second of 136 MW (Abanico). The second plant uses the leakage of the lake to produce electricity (between 20 and 45 m^3/s, depending on the water level at the lake) with a waterfall of only 80 m. These figures reveal the important economic implications linked to the leaks. Apart from the difference in the height of the respective waterfalls, the leak represents an important limitation for the maximum water level that can be achieved in the lake. When the level rises, the flow rate of leakage also increases, and certain equilibrium is attained between the level in the reservoir and the flow rate of the leakage, which is governed by the total water ingress into the lake.

The progressive accumulation of volcanic eruptions along the valley occupied by the River Laja formed Lake Laja. The responsible volcanoes in the vicinity of the lake are Sierra Velluda, Laja and Antuco. The last eruption of the volcano Antuco took place in 1853 and caused a marked increase of the lake's water level by about 30 m. Observations made on the ancient riverbed at the dam site show that the lake was flowing over the volcanic materials of the natural dam before the last eruption. The riverbed was established along the contact of the volcanic materials and the impervious Cretaceous porphyrite of the right side (Fig. 12.18).

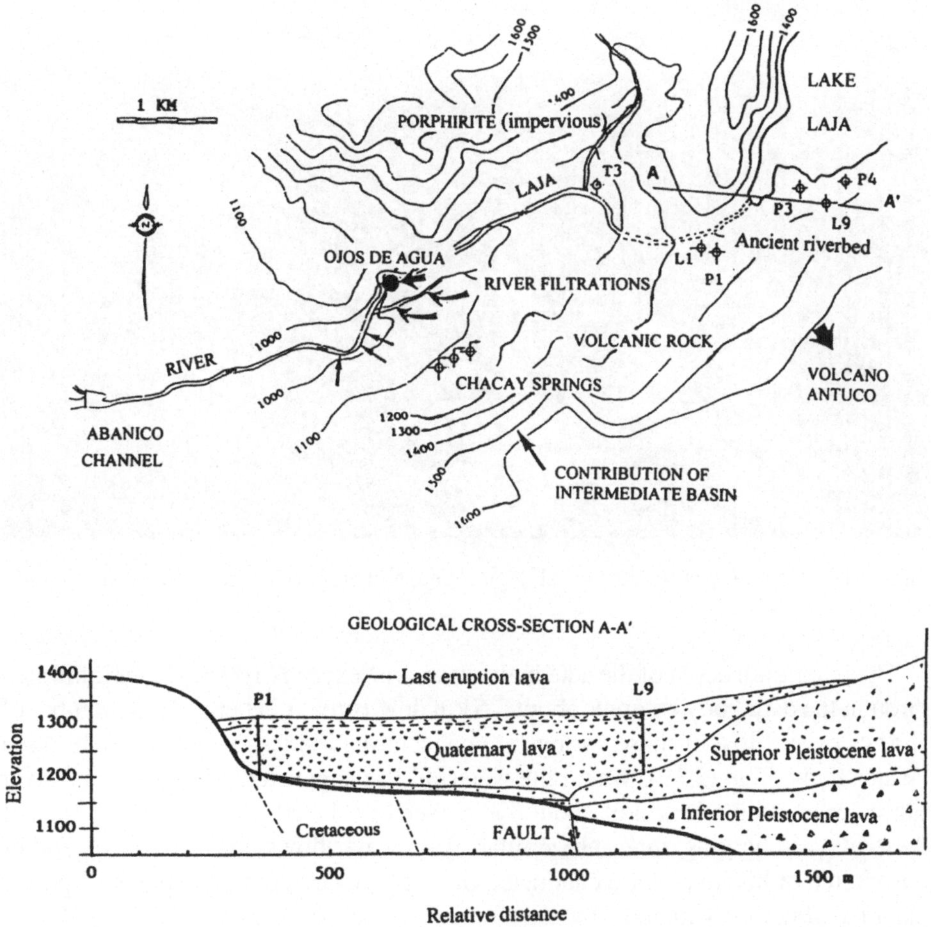

Fig. 12.18. Geology and water points sampled in the study area.

The volcanic materials provided by this last eruption produced a new damming of the lake, and the average water level rose about 30 m. The water stopped flowing over the natural dam, but a high filtration rate is taking place through the lava deposits. Figure 12.18 also shows the location of the most representative water points.

According to the previous information, it seems likely that most of the leakage should take place through the lava layer formed during the last eruption and therefore above the elevation of 1,300 m.a.s.l., which was the elevation of the top of the old dam. The highest permeability should correspond to the zone where the lake water was flowing before this eruption. The contact of the expelled incandescent lava with this water would have produced the rapid disintegration of the deposited materials, and a zone of preferential flow could have been created.

Fig. 12.19. General view of the Ojos de Agua zone, with important groundwater discharges.

The maximum depth of the lake in the dam zone exceeds 100 m. Water infiltrated through the dam body emerges about 3.5 km downstream, where a large number of springs exist. Most of these springs are concentrated in two zones: Ojos de Agua (Water Eyes, Fig. 12.19), where the waterfall has excavated a big hole, and river Filtraciones, with a length of about 800 m (see Fig. 12.20). The total flow rates of these two discharging zones are similar. Both water flows converge at the end of this stretch of 800 m and are conducted through a channel to the Abanico's power plant (see Fig. 5.9, Chapter 5).

A third relatively small water flow is added to the two fluxes previously mentioned, which is formed by the snowmelt deposited at high elevations of the volcano Antuco. These waters form a small lake during the snowmelt period and infiltrate in the lava rocks at an elevation similar to the lake elevation. It emerges at several points of the intermediate basin. As shown in Section 5.3.3, the difference in electrical conductivity between both types of waters allowed the determination of the contribution of this intermediate basin to the total flow diverted to the Abanico's power plant.

The main objective of the study was the localisation of possible preferential infiltration zones at the bottom of the lake, as well as the identification of permeable strata or zones in the dam body to assess the feasibility of undertaking engineering works to reduce the leakage. The following techniques were used to this purpose:

1) Tracer dilution tests at the bottom of the lake for the localisation of areas with preferential infiltration.
2) Tracer tests in boreholes for the localisation of permeable strata and horizons.

Fig. 12.20. View of the River Filtraciones at the waterfall.

3) Study of the distribution of temperature and electrical conductivity in the lake, boreholes and springs.
4) Analysis of major chemical components of water, as well as the contents of deuterium, oxygen-18 and carbon-13.
5) Interconnection experiments between the lake (and boreholes) and springs using artificial tracers.

The results obtained are summarised below.

Tracer dilution tests performed at the bottom of the lake Both the applied technique and the results obtained have been described in Section 10.7.6 (Chapter 10). Figure 10.13 shows that most of the leakage is located at a zone in which the lake was discharging before the 1853 eruption of the volcano Antuco. This coincides, more or less, with the contact between the lava and the porphyritic formation.

Water temperature Figure 5.1 (Section 5.2.1) shows a diagram with the variation of temperature of Lake Laja throughout the year. It is observed that below a depth of 45 m, water temperature is permanently kept constant between 5°C and 6°C.

On the other hand, Fig. 12.21 shows the variation of temperature of the shallow layer of the lake throughout the year (average values have been used), as well as of the two most representative springs, Ojos de Agua and River Filtraciones. It can be observed that water temperature at the lake surface varies between 13°C and about

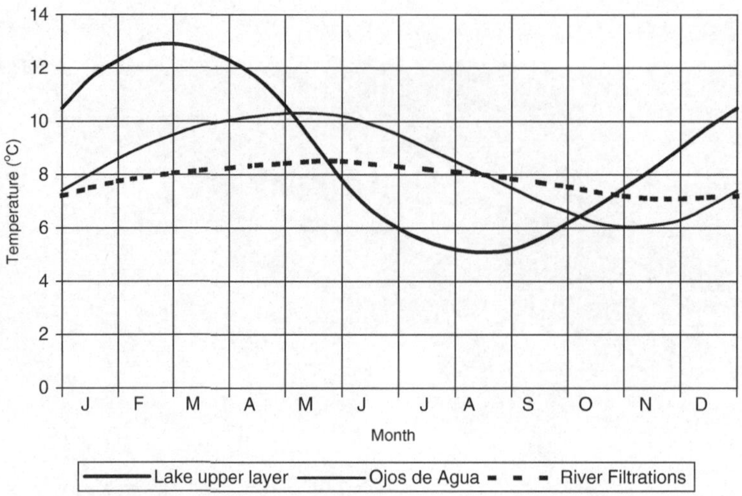

Fig. 12.21. Temperature variation throughout the year at the upper layer of Lake Laja, Ojos de Agua and River Filtraciones.

5°C. For the Ojos de Agua spring, water temperature varies between 10.3°C and 6°C. Obviously, the existence of such thermal variations is a clear demonstration that most of the water that emerges at this spring is formed by the infiltration occurring at the surface layer of the lake. In the case of the River Filtraciones, the lower amplitude of the thermal variation (between 8.5°C and 7.1°C) is explained by a higher contribution of the groundwater flow, which has a constant temperature. Such a flow is partly formed in the intermediate basin (derived from snowmelt of the volcano Antuco) and partly from a deep leakage at the lake, which most probably is taking place through the fault shown in Fig. 12.18.

The annual average temperature of water emerging through the Ojos de Agua spring amounts to 8.4°C. This value is very close to the temperature observed in the upper layer of the reservoir with a thickness of 38 m. This layer is responsible for most of the leakage that emerges at this point, as well as for most of the water emerging at the River Filtraciones. The yearly average temperature of the River Filtraciones is 7.7°C. This lower temperature is explained by the deep leak affecting the lake, as mentioned above.

Figure 12.21 shows that an important time delay exists between the temperature peak observed in the lake and in the springs. The delay is about 2 months for the Ojos de Agua spring and a bit longer for the River Filtraciones. This delay is partly due to the heat exchange between water and rock (see Section 5.2.1, Chapter 5) and partly to the transit time of water from the lake to the springs.

Flows in boreholes Five boreholes were drilled as part of the investigation, whose location is shown in Fig. 12.18. Tracer flow tests were performed in all cases

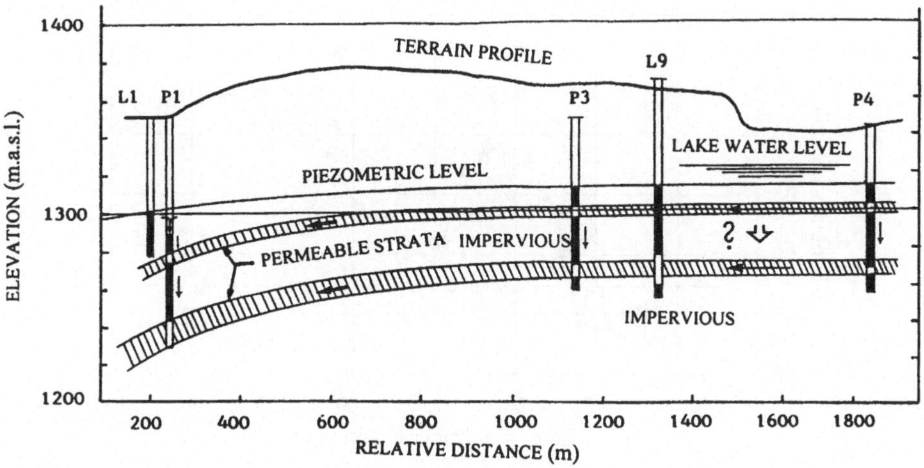

Fig. 12.22. Permeable strata identified at the dam site by means of the performed tracer tests in boreholes.

using ^{131}I as tracer, according to the techniques described in Chapter 8 for the measurement of horizontal and vertical flows. A very slow horizontal flow with a Darcian velocity of about 0.7 cm/d was detected in borehole L1. Boreholes P1, P3 and P4 showed vertical descending flows with very high velocity. In the case of borehole P1, the vertical descending flow was produced by water entering the borehole at a depth above the piezometric level (it was possible to hear water falling along the borehole). An additional water inflow point was detected in this borehole at a depth of 60 m. In borehole L9, a permeable stratum with a small horizontal flow was detected between 95 and 115 m.

The flow tests allowed the identification of two permeable horizons with preferential water flow, as illustrated in Fig. 12.22.

Interconnection experiments Ten interconnection experiments were performed during the implementation of the project. The tracers ^{82}Br, ^{131}I, tritium and uranine were used. Most of these experiments were carried out during 1988 and 1989. Due to space limitations, these experiments are not described in a detailed form. In Section 9.4.2, one of these experiments is described, in which 20 Ci of tritium were injected on the shallow layer of the lake at a point where it was possible to see the infiltration of water through the lava formation. The results obtained are shown in Fig. 9.6. The differences of the areas obtained for different sampling points are explained by the mixing of the labelled lake water with nonlabelled water from the intermediate basin or even from the lake but flowing through a different pathway. The conflict created by the long mean value of the transit time of water between the lake and the springs (about 20 days) has been discussed in Section 9.4.2. Such a long transit time would require the existence of an underground reservoir of about

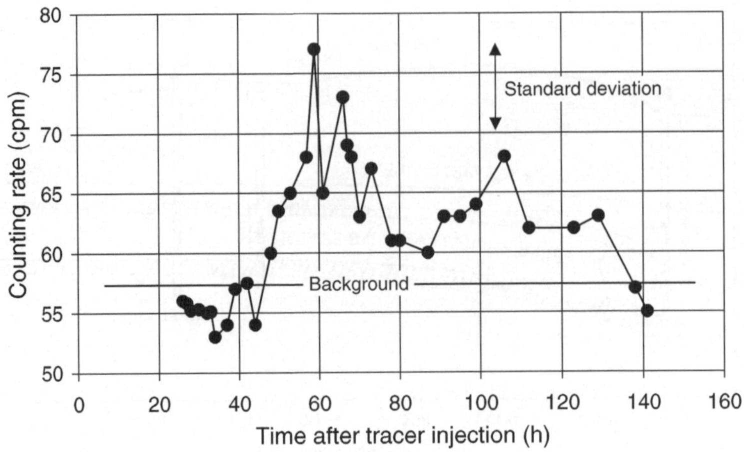

Fig. 12.23. Breakthrough curve obtained at the River Laja (total leakage of Lake Laja) after the injection of 7 Ci of ^{82}Br in the lake. The experiment was carried out by C. Silva and the senior author in 1976.

$45.6 \times 10^6\, m^3$, which is difficult to imagine taking into account the volume of the permeable rock formation existing in the involved area. Similar values of the transit time were obtained in other interconnection experiments.

In 1976, the senior author, together with C. Silva, Chilean Nuclear Energy Commission, carried out an interconnection experiment in which 7 Ci of ^{82}Br were injected in a point of the lake located close to the zone where water was overflowing before the 1853 volcanic eruption. A small fraction of the injected tracer, concretely 1.5%, arrived at the measuring station, which was situated downstream of the confluence points of the Ojos de Agua spring and the River Filtraciones. The tracer was detected by means of a scintillation detector fitted with a NaI(Tl) crystal submerged in water. The breakthrough curve is shown in Fig. 12.23. Two different peaks were obtained, which were respectively related to the two above-mentioned water flows. The respective mean transit times of water were 62 and 110 h, that is, much shorter than the 20-day value obtained for the tritium experiment carried out in 1989.

In 1989, 2 Ci of ^{131}I were injected in borehole P-1. The injection was made at a depth of 60 m, where the upper permeable horizon had been detected by means of flow tests using artificial tracers. However, taking into account that a very rapid vertical descending flow was prevailing, most of the tracer travelled toward the lower permeable horizon located at about 75 m depth and escaped from the borehole at this level. The detection of the tracer was made using the concentration method described in Section 7.6.3, which is based on the coprecipitation of the ^{131}I with nonradioactive AgI. The strange breakthrough curve shown in Fig. 12.24 was obtained at the sampling station located downstream of the confluence of the Ojos de Agua spring and the River Filtraciones. A small fraction of the tracer

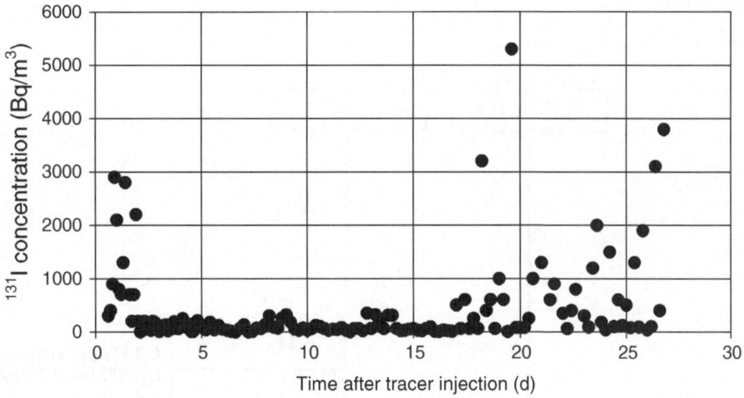

Fig. 12.24. Breakthrough curve obtained at the River Laja (total leakage of Lake Laja) after the injection of 2 Ci of ^{131}I in borehole P1.

arrived at the sampling station between 1 and 2 days after the injection. During the next 18 days, no tracer was detected at this station. Between 18 and 27 days after the injection, small tracer concentrations were detected with an irregular distribution. These results are in good agreement with those obtained for both the ^{82}Br and the tritium experiments. A small fraction of the injected tracer escaped from the borehole through the upper permeable horizon. The rest of the tracer was incorporated into a deep flow, with 20 days as the mean transit time, and its detection was strongly hindered by the low half-life of ^{131}I (8 days) and the high dilution rates.

Conclusions The results obtained in the complete study are summarised in the flow diagram presented in schematic form in Fig. 12.25. Water temperature values measured in March 1989 are also shown in the same figure. According to this scheme, the downstream water flow conducted to the Abanico's power plant is formed by the three following components:

1) Water from the upper layer of the lake, above the elevation of about 1,300 m.a.s.l.
2) Water infiltrated at the bottom of the lake, most probably through the fault marked in Fig. 12.18.
3) Flow from the intermediate basin mainly formed by snowmelt in areas close to the Antuco volcano.

The temperature of water of the first flow component varies throughout the year between 6°C and 10.5°C, approximately. Water from the second flow component shows a constant temperature of about 5.5°C. The third component shows a more or less constant temperature of 12.1°C (average value).

Temperature and electrical conductivity values obtained at the different springs that form the River Filtraciones can be explained only if the existence of a

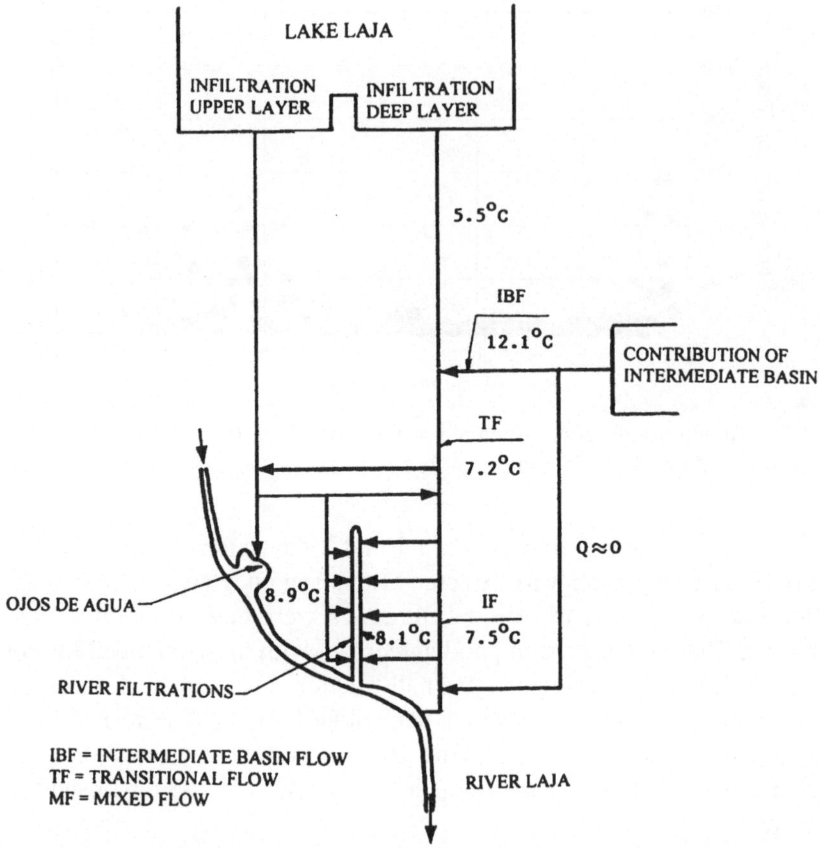

Fig. 12.25. Schematic diagram illustrating the origin of the water emerging at Ojos de Agua and River Filtraciones. Average temperature values of the different water flows are indicated.

"transition" or intermediate flow is admitted. This transitional flow has to be formed by the confluence of the flow generated in the intermediate basin (12.1°C and 450 μS/cm) with the flow related to the deep infiltration at the bottom of the lake (5.5°C and 50 μS/cm). The contribution of this transitional flow to the Ojos de Agua spring is very small but sufficient to explain the small rise in conductivity observed in this spring with respect to the lake water. Most of the water from this transitional flow discharges to the River Filtraciones because it is located on the left side of the river, as in the case of the intermediate basin.

Groundwater flow from the upper layer of the lake arrived at the springs (both Ojos de Agua and River Filtraciones) with an average transit time varying between 2 and 5 days, depending on the water level at the lake. On the contrary, the deep flow arrives at these springs with an average transit time that can vary from 15 to 25 days, also depending on the lake water level. This deep flow feeds both the Ojos de Agua springs and the River Filtraciones.

A thorough analysis of the results allowed the elaboration of the water balance of the leak presented in Table 12.5, in which the contribution of the three above-mentioned groundwater flows are given. This contribution is valid for the conditions prevailing both in March 1988 and in March 1989. The water balance is based exclusively on the electrical conductivity values and on the results obtained in two interconnection experiments carried out by injecting artificial tritium in the lake. The available temperature values have not been used for the balance because this parameter cannot be considered of a conservative nature. The probable uncertainties for the involved parameters used in the calculations are also shown.

The results of the leakage balance show, as expected, that most of the water emerging at the springs is formed by the infiltration taking place in the upper layer of the lake, above the elevation of 1,300 m.a.s.l. The contribution of the deep leak must be on the order of 6 m^3/s. The variation of this contribution with the water level at the lake is relatively small. For example, its value was about 6.4 m^3/s, for a total leakage of 36 m^3/s (March 1998) or 5.7 to 26 m^3/s (March 1989). This small variation is proof for the deep origin of the leakage. The changes in water level are small (a few metres) compared with the depth at which this leakage originates (about 100 m).

The contribution of the intermediate basin is about 2 m^3/s, and obviously, this value is not affected by the water level at the lake. The variation of this contribution throughout the year is also small because of the high capacity of the saturated rock formation in which these waters are recharged.

Unfortunately, the study could not be finished for reasons beyond the control of the team. The precise and irrefutable localisation of the infiltration area responsible for the rapid flow connecting the upper layer of the lake to the springs is missing. The tracers injected in the lake at points not situated inside this area are always incorporated into the deep flow with some delay and long transit time.

Obviously, the identification of this area of preferential infiltration is crucial for the evaluation of the feasibility of undertaking any work to decrease the magnitude of the leakage. A decrease of just 10% in the flow rate of the leakage would represent an important economic benefit due to the high waterfall of the El Toro's power plant. In the opinion of the senior author, an important part of the leakage is concentrated in a small area located close to the contact of the porphyritic rock formation and the lava, above the elevation of 1,300 m.a.s.l. Therefore, is very likely that an important fraction of the leakage can be stopped at a reasonable cost.

12.6 THE COLBÚN RESERVOIR, CHILE

Generalities The Colbún reservoir is located on the River Maule at a distance of 260 km south of Santiago, Chile. Its storage capacity is about $1,500 \times 10^6$ m^3. The reservoir is used for hydropower generation and irrigation purposes. Three dams

Table 12.5. Water balance of the leaks affecting the Lake Laja. Integers shown in the first column (row) are used for the calculation of the detailed water balance of each flow component. When the indication (1−) is included, the number has its real value.

Row	Water point/parameter	March 1988	March 1989
	Experimental data		
	Interconnection experiments using tritium		
1	Area of breakthrough curve at Ojos de Agua (OA)	$4.28 \pm 0.13\ \mu Ci \cdot d \cdot m^3$	$10.59 \pm 0.21\ \mu Ci \cdot d \cdot m^3$
2	Area of breakthrough curve at the River Filtraciones	$1.57 \pm 0.08\ \mu Ci \cdot d \cdot m^3$	$2.76 \pm 0.08\ \mu Ci \cdot d \cdot m^3$
	Conductivity values ($\pm 0.5\%$ μs/cm)		
3	Lake Laja	41.3	42.0
4	Ojos de Agua (mean value)	44.5	51.1
5	Springs at River Filtraciones (SRF) (mean value)	100.6	112.0
6	River Filtraciones at the mouth	78.0	88.5
7	River Laja downstream the River Filtraciones	65.2	74.7
8	Chacay springs (mean value)	454.0	476.0
	Temperature ($\pm 0.2°C$) (only for comparison)		
9	Ojos de Agua	9.6	8.9
10	River Filtraciones at the mouth	8.7	8.1
11	Springs at River Filtraciones (SRF)	8.1	7.5
	Water balance		
12	Total leakage (m^3/s)	36	26
	Contribution of River Filtraciones (m^3/s)		
13	$6f + 4(1 - f) = 7;\ f = 0.618 \pm 0.014\ (0.631 \pm 0.015)$	$0.618 \times 36 = 22.2 \pm 0.55$	$0.631 \times 26 = 16.4 \pm 0.59$
14	Contribution of Ojos de Agua (12–13) (m^3/s)	13.8 ± 0.66	9.6 ± 0.64
	Contribution of SRF to River Filtraciones (m^3/s)		
	Conductivity:		
15	$5f + 4(1 - f) = 6;\ f = 0.597 \pm 0.01\ (0.614 \pm 0.01)$	$0.597 \times 22.2 = 13.3 \pm 0.52$	$0.614 \times 16.4 = 10.1 \pm 0.40$

	Temperature:		
16	$11f + 9(1 - f) = 10;\ f = 0.6\ (0.57)$	$0.6 \times 22.2 = 13.3$	$0.57 \times 16.6 = 9.3$
	Contribution of OA-flow to SRF-flow (m^3/s)		
17	$f = 2/1;\ f = 0.367 \pm 0.022\ (0.261 \pm 0.009)$	$0.367 \times 13.3 = 4.9 \pm 0.35$	$0.261 \times 10.1 = 2.6 \pm 0.14$
18	Flow-rate of transitional flow (T-flow) (16–17) (m^3/s)	$13.3 - 4.9 = 8.4 \pm 0.63$	$10.1 - 2.6 = 7.5 \pm 0.42$
	Conductivity of T-flow (μS/cm)		
19	$4(2/1) + K(1 - (2/1)) = 5;\ K = 133.1 \pm 4.9\ (133.5 \pm 1.9)$	133.1 ± 4.9	133.5 ± 1.9
	Contribution of IB-flow to T-flow (m^3/s)		
20	$8f + 3(1 - f) = 19;\ f = 0.222 \pm 0.012\ (0.211 \pm 0.0045)$ (20b)	$0.222 \times 8.4 = 1.9 \pm 0.17$	$0.211 \times 7.5 = 1.6 \pm 0.1$
	Contribution of T-flow to OA-flow (m^3/s)		
21	$19f + 3(1 - f) = 4;\ f = 0.035 \pm 0.0038\ (0.1 \pm 0.0042)$	$0.035 \times 13.8 = 0.5 \pm 0.06$	$0.1 \times 9.6 = 1.0 \pm 0.08$
22	Contribution of IB-flow to OA-flow (21 × 20b) (m^3/s)	$0.5 \times 0.222 = 0.1 \pm 0.01$	$1.0 \times 0.211 = 0.2 \pm 0.07$
	Total balance		
	Contribution of the upper layer leakage		
23	$14 + 17 + (13–15)\ m^3/s$	27.6 ± 1.1	18.5 ± 1.0
24	Contribution of the intermediate basin ($20 + 22\ m^3/s$)	2.0 ± 0.2	1.8 ± 0.1
25	Contribution of the deep leakage ($12 - 23 - 24\ m^3/s$)	6.4 ± 1.2	5.7 ± 1.0

IB = Intermediate basin.
Transitional flow (T-flow) = flow formed by the mixing of the deep infiltration flow and IB-flow.
d = day. 20b corresponds to the value of fraction f.

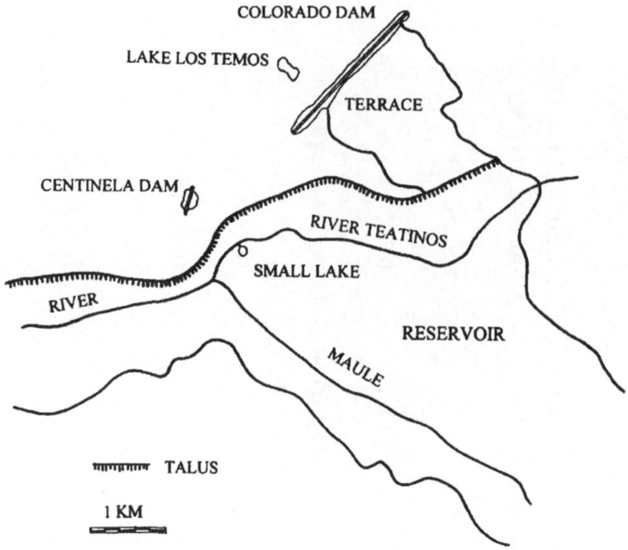

Fig. 12.26. Schematic map illustrating the study area in the vicinity of the Colorado dam.

were constructed. The leakage problem is located at the Colorado dam, which is not the main dam. The construction of the Colorado dam was necessary to prevent water from escaping along a glaciofluvial valley, at the end of which the city of Talca is located.

The Colorado dam is a homogeneous earth-filled dam with a height of 29 m and a length of 2,600 m (see Fig. 12.26). The dam was constructed on a terrace of Quaternary materials of different origins, whose thickness reaches about 300 m. The elevation difference between the terrace and the valley of the River Maule is about 40 m. A small river, called River Teatinos, was running at the bottom of the valley, as shown in Fig. 12.26. The Quaternary terrace ends in a talus of about 20 m.

Description of the problem The leak appeared during the first filling, when the water level at the reservoir reached the elevation of 409.5 m.a.s.l. Water initially emerged in a small lake called Los Temos (see Fig. 12.26), which corresponds to the lowest elevation of the dam site. When the water level reached the elevation 419.8, the leak flow rate reached 800 l/s, and at the maximum elevation in the reservoir, a value of 11 m^3/s was attained. Due to the high risk that this leakage could represent for the stability of the dam, the engineers responsible for the dam management immediately initiated an intensive drilling campaign of relief wells. Simultaneously, a thick gravel and sand layer was deposited on the ground to increase its stability. About 70 relief wells were drilled and a drainage network was constructed to collect the emerging water from the dam site. The location of most of these relief wells is shown in Fig. 12.28 below.

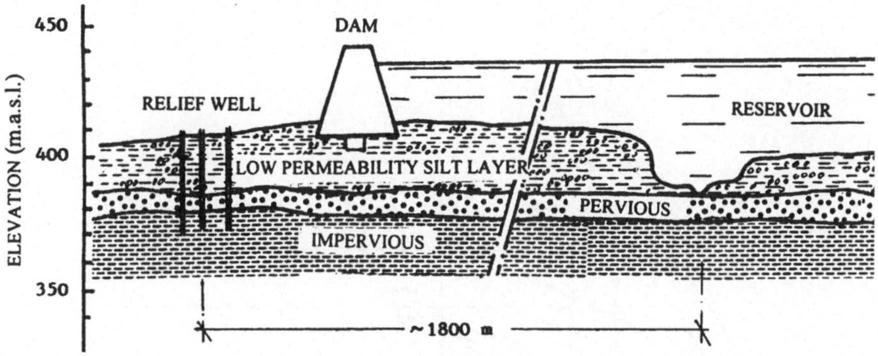

Fig. 12.27. Schematic diagram illustrating the origin of the leaks affecting the Colorado dam during the investigation carried out in the Colbún reservoir, Chile.

Hypothesis on the origin of the leakage The results obtained during the experimental campaigns were used to elaborate a hypothesis about the precise origin of the leakage (Plata et al., 1989a). This hypothesis is presented here because it will help in understanding the results obtained. The hypothesis is illustrated in a schematic form in Fig. 12.27. A gravel and sand layer with very low clay content was identified on the ground and below the dam site at an elevation of about 380 m.a.s.l. A low permeability feldspathic silt layer covers this permeable layer. The River Teatinos excavated its bed by crossing this layer of fine materials and excavating down to the top of the gravel layer, which could not be eroded due to the coarse grain size. Reservoir water was transmitted through this permeable layer and emerged at the point showing the lowest resistance, which coincides with the area close to the Lake Los Temos.

Prior to the dam construction, groundwater in the area close to the dam site was discharging in the River Teatinos. This river showed a permanent water flow even during extreme dry periods due to the discharge of groundwater. A small permanent lake, fed by water of the same origin, existed at the bottom of the valley at a short distance from the River Teatinos, as shown in Fig. 12.26. Frequent landslides occurred at the foot of the talus during the rainy periods as a result of groundwater discharge. The same pathway used for groundwater to discharge in the bottom of the valley was used for the reservoir water to escape from the reservoir once a sufficient elevation was attained. The existence of this layer of gravel and sand was known by the technicians responsible for the dam construction, but it was underestimated from the point of view of a potential source for a significant leak.

Deuterium contents Shortly after the appearance of the leakage, technicians from the Chilean Nuclear Energy Commission started a program for the measurement of deuterium in samples collected at the emerging water points and the reservoir (the available mass spectrometer was not ready for the measurement of oxygen-18 at

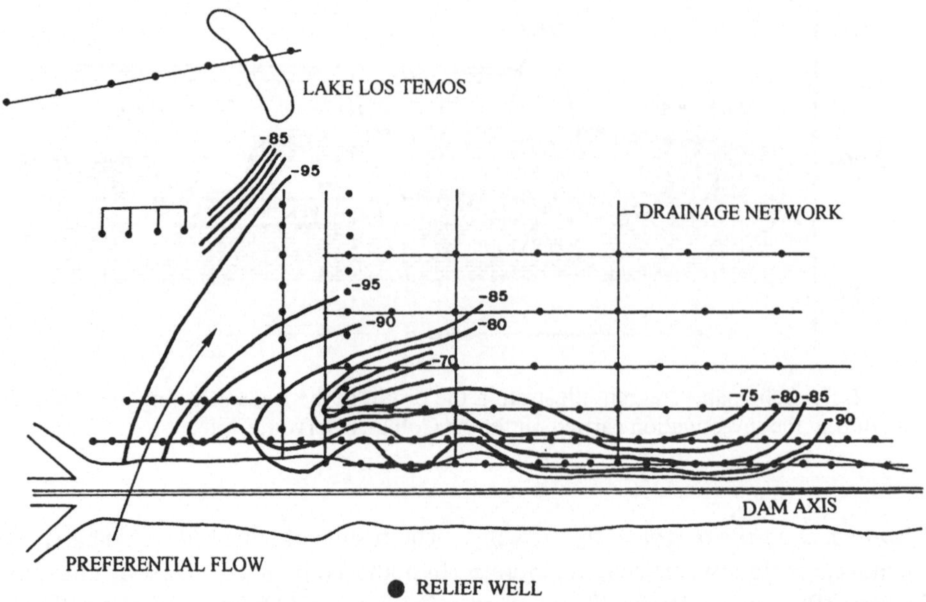

Fig. 12.28. Distribution of the δD values (expressed in ‰) of groundwater emerging in the relief wells drilled at the dam site. Each point represents a well.

that time). The results obtained showed that the δD values of the emerging groundwater through the relief wells varied from −70‰ and −96‰. Several hundred water samples were analysed. The typical distribution of the δD values is shown in Fig. 12.28. As shown later, the more negative δD values corresponded to the water flow fed by the reservoir.

When the senior author started the study at the beginning of 1985, one of the first actions undertaken was the implementation of a sampling program for chemical and δD analyses of all water points close to the dam site. The reason for this sampling was to characterise the chemical and isotopic compositions of the locally recharged groundwater not related to the leakage. Thirty water samples were analysed. The mean δD value of these samples was −71.1‰, with a mean standard deviation of 1.6‰. On the contrary, the δD values obtained for the reservoir water ranged from −95‰ to −96‰.

This relatively large contrast of δD values in the reservoir and the local groundwater permitted the determination, for each relief well, of the mixing proportions of both types of waters. Due to the fact that the piezometric level of the local aquifer was higher than the water level at the reservoir, the aquifer water was forced to find new pathways in the underground. The presence of the reservoir impeded its discharge at the bottom of the valley, as happened prior to the dam construction. This fact explained the existence of some relief wells with vertical descending flows. The aquifer water was flowing through these wells before being incorporated into

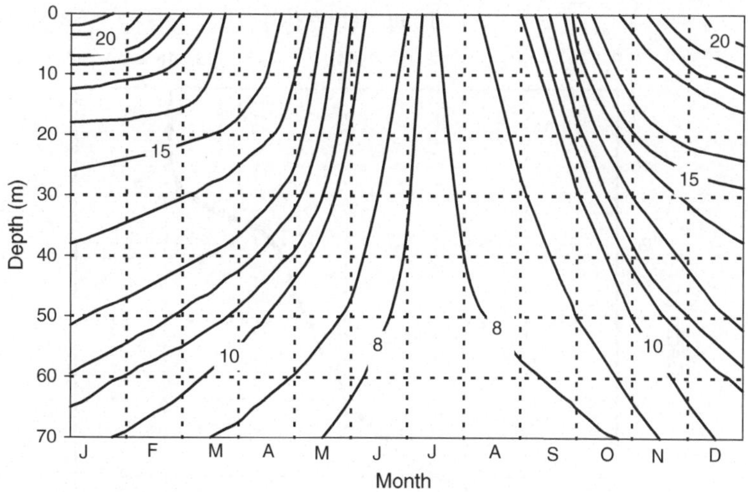

Fig. 12.29. Typical seasonal variation of temperature of the impounded water in the Colbún reservoir, Chile (values in °C).

the flow of the emerging waters fed by the reservoir. Of course, most of the relief wells showed vertical ascending flow.

The calculations made for the contribution of the local groundwater flow to the total water volume emerging at the relief wells provided the following values: 587, 664, 557, 734 and 649 l/s, respectively, for a total leakage flow rate of 1.2, 3.9, 2.2, 3.8 and 5.4 m^3/s. The mean contribution was 658 l/s. Obviously, the contribution of the local groundwater does not depend on the water level at the reservoir.

Temperature measurements Water temperature proved to be an excellent tracer for the leakage study. Figure 12.29 shows the annual temperature distribution in the reservoir. As expected, the reservoir becomes a well-mixed water body at the end of June due to the progressive cooling of the upper layer (density effect). Figure 12.30 shows a vertical profile of temperature measured in March 1986. This profile shows that water temperature at the deepest 10-m layer of the reservoir ranged from 9.5°C to 12°C. The temperature distribution of the water that was emerging at this time through the relief wells is shown in Fig. 12.31. In the same way, the temperature distribution measured at the bottom of the reservoir at the same time is shown in Fig. 12.32.

The comparison of these results leads to the conclusion that most of the water emerging at the relief wells infiltrates at the deepest layer of the reservoir. Temperatures lower than 11°C were present in the reservoir only below 375 m.a.s.l. Such a low temperature only existed within the investigated zone in a small area located at the confluence of the River Teatinos and the River Maule and along a stretch of the bed of the River Teatinos of about 400 m upstream of this point, as shown

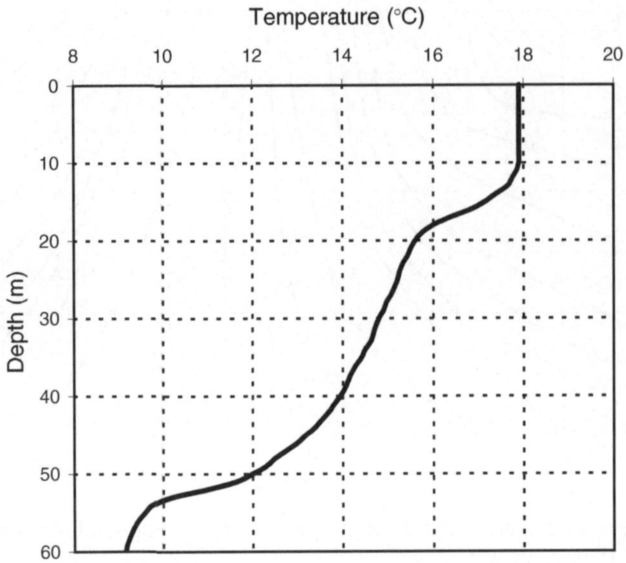

Fig. 12.30. Vertical profile of temperature in the Colbún reservoir, March 1988.

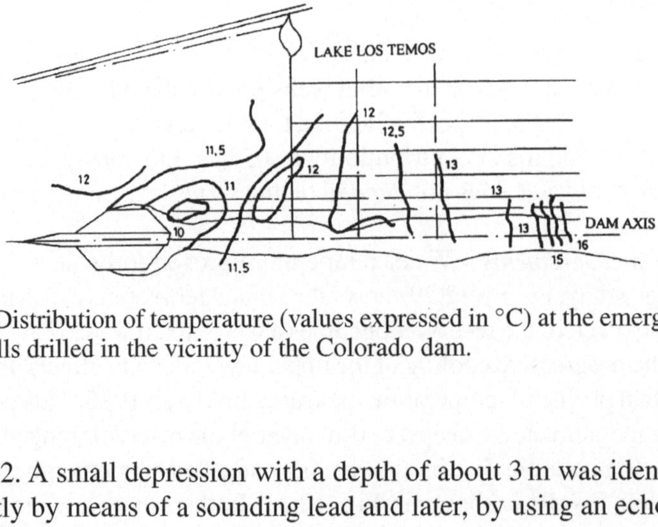

Fig. 12.31. Distribution of temperature (values expressed in °C) at the emerging waters in the relief wells drilled in the vicinity of the Colorado dam.

in Fig. 12.32. A small depression with a depth of about 3 m was identified in this stretch, firstly by means of a sounding lead and later, by using an echo sounder.

The relief wells showing a temperature of 11°C and below were those presenting δD-values close to −95‰, that is to say, the wells where pure reservoir water was emerging. These wells also showed the highest flow rates of the emerging waters. Emerging waters with a temperature higher than 11°C were formed through the mixing of reservoir water with locally recharged groundwater, the temperature of which was around 18°C. However, the contribution of other leaks from the top of the terrace, which was located immediately upstream of the dam, could not be fully disregarded. The temperature of this water was also approximately 17°C.

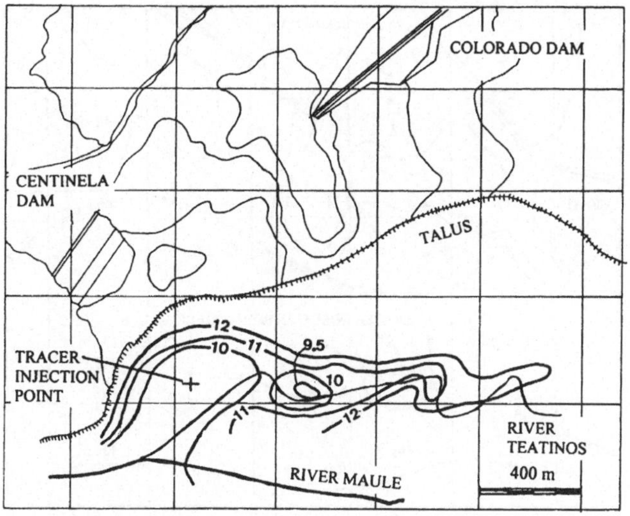

Fig. 12.32. Distribution of temperature of water at the bottom of the reservoir (values in °C).

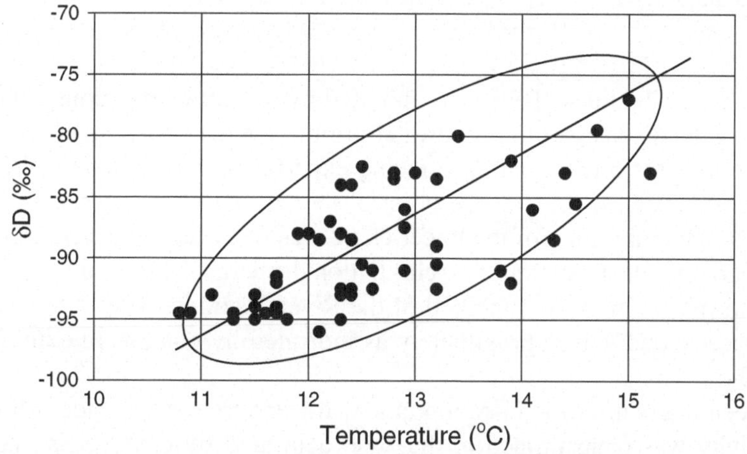

Fig. 12.33. Relationship between the isotopic deviation δD and temperature of the water emerging through the relief wells.

Figure 12.33 shows the correlation between the water temperature and the δD-values, which supports the previous statement. The dispersion of the points was mainly due to the nonconservative character of temperature.

Interconnection experiment between the reservoir and the emerging groundwater using ^{131}I as tracer Temperature data clearly indicate that the leak, or at least most of it, could only be taking place from the deepest areas of the reservoir located

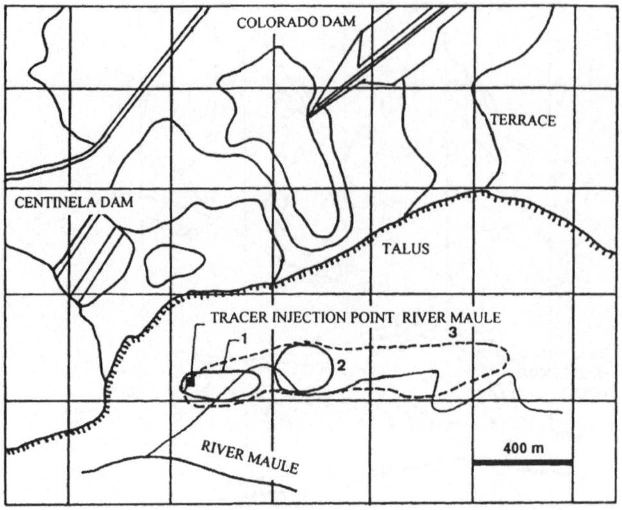

Fig. 12.34. Migration of the tracer cloud of the ^{131}I injected close to the bottom of the reservoir. Curve 1 was measured 3.5 h after the injection. Curve 2 was obtained 18 h after the injection. Curve 3 represents the maximum spread of the injected tracer. Most of the tracer was infiltrated at the zone represented by curve 2.

along the River Teatinos. This was supported by the above-mentioned statements concerning the discharge of groundwater prior to the dam construction.

To confirm this crucial point, a tracing experiment was undertaken, in which 4.7 Ci of ^{131}I were injected at the point of the bottom of the reservoir shown in Fig. 12.34. The migration of the tracer cloud was followed for several hours after the injection by means of NaI(Tl) scintillation detectors. This cloud moved to the east, following more or less the bed of the River Teatinos. After 3.5 h, the shape of the tracer cloud was approximately as indicated by contour line number 1 in Fig. 12.34.

Eighteen hours after the injection, a new tracer survey was made. Most of the radioactivity was concentrated at the zone delimited by contour line number 2 in Fig. 12.34, which coincides with the zone where the bottom depression had been identified 2 days before and where the lowest temperature values were found. Two additional radioactivity surveys were carried out, and similar results were obtained. Most of the materials of the bottom of the reservoir labelled with ^{131}I were concentrated at the above-mentioned area. It became clear that most of the tracer was infiltrated in the ground at this zone.

The recovery of the tracer at the emerging water was determined using the measuring technique based on the coprecipitation of the ^{131}I with nonradioactive AgI, which has been described in Section 7.6.3. The tracer was detected in the relief wells about 3 days after the injection. The breakthrough curve shown in Fig. 12.35 was obtained, on which the net counting rate (counts per minute, cpm)

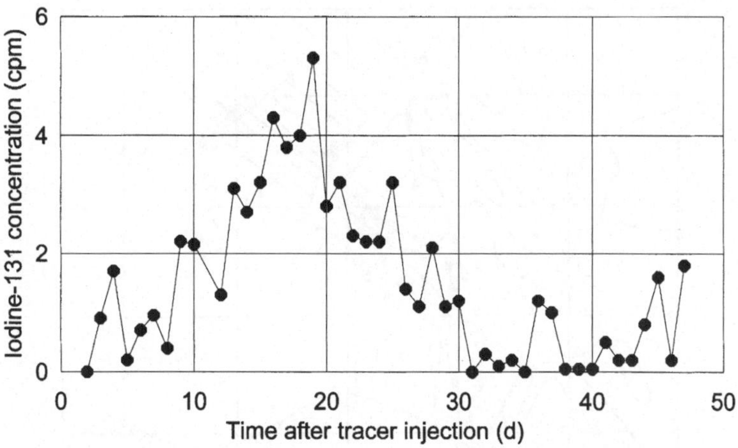

Fig. 12.35. Breakthrough curve measured at the emerging water representing the total flow after the injection of 4.7 Ci of ^{131}I at the point of the reservoir indicated in Fig. 12.34.

of the measured samples is plotted as a function of time. The background of the measuring system was kept very low (about 15 cpm) by using a shielding of the scintillation detector made with water and by using a measuring window covering only the photopeak of the gamma rays of 0.36 MeV of the ^{131}I. To get a low statistical counting error, the samples were measured for 30 to 60 minutes. All water samples were collected at a point where the total leakage was concentrated.

The large dispersion undergone by the tracer during its migration, first at the reservoir bottom and later through the rock formation, was responsible for the observed high dilution. Apart from this dilution, the low tracer concentration was also derived from the decay of the ^{131}I. The long transit time of the tracer between the reservoir and the relief wells was due to the fact that during the experiment, the water level at the reservoir was very low and the total flow rate of the leak was only about 1 m^3/s.

Interconnection experiments between the boreholes and the relief wells To investigate the flow pattern at the dam site, 12 interconnection experiments using fluorescent tracers (uranine and rhodamine B) were performed. The tracers were injected in a set of boreholes drilled at the dam site, which are numbered 1 to 9 in Fig. 12.36. All experiments except one gave a positive result. The tracer injected in a borehole was detected in a single relief well, indicating therefore the local flow direction. The results obtained are shown in Fig. 12.36. The marked lines indicate the flow direction. It can be observed that the flow lines on the left abutment intercepted, more or less, the area of the reservoir where the tracer ^{131}I infiltrated. These flow lines also corresponded to the highest emerging flow rate. However, the direction of the flow lines from the left to the right abutment becomes more and

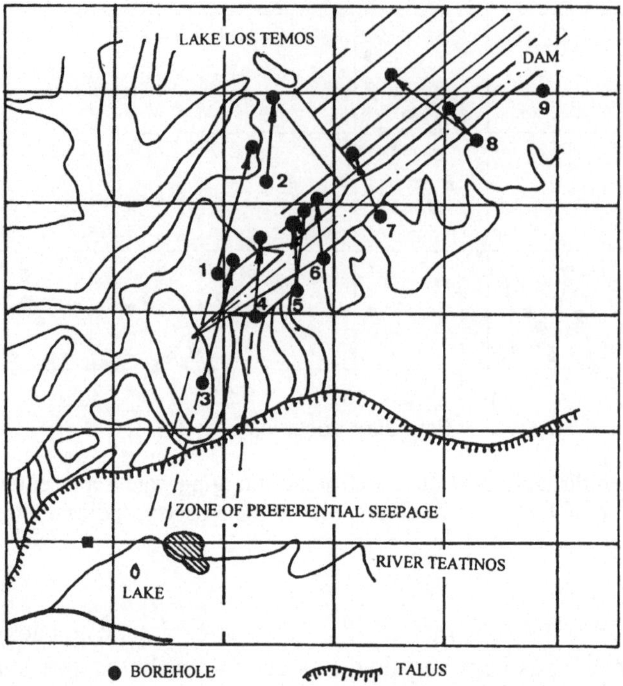

Fig. 12.36. Results of the 10 interconnection experiments carried out by injecting fluorescent tracers in boreholes of the dam site. The arrows indicate the flow direction and the relief wells at which the tracer was detected.

more perpendicular to the dam axis, suggesting that some leakage is also taking place in other points of the River Teatinos located at even longer distance upstream of the point of preferential tracer infiltration. However, the flow lines on the central and right parts of the dam could also be interpreted as leakage coming from the terrace located immediately upstream of the dam.

Flow tests in relief wells In April 1987, the water level in the reservoir experienced a significant decrease, and as a result, all relief wells situated along a line parallel to the dam axis ceased to discharge water. Taking advantage of this circumstance, flow tests were performed using ^{131}I as a tracer, according to the techniques described in Chapter 8 for the measurement of horizontal and vertical flows.

The results obtained with these flow tests showed that in spite of the fact that water was not emerging through the mouths of the boreholes, vertical ascending flows with very high velocity were present in most of the wells. The velocity of the ascending flow multiplied by the cross-section of the boreholes provided flow rates up to 25 l/s. The ingress of water to the wells was taking place at a depth corresponding to the elevation of the gravel horizon shown in Fig. 12.27. The outlet was taking place through the gravel layer added to the terrain after the appearance of the leaks to increase its stability. The casing of the relief wells was perforated with slots throughout the whole length. In eight of the relief wells, the velocity of the vertical ascending flow was so high that its measurement could be performed directly using a propeller current meter.

The flow tests were performed in 32 wells distributed all along the dam site. The results obtained with the tracer technique were in good agreement with those obtained using the propeller current meter. These results allowed the localisation of the permeable stratum responsible for the leakage. It was found that the average thickness of this stratum was only a few metres, and it was located between the elevations 378 and 385 m.a.s.l. At the zone close to the right abutment, this stratum was not completely horizontal. It was dipping slightly upward.

The flow tests for the measurement of vertical flow also showed that the drainage system constructed for the evacuation of the leak from the dam site was not fully effective. An important fraction of the water flowing at the bottom of the relief wells was not collected by the drainage system but continued flowing through the rock formation along the valley. Tracer injections made at the deepest points of the water column showed that the tracer never arrived at the top of the water column. However, when the detector was lowered to the injection point, the tracer was not there. It had abandoned the well immediately after the injection due to the horizontal flow prevailing at this depth. In other words, only the upper stretch of the permeable layer was contributing to the vertical ascending flow. This result meant that the total leakage was much higher than the value measured at the point where all emerging waters were collected. Obviously, the underground flow escaping from the dam site could emerge at any point along the valley. It happened shortly after conducting the mentioned flow tests. At a distance of about 5 km from the dam, several springs appeared, reaching, in a short period of time, a flow rate of 1 m^3/s.

REFERENCES

Florkowski, T. and Wotjas, J., 1979. Model experiment for tritium balance of surface water reservoir including molecular exchange. *Nukleonika*, 24(12/79), 1247–1252.

International Atomic Energy Agency (IAEA), 1983. *Guidebook on Nuclear Techniques in Hydrology*, Technical Reports Series No. 91, IAEA, Vienna, 441 pp.

Plata, A., Araguás, L., Avilés, J. and Peña, R., 2001. Relación entre el lago de Managua (Nicaragua) y las aguas subterráneas de su entorno. *Ing. Civil*, 121, 127–137 (in Spanish).

Plata, A., Gonfiantini, R. and López, A., 1994. Assessment of contamination risks of Asososca Lake, Nicaragua. In T. Suokko and J. Soveri (Eds.), *Future groundwater resources at risk (Proc. of the Helsinki Conference, June 1994)*. IAHS Publ. no. 222, Helsinki, pp. 177–188.

Plata, A. and Iragüen, E., 1992. Leakage study at lake Laja, Chile. In Hötzl and Werner (Eds.), *Tracer hydrology*. Balkema, Rotterdam, pp. 387–396.

Plata, A., Iragüen, E. and Heresi, N., 1989a. Estudio de las filtraciones de la presa "El Colorado" con técnicas isotópicas. In *Isotope hydrology investigations in Latin America*. International Atomic Energy Agency-TECDOC-502, pp. 169–189 (in Spanish).
Plata, A., Pérez, E. and Díaz, M.F., 1997. *Estudio de filtraciones en el embalse de Arquillo (Teruel)*. Internal report, CEDEX, Ministerio de Fomento, Madrid, 52 pp. (in Spanish).
Plata, A., Torres, A., Perez, O., Saint-Hilaire, J. and Bueno, E., 1989b. *Estudio de las filtraciones de la presa de Sabaneta, República Dominicana*. Final report IAEA, Vienna, Project DOM/8/002, 103 pp. (in Spanish).
Slattery, M.W. and Ingraham, N.L., 1994. Detritiation of water by isotopic exchange: experimental results. *Environ. Sci. Technol.*, 29, 1417–1421.

ANNEX 1

Calculation of under-dam seepage in case of isotropic granular medium

The under-dam seepage in the case of isotropic granular medium can be computed using a simple procedure assuming that the boundaries of the rock formation are well-defined. The method is based on the construction of a flow-net. The simplest case is illustrated in Fig. A1.1A, in which the permeable formation is embedded between two impervious layers considered of infinite dimension. Therefore, flow lines must be parallel to the contact surfaces and the equipotential lines must be perpendicular to the flow lines. The area between two flow lines is called a flow tube. If the distance between consecutive flow lines is identical, the flow rates through the different flow tubes are the same.

Let's consider the flow at the sector ABCD of Fig. A1.1A. Let's call *ds* and *dm*, respectively, the distances AB and BC, and *dh* the drop in the hydraulic head between AD and BC. According to Darcy's law, the flow rate through a unit of the cross-section is given by the following equation:

$$dQ = K\,\frac{dh}{ds}\,dm \tag{1}$$

where K is the permeability or hydraulic conductivity.

If the flow-net is constructed with a square form, *ds* is equal to *dm* and the previous equation is transformed into the following one:

$$dQ = K \cdot dh \tag{2}$$

For a system with m flow tubes, the total flow rate will be:

$$Q = m \cdot K \cdot dh \tag{3}$$

If the drop in hydraulic head referred to the whole area considered is H and the flow-net contains n cells equal to the cell defined by ABCD, $H = n \cdot dh$, equation (3) is transformed into:

$$Q = \frac{m \cdot K \cdot H}{n} \tag{4}$$

For example, in the case illustrated in Fig. A1.1A, $m = 3$, $n = 6$ and $H = 100-40 = 60$. Therefore, for a value of the hydraulic conductivity $K = 10^{-4}$ m/s, the flow

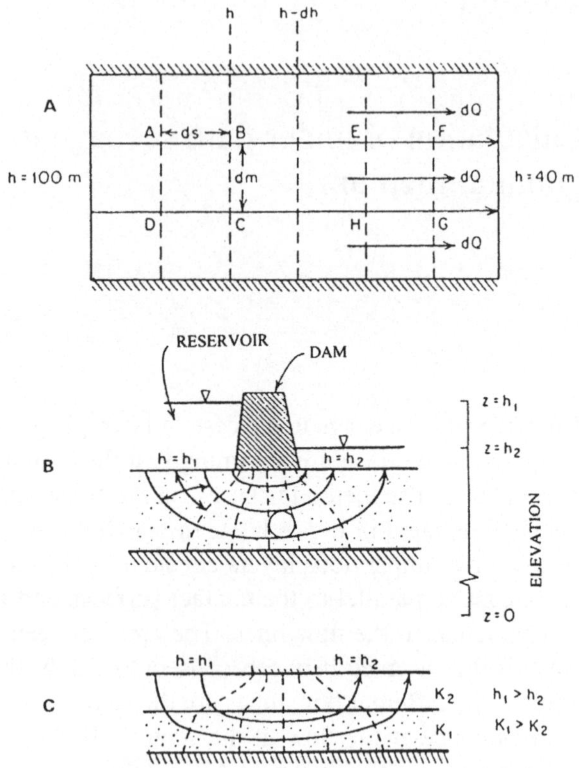

Fig. A1.1. Illustration of under-dam seepage.

rate is equal to $Q = 3 \times 10^{-3}$ m^3/s per metre of the cross-section perpendicular to the flow-net.

Obviously, the flow-net showed in Fig. A1.1A does not exist in the case of real dams. In the case of a dam constructed over an isotropic granular medium, impervious contacts of infinite dimension do not usually exist, and the flow-net has the form shown in Fig. A1.1B. The borders of the squares are not defined by straight lines but by curves. In this case, the flow rate through the permeable stratum is computed in the same way as described for the previous case but using the value $m = 3.5$ instead of $m = 3$.

Figure A1.1C shows the flow-net when two permeable layers are present under the dam, each of them characterised by a different hydraulic conductivity. In such case, the solution is more complicated but can be computed.

Polubarinova (1962) described a simple method for the calculation, based on the previous theory developed for the determination of under-dam seepage in the case of isotropic medium. The method is illustrated in Fig. A1.2, where q represents the seepage flux per metre of the dam length (l). The diagram shows the relationship between the parameter $q/(K \cdot \Delta H)$ and the ratio x/b, with x being the dam width at ground level and b the thickness of the permeable formation under the dam.

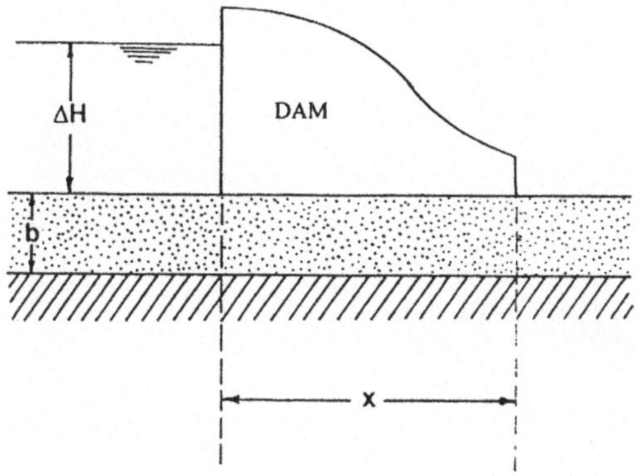

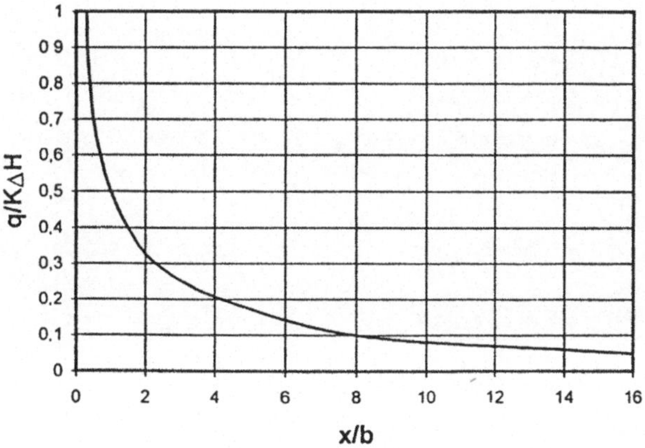

Fig. A1.2. Variation of the parameter $q/K\Delta H$ with respect to the ratio x/b.

For example, in the case of a dam with a width of 50 m and a thickness of the permeable formation of 20 m, the value of the ratio x/b is equal to 2.5. The corresponding $q/(K \cdot \Delta H)$ value, given by the diagram of Fig. A1.2, is 0.3. If $\Delta H = 50$, the value of q is given by $q = 0.3 \times 50K = 15K$. For a value of K equal to 10^{-5} m/s, a value of q equal to 0.15 l/s per metre of the dam length is obtained. For a dam with a length of 100 m (1 = 100 m), the seepage flow rate would be 15 l/s. Obviously, all these calculations are only a preliminary approximation.

REFERENCE

Polubanirova, K., 1962. *Theory of groundwater movement.* Princeton University Press, Princeton, NJ, USA.

ANNEX 2

Measurement of open flows using tracers

1 INTRODUCTION

The measurement of the flow rate of surface flows is a basic requirement for many hydrological studies, as well as for the planning and design of hydraulic structures, such as dams. The main methods available nowadays for this task are:

1) Overflow devices.
2) Methods based on the use of calibrated current meters.
3) Methods based on the use of artificial tracer.

The first method is mostly used for the measurement of small flow rates, for example, below a few cubic metres per second. The gauging station, when using this method, can be equipped with automatic level-meters, allowing continuous monitoring of the flow rate. This type of gauging stations is very commonly used for the continuous measurement of the flow rate of springs, such as those generated by a leak from a reservoir. The accuracy of this measuring technique is relatively small (between 5% and 10%, depending on the prevailing conditions). Apart from that, the overflow device can only be used for a specific flow rate range.

The second method is the most commonly used. However, it can involve serious difficulties when the cross section of the flow is very irregular or when dealing with flows of high velocity and turbulence.

The method based on the use of artificial tracers can be considered a complementary and alternative technique. However, in some cases, this technique can be the most appropriate and easiest to apply.

2 METHODS

Two methods are used for the measurement of flow rate in rivers and open channels using artificial tracers (Plata, 1972). These are the following:

1) *Continuous injection method*, in which a tracer solution of a known concentration is continuously injected into the flow for an appropriate period of time (usually, less than one hour) and at a constant rate (Joly, 1922; International Organization for Standardization, 1974).

2) *Point injection method*, in which a known mass of tracer is instantaneously injected in a point of the flow, and the integral of the passage or breakthrough curve is detected and analysed at a control station located downstream of the injection point.

3 THE CONTINUOUS INJECTION METHOD

3.1 Fundamentals of the method

As mentioned above, in this method, a tracer solution of a known concentration is continuously injected at a constant flow rate in a point of the measuring flow for a determined period of time. The permanent and steady state conditions are fulfilled by the flow. Under these circumstances, the tracer concentration measured at the downstream control point, located at a distance from the injection point far away enough to ensure good mixing of the tracer, will vary as a function of time according to curve number 1 shown in Fig. A2.1. After a certain period of time, a "plateau" or stretch of the passage curve with a constant concentration is achieved. The duration of this plateau will depend on the duration of the tracer injection, as well as on the distance and flow dispersivity between the injection and the measuring stations (Clayton and Smith, 1960, 1964).

The continuous injection method can be considered equivalent to many successive point injections of equal mass made at regular time intervals. Each individual injection will generate in the measuring station a breakthrough curve similar to the one indicated as number 2 in Fig. A2.1. The plateau at the measuring station will be established at the moment when the breakthrough curve corresponding to the first point injection has been completed. Obviously, the establishment of the plateau may require a long period of time if significant stagnant water exists in the stretch between the two stations. The presence of zones of stagnant water will involve a marked increase in the duration of the injection and, therefore, of the mass of tracer to be injected. For this reason, a reach of the flow should be selected in which the presence of zones of stagnant waters is minimal.

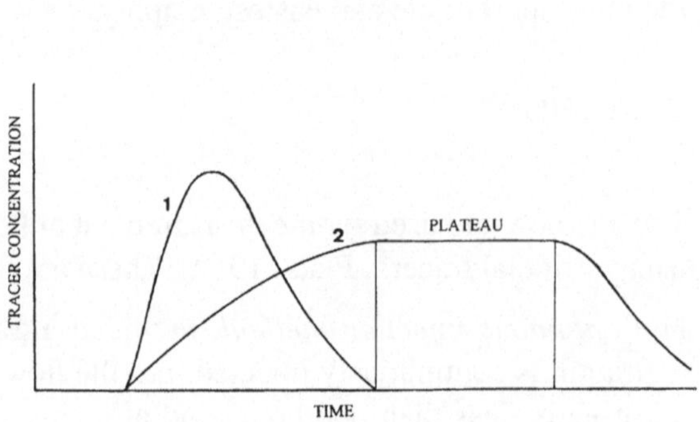

Fig. A2.1. Typical breakthrough curves obtained in the case of a point injection of a tracer (1) and a continuous injection at a constant flow rate (2).

The tracer concentration corresponding to the plateau of the passage curve will decrease at increasing flow rates because of the parallel increase of the tracer dilution. Let's call C_1 and C_2, respectively, the tracer concentrations of the injected solution and the mean value of samples corresponding to the plateau. The dilution factor will be given by the ratio C_1/C_2. If Q is the measuring flow rate and q the flow rate of the tracer injection, the following equation can be written:

$$\frac{Q}{q} = \frac{C_1}{C_2} \tag{1}$$

The concentration C_2 is determined by collecting several water samples from the surface flow when the plateau is established.

3.2 Basic conditions

The continuous injection method requires the following conditions to be fulfilled:

1) Flow to be gauged must be stationary and steady state.
2) Good mixing condition must be attained, which in this case means to have the same tracer concentration at any point of the cross section at the sampling station during the duration of the plateau. To check this condition, water samples collected from different points of this cross section have to be measured.
3) The duration of the injection has to be long enough for the establishment of the plateau. To reduce the amount of tracer to be used, the distance between the injection and the sampling stations should be chosen to be as short as possible, assuming that both conditions are fulfilled. Increasing distances between the stations involved in the gauging require longer duration of the injection to establish the plateau and, therefore, higher tracer consumption.
4) No significant tracer losses must exist between the injection and the sampling stations. In case of tracer losses due to branching of the main flow, the injection has to be performed at a point located upstream at a distance from this branching point long enough to fulfil the good mixing condition.
5) If a tributary joins the main flow between the injection and the sampling stations, the good mixing and plateau conditions have to be fulfilled for the stretch downstream of the confluence of both flows. The measured flow rate will then correspond to the sum of these two flows.

The tracer can be injected using a constant flow rate pump or the constant level cell described in Section 4.3. For example, a synchronic pump fitted with a motor can be used, whose turning velocity is governed by the frequency of the electrical direct current.

4 THE POINT INJECTION METHOD

4.1 Fundamentals of the method

In this method, a known mass M (or activity A, if a radioactive tracer is used) is injected in a point of the flow. The tracer concentration at any point downstream of the injection point will vary as a function of time according to a bell-shaped breakthrough curve similar to curve number 2 shown in Fig. A2.1. The tracer balance at the measuring or sampling station can be written in the following way:

$$M = \iint \left[dQ \int_0^{\infty} C(t)\, dt \right] dS \tag{2}$$

where $\int C(t)dt$ represents the area of the breakthrough curve and dS is a flow tube of the cross section at the measuring station through which a flow rate dQ is flowing. If this integral is kept constant for the entire measuring cross section and the flow rate Q is also kept constant during the passage on the tracer cloud, the previous equation is simplified in the following form:

$$M = \int_0^{\infty} C(t)\, dt \iint dQ\, dS = M \int_0^{\infty} C(t)\, dt \text{ and therefore:}$$

$$Q = \frac{M}{\int C(t)\, dt} \tag{3}$$

If the average tracer concentration of the breakthrough curve is designated by C_m, the integral expression $\int C(t)\, dt$ can be replaced by the product $C_m \cdot t$ and equation (3) is modified in the following form:

$$Q = \frac{M}{C_m \cdot t} \tag{4}$$

It is important to observe that the value of C_m can be referred to any period of time with the only condition that the full breakthrough curve is covered. The longer is the time, the lower will be the concentration, but the product $C_m \cdot t$ will remain constant because water collected before and after the time interval corresponding to the breakthrough curve will not contribute with additional mass of tracer.

4.2 Basic conditions

This method requires that all conditions mentioned in Section 3.2 for the constant rate injection method be fulfilled, except condition 3. The good mixing condition requires, in both methods, that any tracer molecule injected in the flow has the same probability to cross, at the measuring or sampling station, an element or flow tube with cross section dS. It can be proved that the condition $\int C(t)\, dt = constant$ is necessary and sufficient to fulfil the good mixing condition (Alquier et al., 1970). To check that the good mixing condition has been fulfilled in a particular case, the value of the expression $\int C(t)\,dt$ must be determined at two or three different points of the measuring or sampling station.

4.3 Methods for determining the value of the integral $\int C(t)\,dt$

Obviously, the measuring technique for the tracer injected in the current will depend on the type of tracer. However, independently, the value of the expression $\int C(t)\,dt$ can be obtained by means of the procedure described below.

1) Sampling at a constant flow rate In this method, water is collected at the sampling station with a constant flow rate during the full period of passage of the tracer cloud. The collected water is placed in a single container, and the average tracer concentration C_m is measured. The flow rate is obtained using equation (4). The time t corresponds to the period in which the sampling took place.

Basic conditions of the method are that the sampling flow rate is kept constant during the whole period of time and that the sampling duration covers the entire breakthrough curve. Obviously, the fraction of tracer collected during pumping is equal to the flow fraction collected, that is to say:

$$\frac{Q}{q} = \frac{M}{m} \tag{5}$$

where q is the sampling flow rate and m the mass of the collected tracer.

Several authors (Baonza and Plata, 1967a, b; Dinçer, 1967; Guizerix et al., 1964) have used a sampling method based on the use of a constant level cell such as the one shown in Figs. A2.2 and A2.3. River water is pumped through this cell using, for example, a small peristaltic pump. The cell provides a constant flow rate, the value of which depends on the diameter d of the lower disc and the height of the water column. The flow rate provided by the pump is slightly higher than in the previous case so that the level at the cell is kept constant all the times. The excess of water is poured away. This constant level cell can also be used for the tracer injection when using the constant rate injection method previously described.

2) Sampling at regular time intervals For more than 20 years, this has been the method commonly used by the authors for the measurement of small and medium flow rates using tracer techniques. The integration of the breakthrough curve is carried out by collecting from the flow small water samples of equal volume at regular time intervals. For example, samples of 50 ml collected every 15, 30 or 60 s, depending on the expected duration of the breakthrough curve. The volume of the samples does not need to be known. The only conditions to be fulfilled are that all samples have equal volume and the time interval between consecutive samples is always the same. These samples are mixed together in a 5- or 10-l plastic bottle. If the time intervals between samples are kept short relative to the duration of the breakthrough curve, the average tracer concentration in the container where

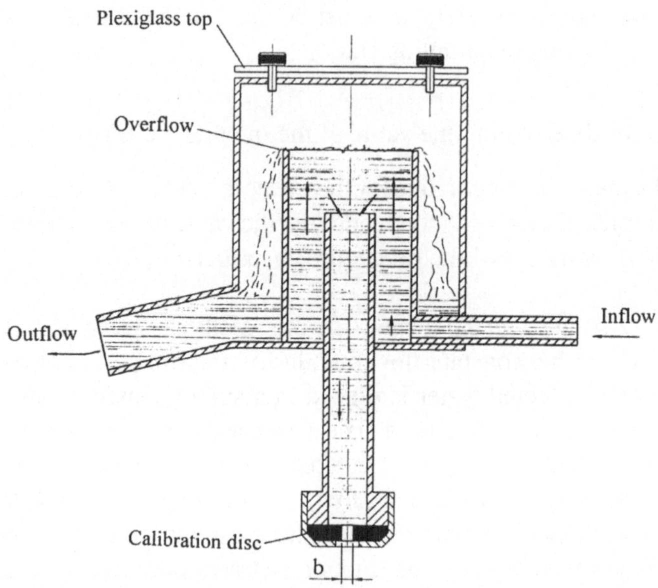

Fig. A2.2. Illustration of a constant level cell.

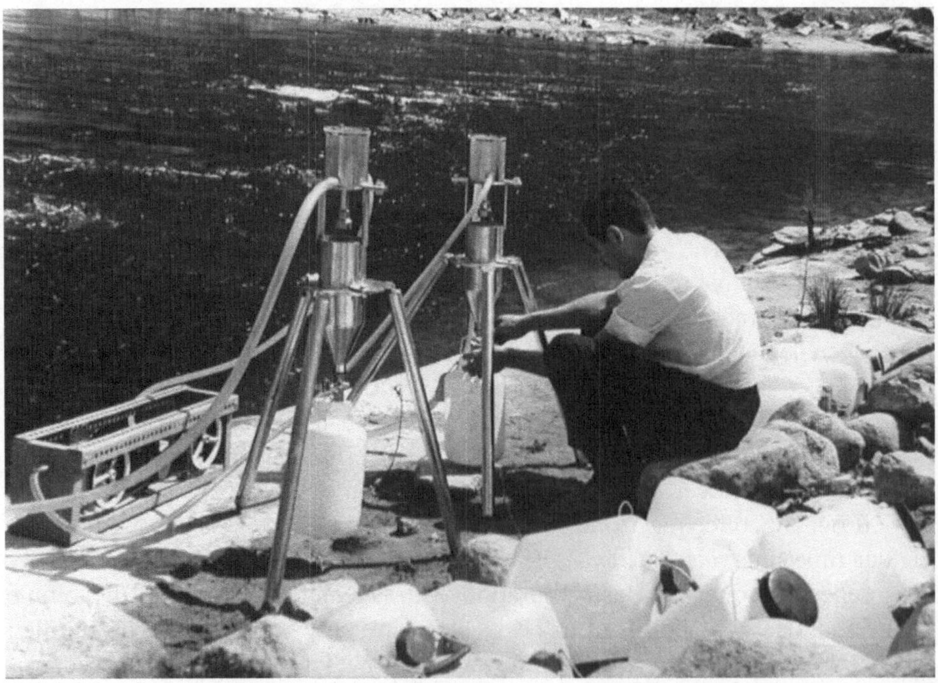

Fig. A2.3. Collecting water samples from a river at a constant flow rate, using constant level cells and a peristaltic pump.

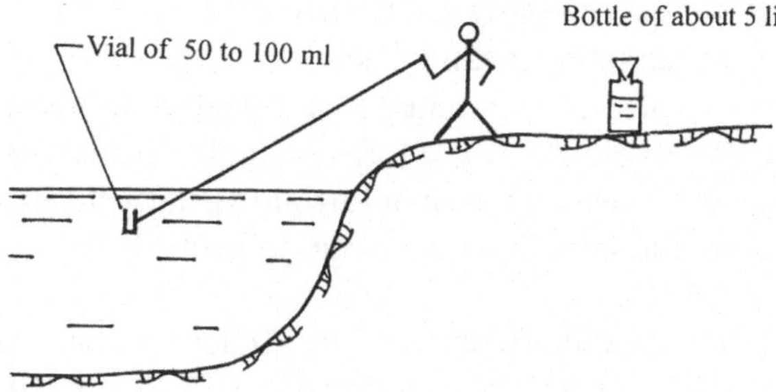

Fig. A2.4. Illustration of the flow rate measurement technique based on the collection of discrete water samples of equal volume at regular time intervals.

Fig. A2.5. Measuring the flow rate at the discharge of the Sermo reservoir, Indonesia, using the method based on the collection of discrete water samples of equal volume at regular time intervals.

all samples are collected is the same as the one obtained with the continuous sampling method described above. The main advantage of this procedure is the great simplicity of the method. The samples are collected with small plastic bottles or vials fixed to the end of a pole or stick as shown in Figs. A2.4 and A2.5. No pumps are necessary to conduct the flow gauging. Obviously, to check the fulfilment of the good mixing condition, the sampling has to be performed at two or three different points along the cross section.

The method requires the sampling to be performed covering the whole breakthrough curve. Usually, sampling is started before the tracer arrives at the sampling station and is finished when it is assumed that a negligible tracer concentration remains in the flow. If a dye tracer is used (for example, rhodamine B or WT), the moments of starting and finishing the sampling are visually determined. If a non-coloured tracer is used, the sampling control has to be made by submerging in the flow an appropriate detector.

Very frequently, when dealing with small flow rates or flows with high turbulence, the amplitude of the breakthrough curve is very small. The tracer passes through the sampling station in a matter of minutes and the operator does not have enough time to make a representative sampling. In such a case, the tracer injection is not injected instantaneously but slowly, for example, over a period of several minutes. Obviously, the way that the tracer is injected does not play any role. The integral of the breakthrough curve will be always the same because it will depend only on the mass of injected tracer and the flow rate.

It is important to mention that in many cases the mixing of the tracer is facilitated if it is distributed over the entire cross section of the flow. Of course, this is not true from a theoretical point of view but it is from a practical point of view. For example, if the sampling is made at three points of the flow, the dispersion of the individual flow rate values obtained with the three samples can be much lower with this multipoint injection method.

The method described above, as well as the one previously described, allow the utilisation of tracer enrichment techniques to reduce the amount of tracer needed for the flow rate measurement. This option may be important when dealing with very high flow rates. The enrichment method will depend on the type of tracer. In the case of a fluorescent dye tracer, the enrichment can be done using the techniques described in Section 7.5.3. If using ^{131}I as a tracer, the method described in Section 7.6.3 based on the coprecipitation of the radioisotope in the form of AgI can be applied.

It is also important to mention that both methods do not require the determination of the absolute tracer concentration in the collected water sample. It is only necessary to determine the relative concentration with respect to the tracer solution injected. Usually, the tracer to be injected is dissolved or diluted in a known water volume. The measurement of the tracer concentration in this solution requires a dilution process of an aliquot to get a solution with the appropriate concentration. The concentration of this diluted solution is compared with the composite water sample collected from the flow. Both the dilution factor and the volume of injected solution have to be known.

3) Total count method This method can be applied when a gamma ray emitter is used as a tracer, for example, ^{131}I (Hull, 1958, 1962a, b; Hull and Macomber, 1958). The detection of the tracer is made in place by means of a submerged scintillation detector. The value of the expression $\int C(t)\,dt$ is determined by summing up the pulses produced by the detector in a digital scaler during the passage of the tracer cloud through the measuring station. The detector can be submerged in the measuring flow with the required saturation geometry or can be placed in the centre of a container in which a small flow is diverted from the main flow using a pump. The detector must be previously calibrated in the laboratory so that the counting rate (counts per minute, cpm) can be transformed into concentration units ($\mu Ci/m^3$). If ^{131}I is used as a tracer, the saturation geometry is obtained by placing the detector in the centre of a container large enough to ensure that the detector has in all directions at least 45 cm of water spiked with the radioactive solution. This means that the counting rate obtained for the same tracer concentration will also be the same, independent of the thickness of the calibrated solution, assuming it is higher than 45 cm. Obviously, the place where the detector is submerged in the river must have at least 45 cm of water in all directions. If the measurement is made with the detector placed inside a derivation container, the calibration is made by filling this container with a radioactive solution of a known concentration. In both cases, a calibration factor is obtained, which is expressed in cpm per microcurie per litre of concentration ($\text{count} \cdot \text{minute}^{-1} \cdot \text{litre} \cdot \mu\text{Ci}^{-1}$).

Obviously, the total number of pulses, N, accumulated by the scaler during the passage of the tracer cloud will be directly proportional to the injected tracer activity A and inversely proportional to the flow rate Q, so that the following equation can be applied:

$$N = \frac{A}{Q} \cdot F \tag{6}$$

where F is the above-mentioned calibration factor.

5 COMPARISON OF BOTH METHODS

Experimental difficulty The continuous injection method requires the use of a relatively sophisticated injection system, whereas the sampling technique is very simple. The contrary occurs in the case of the instantaneous injection method. When using a radioactive tracer, the first method may create some radiological protection problems because a radioactive solution with a high concentration has to be handled throughout the long period of time that is required for the injection.

Amount of tracer needed Let us call C_{min} the minimum tracer concentration that can be measured with the required accuracy. The mass of tracer M that has to be injected in the surface flow to reach this average concentration with the instantaneous injection method is given by the following equation (see equation (4)):

$$M_1 = Q \cdot C_{min} \cdot t_1 \tag{7}$$

where t_1 is the passage time of the tracer cloud.

In the case of the continuous injection method, a volume V of tracer solution has to be injected, which is equal to $q(t_1 + t_2)$, with t_2 being the duration of the plateau. Therefore, a mass of tracer $M_2 = V \cdot C$ has to be injected. Therefore, the following expression can be written:

$$\frac{Q}{q} = \frac{M_2}{q(t_1 + t_2) \cdot C_{min}} \quad (8)$$

Then the ratio M_1/M_2 is given by the following expression:

$$\frac{M_1}{M_2} = \frac{t}{t_1 + t_2} \quad (9)$$

This expression proves that a higher mass of tracer is required for the continuous injection method.

Accuracy The continuous injection method offers some advantages from the point of view of the accuracy that can be obtained. This method allows a better check of the fulfilment of the good mixing condition. A large number of water samples can be collected from the river at different times to check or verify the establishment of the plateau and at different points of the cross section during the duration of the plateau to check the good mixing condition of the tracer. The average tracer concentration of all these samples will provide the figures needed to calculate the flow rate of the stream. The dispersion of such values will provide information regarding the fulfilment of the good mixing condition as well as the slope of the plateau. Only samples corresponding to a slope equal to or close to zero, that is, showing a constant tracer concentration, should be accepted. Apart from that, the average value of the parameter q (injection flow rate) can be very precisely determined by weighting the injection solution both at the beginning and at the end of the experiment.

6 GOOD MIXING DISTANCE

The minimum distance between the injection and the measuring or sampling stations that is necessary to fulfil the good mixing condition in the case of open flows is strongly influenced by the hydraulic and geometrical characteristics of the flow in this reach. As a general rule, a reach with high turbulence should be selected. The presence of narrow stretches and small waterfalls between these stations will facilitate the mixing of the tracer with the flowing water. The contrary can be said in areas with stagnant waters. Moreover, the injection technique plays an important role, as mentioned below.

Obviously, the shorter the good mixing length, the smaller the amount of tracer to be injected in the flow. However, very frequently, logistic conditions, such as the access to the river, presence of a bridge, etc. play an important role in the definition of the reach where the measurement is to be performed. In other cases, the flow measurement has to be carried out in a specific reach due to hydrological reasons.

The fulfilment of the good mixing condition can be checked only experimentally. When measurements are to be repeated in the same reach, the results obtained in previous measurements can assist in ensuring that the correct method is used. In the successive campaigns, measurement can be performed using a single measuring or sampling point in the cross section.

Several equations have been proposed for the calculation of the good mixing distance. However, these equations provide only a preliminary estimate. One of these equations, based on experimental data, was proposed by Hull (1958). The good mixing length L is given by the following equation:

$$L = k \cdot Q^{1/3} \tag{10}$$

where k is a constant; a value of 50 is used for k if the injection is made in the centre of the flow, and 200 is used if it is made at the bank. This equation underestimates the value of the good mixing length, which, on the other hand, can not be exclusively computed from the flow rate.

A second equation suggested by Guizerix and Florkowski (1983) is the following:

$$L = 10\sqrt{\frac{b^3}{h}} \tag{11}$$

where b and h are, respectively, the average values of the width and depth of the flow in the selected reach. The main problem, in this case, is the determination of these parameters.

Finally, the Rimar equation has been suggested, which, in spite of being an empirical equation, has, to some extent, hydraulic justification:

$$L = \frac{0.13 \cdot b^2 \cdot C \cdot (0.7C + 2\sqrt{g})}{g \cdot h} \tag{12}$$

where C is the Chezy coefficient, the value of which ranges from 15 to 50:

$$C = \sqrt{\frac{8g}{\lambda}} \tag{13}$$

where λ is the coefficient of the universal formula for the loss of hydraulic head per unit of length (Δh) given by:

$$\Delta h = \frac{\lambda - v}{8R \cdot g} \tag{14}$$

with v being the mean velocity of the flow, g the acceleration due to gravity and R the hydraulic radius.

Some authors (Guizerix and Florkowski, 1983; Clayton and Smith, 1964) have found that this equation underestimates the value of L when the width of the flow is above 50 m.

In the case of rivers, the vertical mixing of the tracer is produced more rapidly than the transversal mixing. This is the reason why the good mixing length strongly depends on the width of the flow channel. The mixing length is much shorter when the injection is made in the centre of the flow than on the shores. Good results can be obtained, even with distances between the two stations shorter than the one defined by the good mixing length, when the tracer is distributed along the flow width, especially if higher tracer amounts are injected in the points with higher flow velocities (Baonza and Plata, 1967a, b). Of course, the good mixing length does not depend on the injection method. Nevertheless, practical experience shows that the differences between the flow rate values computed from the data collected at different sampling or measuring points are much smaller when this multipoint injection method is applied.

7 TRACERS

Many tracers have been used for flow rate measurement in open flows (Dumas, 1943, 1953; Guizerix and Florkowski, 1983; Guizerix et al., 1964). For many years, rhodamine WT has been the most commonly used tracer by the authors. The fact of being supplied in a liquid form facilitates the easy and precise measurement of the volume of tracer solution to be injected. This tracer can be used in a wide range of flow rates, from a few l/min to about 500 m^3/s. Uranine is not recommended due to its high photodecomposition.

Table salt can be used in the form of a saturated solution or even in a solid form to measure flow rates up to about 200 l/s. The measurement can easily be performed using a conductivity meter. If a saturated salt solution is used, the mass of salt is determined by measuring the electrical conductivity of calibration samples prepared by diluting in distilled water an aliquot of this solution with a factor close to 100. The net conductivity of this sample will be between 3 and 5 mS/cm. If solid table salt is used, the calibration is made by measuring the conductivity of a solution prepared by dissolving a known mass of salt into a fixed volume (for example, 5 g of salt dissolved in 1 l of distilled water).

Artificial tritium in the form of tritiated water is used in exceptional cases for the measurement of very high flow rates (above 500 m^3/s), when the conventional methods cannot be applied. In fact, the authors never used tritium when the only objective was to determine the flow rate. Only when other flow parameters are necessary, such as transit time, dispersion parameters, water losses, etc., and no other possibility exists is the use of tritium considered.

Sodium dichromate has been used in the past for flow rate determination up to about 50 m^3/s (André, 1960). The characteristics and measuring procedure of this tracer have been described in Chapter 7. Nowadays, the utilisation of this tracer has been abandoned due to its high toxicity.

8 TRACER AMOUNT

The tracer amount necessary for flow rate measurement can easily be calculated by using an inverse computing procedure, as shown below. The calculation requires the estimation of the value of the flow rate to be measured as well as the passage time of the tracer cloud at the measuring or sampling station. Both estimations do not create any problem for technicians familiar with this type of work. Only a rough estimation is necessary. The duration of the passage or breakthrough curve can even be estimated using the dispersion coefficients for open flow existing in the literature and assuming a breakthrough curve with a Gaussian shape. The calculation procedure is better illustrated by means of two examples.

Example 1 Let us suppose that the flow rate to be measured is about 100 m^3/s and the duration of the breakthrough curve will be about 20 min. The flow rate determination will be made using rhodamine WT as a tracer, and the method based on the collection of discrete water samples of equal volume at regular time intervals is going to be used (equation (4)). To be sure that the sampling will cover the full passage of the tracer cloud, a sampling time of 30 min (1800 s) will be adopted.

On the other hand, the available fluorometer allows the precise measurement of a concentration of rhodamine WT equal to 10 μl/m^3. According to equation (4), the volume of rhodamine WT to be injected will be:

$$V = 100 \times 10 \times 1800 = 1.8 \times 10^6\ \mu\text{l} = 1.8\ \text{l}$$

Example 2 Let us calculate the maximum flow rate that can be measured using 50 kg of table salt as a tracer, using the same method of the previous example. The tracer will be directly injected in the flow in solid form to avoid the need for handling a high volume of tracer solution. A calibration sample is prepared by dissolving, for example, 2.3115 g of salt in a volume of one litre, using distilled water. The electrical conductivity of this solution is measured using the same conductivity meter to be used for the measurement of the water samples collected from the river in order to keep the measuring error low. A conductivity value of 3.45 mS/cm is obtained. The conductivity of the water used for the preparation of the calibration sample was 12 μS/cm. Therefore, the net conductivity of this sample amounts to $3.45 - 0.012 = 3.438$ mS/cm.

The duration of the breakthrough curve is estimated to be only 12 min (720 s) because of the small value of the flow rate.

Let's suppose that the electrical conductivity of the flow to be measured is 350 μS/cm. Therefore, the electrical conductivity of the river water sample equal to 1 mS/cm (1000 μS/cm) is considered acceptable. The net conductivity of this water sample will be 1 − 0.35 = 0.65 mS/cm, and the salt concentration will be:

$$0.65 \times 2.3115/3.438 = 0.437\ \text{g/l}.$$

Under these conditions, the maximum measurable flow rate will be:

$$Q = (50 \times 10^3)/(0.437 \times 720) = 158.9\ \text{l/s}$$

REFERENCES

Alquier, M., Gruat, J., Courtois, G. and Sauzay, G., 1970. La notion de bon melange dans l'eau, emploi de traceurs vitesse debitante et vitesse du centre de gravite d'un nuage de traceurs dans l'etude d'un ecoulement. In *Isotope Hydrology 1970*. International Atomic Energy Agency, Vienna, pp. 441–462 (in French).

André, H., 1960. Méthode chimique de dilution. Procédé par intégration. *La Houille Blanche, no. special B*, 833–843 (in French).

Baonza, E. and Plata, A., 1967a. Aplicación de los isótopos radiactivos a la medida de caudales superficiales de agua. *Revista de Obras Públicas*, 1199–1216 (in Spanish).

Baonza, E. and Plata, A., 1967b. Informe sobre el estudio y ensayo de las aplicaciones de los isótopos radiactivos a la medida de caudales superficiales de agua. *CEDEX, GANOP, Internal report, EN-7 y EN-17*, 174 pp. (in Spanish).

Clayton, C.G. and Smith, D.B., 1960. The accurate measurement of turbulent flow in pipes using radioactive isotopes—the isotope dilution method. *Proc. of a Symposium on Flow Measurement in Closed Conduits*, H.M. Stationery Office, Glasgow, Scotland.

Clayton, C.G. and Smith, D.B., 1964. A comparison of radioisotope methods for river flow measurements. In *Radioisotopes in Hydrology*. International Atomic Energy Agency, Vienna, pp. 1–24.

Dinçer, T., 1967. Application of radiotracer methods in streamflow measurements. In *Isotopes in Hydrology.* International Atomic Energy Agency, Vienna, pp. 93–110.

Dumas, H., 1943. *Jaugeages de cours d'eau par la méthode chimique*. International Association for Hydraulic Structures Research, troisième réunion, 5, 6 and 7 September (in French).

Dumas, H., 1953. La méthode chimique pour la mesure du débit des cours d'eau. *La Houille Blanche*, Oct.–Nov. 1952, juin-juillet 1953 (in French).

Guizerix, J. and Florkowski, T., 1983. Streamflow measurements. *Guidebook on nuclear techniques in hydrology*, Technical Reports Series No. 91, International Atomic Energy Agency, Vienna, pp. 65–79.

Guizerix, J., Granclèment, G., Hours, R., Andre, H., Wolf, R. and Perez, R., 1964. Les mesures de débits effectuées en France a l'aide de traceurs radioactifs par la méthode d'intégration. In *Radioisotopes in Hydrology*. International Atomic Energy Agency, Vienna, pp. 255–282 (in French).

Hull, D.E., 1958. The total-count technique: A new principle in flow measurement. *Int. J. Applied Radiat. Is.*, 4, 1–15.

Hull, D.E., 1962a. Total-count flow measurements in water resources research. *Am. Nucl. Soc. Trans.*, 5, 273–274 (summary only).

Hull, D.E., 1962b. Dispersion and persistence of tracer in river flow measurement. *Int. J. Appl. Radiat. Is.*, 13, 63–70.

Hull, D.E. and Macomber, M., 1958. Flow measurements by the total count method. In *Proc. 2nd U.N. Intern. Conf. on the Peaceful Uses of Atomic Energy*, 19, 324–332.
International Organization for Standardization, 1974. *Liquid flow measurement in Open Channels—Dilution methods for measurements of steady flow*. ISO-555 (1973), ISO-555/II (1974), 18 pp.
Joly, J., 1922. On a new method of gauging the discharge of rivers. *R. Dublin Soc. Sci. Proc.*, 16, 489–491.
Plata, A., 1972. *Isótopos en hidrología*. Editorial Alhambra, Madrid, 328 pp. (in Spanish).

ANNEX 3

Isotopic balance of lakes and reservoirs

1 THE CRAIG–GORDON EVAPORATION MODEL

As mentioned in Chapter 6, during evaporation of water from surface water bodies, the water vapour released to the atmosphere is depleted in the heavy molecules DHO and $H_2{}^{18}O$ because of their lower vapour pressure with respect to the more abundant light molecules $^{1}H_2{}^{16}O$. The evaporation process is rather complex. The model proposed by Craig and Gordon (1965), which has been adopted by most investigators, assumes that the evaporation is produced according to the following four stages:

1) Water vapour in equilibrium with the liquid phase is first released from the surface of the evaporating water body. An almost saturated sublayer is formed in the interface of air–liquid. The vapour produced in this way is depleted in the heavy molecules according to the fractionation factor for equilibrium conditions α_e. Therefore, we can write:

$$\alpha = \frac{R_i \, \textit{in the liquid phase}}{R_i \, \textit{in the vapour phase}} \tag{1}$$

   where R is the ratio between molecules containing heavy and light isotopes and the subscript i defines the relevant heavy molecule.
2) Water vapour escapes from this sublayer and reaches a second sublayer of the atmosphere, where the vertical migration of the water molecules is governed by the molecular diffusion according to a laminar flow. The vertical transport of the water molecules inside this second sublayer produces an additional depletion of the isotopic composition of water due to the fact that the diffusion coefficients of the heavy molecules (D_i) are lower than the diffusion coefficient of the light molecules (D_l). It has been proven that the isotopic depletion of natural water bodies derived from this process is approximately given by the ratio $(D_i/D_l)^{1/2}$.

This fractionation process is called kinetic fractionation and can be expressed by the following equations:

$$\Delta\varepsilon = \varepsilon_k(1-h); \quad \text{being} \quad \varepsilon_k = \sqrt{\frac{D_i}{D_l}} - 1 \tag{2}$$

where h is the relative humidity of the atmosphere at the beginning of the zone with turbulent flow.

3) Water vapour moves to a zone of the atmosphere where turbulent flow dominates and is mixed with a vapour fraction formed in other regions. No isotopic fractionation is produced in this zone.
4) A fraction of the water vapour from the zone with turbulent flow also interacts with the molecular diffusion layer and condenses on the water surface. This process is designed as a molecular exchange between the water body and atmosphere.

2 ISOTOPIC COMPOSITION OF THE NET EVAPORATION FLUX

According to the Craig–Gordon evaporation model, the net flux of vapour crossing the molecular diffusion layer, that is to say, the net evaporation E, obeys the Fickian diffusion law. Therefore, it is proportional to the difference between the vapour concentration at the lower sublayer (air saturated sublayer) and that of the lowest part of the zone with turbulent flow. Therefore, it can be written:

$$E = (C_S - C_T)/\rho = C_S(1-h)/\rho \tag{3}$$

where ρ is a resistance coefficient, C_S is the saturation concentration of water vapour at the interface of air–liquid (lower sublayer) and C_T is the concentration of water vapour in the turbulent flow zone. The relative humidity is given by the ratio $h = C_T/C_S$.

In the same way, for molecules containing heavy isotope species, as indicated by the corresponding subscript i, it can be written:

$$E_i = (C_S R_S - C_T R_T)/\rho_i = C_S\left(\frac{R_L}{\alpha} - h \cdot R_T\right)\Big/\rho_i \tag{4}$$

where R_L, R_S and R_T are, respectively, the isotopic ratios (D/H or $^{18}O/^{16}O$) for the upper layer of the water body, for the lower sublayer saturated in water vapour, and for the turbulent zone of the atmosphere. The term α is given by the ratio $\alpha = R_L/R_S$, which corresponds to the fractionation factor between the liquid and vapour phases under equilibrium conditions.

Let us consider the case of a closed water body, that is to say, with neither inflow nor outflow, in which the only existing water losses are produced by evaporation. In this case, we have $E = dN_l/dt$ and $E = dN_h/dt$, where N_l and N_h are, respectively, the number of light and heavy molecules existing in the liquid phase. The quotient between equation (4) and equation (3) gives:

$$\frac{dN_h}{dN_l} = R_E = \frac{(R_L/\alpha) - h \cdot R_T}{(1-h)(\rho_i/\rho)} \tag{5}$$

where R_E represents the isotopic composition of the net evaporating flux. The term ρ_i/ρ is approximately equal to the term $(D_i/D)^{1/2}$.

3 ISOTOPIC COMPOSITION OF AN EVAPORATING CLOSED LAKE

As mentioned above, the isotopic ratio for the heavy molecule *i* forming part of a closed water body is given by $R_i = N_i/N$. Therefore, we can write $dN_i = N \cdot dR$. Let us call now N_0 the total number of water molecules initially existing in the evaporating water body. The fraction of the remaining water after evaporation over a certain period of time is given by $f = N/N_0$ and $dN = N_0 \cdot df$. Combining these two differential equations, we can write:

$$\frac{dN_i}{dN} = \frac{N \cdot dR_i}{N_0 \cdot df} = \frac{dR_i}{d(f/f)} = \frac{dR_i}{d(ln\, f)} \tag{6}$$

Introducing the kinetic fractionation term given by equation (2) and introducing the δ notation ($\delta_i = (R_i - R_{SMOW}) - 1$), the following expression is obtained:

$$\frac{d\delta_i}{d(ln\, f)} = \frac{h(\delta_i - \delta_a) - (\delta_i + 1)(\Delta\varepsilon + \varepsilon\alpha)}{1 - h + \Delta\varepsilon} \tag{7}$$

with $\varepsilon = \alpha - 1$.

The previous equation gives the variation of the isotopic composition of the lake water undergoing evaporation as a function of the remaining fraction of water in the lake. If the evaporation conditions remain constant, that is to say, if the values of h, δ_a, $\Delta\varepsilon$, ε and α remain constant during the observation time, equation (7) can be integrated, and the following equation is then obtained:

$$\delta_i = \left(\delta_0 - \frac{A}{B}\right) f^B + \frac{A}{B} \tag{8}$$

where δ_0 is the initial isotopic composition of the water and A and B are given by the following expressions:

$$A = \frac{h\delta_a + \Delta\varepsilon + \varepsilon/\alpha}{1 - h + \Delta\varepsilon} \tag{9}$$

$$B = \frac{h - \Delta\varepsilon - \varepsilon/\alpha}{1 - h + \Delta\varepsilon} \tag{10}$$

Equation (8) shows that the ratio A/B represents the final isotopic composition of the evaporating water body, that is to say, the isotopic composition when f tends to zero. Therefore, A/B represents the maximum possible isotopic enrichment of water. It is important to observe that the ratio A/B does not depend on the initial isotopic composition of water.

The different ways of estimating the kinetic enrichment factors have been presented in Chapter 6. However, many authors still use the equations cited in Gonfiantini (1986), in which the kinetic enrichment factor $\Delta\varepsilon$ can be approximately calculated using the following equations:

$$\Delta\varepsilon^{18}\text{O} = 14.2(1-h) \text{ (in per mil)} \quad (11)$$

$$\Delta\varepsilon\text{D} = 12.5(1-h) \text{ (in per mil)} \quad (12)$$

The fractionation factor α for equilibrium conditions can be calculated, according to Bottinga and Craig (1969) and Majoube (1971), respectively, by means of the following equations:

$$10^3 ln(\alpha^{18}\text{O}) = 2.644 - 3.296 \times 10^3 T^{-1} + 1.534 \times 10^6 T^{-2} \quad (13)$$

$$10^3 ln(\alpha\text{D}) = 52.612 - 76.248 \times 10^3 T^{-1} + 24.844 \times 10^6 T^{-2} \quad (14)$$

These two equations provide similar values to those cited in Chapter 6 for the estimation of the same parameter.

4 LAKE BALANCE

The water balance of a lake can be expressed by the following equation:

$$\frac{dV}{dt} = I - Q - E \quad (15)$$

where V is the water volume of the lake, I is the water inflow, Q is the water outflow or discharge and E is the evaporation rate.

For a heavy molecule with isotopic deviation δ_i, the balance is as follows:

$$\frac{V\delta_L + \delta_L dV}{dt} = I\delta_I - Q\delta_Q - E\delta_E \quad (16)$$

The isotopic composition of the water inflow is given by:

$$\delta_I = \frac{\Sigma\delta_i I_i}{\Sigma I_i} \quad (17)$$

All water inflows have to be included, that is to say, precipitation, runoff and underground inflow.

In the case of a well mixed lake, the isotopic composition of the water outflow is equal to that of the lake δ_L, that is to say $\delta_Q = \delta_L$.

On the other hand, if equation (5) is modified by introducing both the δ-notation and the kinetic enrichment factor $\Delta\varepsilon$, and the isotopic deviation of the atmospheric vapour is designed by δ_a, the isotopic composition of the water vapour escaping from a well mixed lake is given by:

$$\delta_E = \frac{1}{1 - h + \Delta\varepsilon} - \left(\frac{\delta_L - \varepsilon}{\alpha} - h\delta_a - \Delta\varepsilon\right) \tag{18}$$

Taking into account this equation, equation (16) can be written in the following way:

$$\frac{Vd\delta_L + \delta_L dV}{dt} = I\delta_I - Q\delta_Q - \frac{E}{1 - h + \Delta\varepsilon}\left(\frac{\delta_L - \varepsilon}{\alpha} - h\delta_a - \Delta\varepsilon\right) \tag{19}$$

If the volume of the water body during the observation period is kept at least approximately constant, $dV/dt = 0$ and $I = Q + E$. The fraction of the water inflow undergoing evaporation in the lake is given by x, $E/I = x$ and $Q - 1 = 1 - x$. Taking into account that the residence time of water in the lake is given by $\tau = V/I$, equation (19) can be transformed in the following way:

$$\frac{d\delta_L}{dt} = -[(1 + Bx)\delta_L - \delta_i - Ax]/\tau \tag{20}$$

where A and B are the constants given by equations (9) and (10).

In the case of a lake with very long residence time of water, it can be assumed that the parameters δ_i, δ_a, τ, h, α, ε and $\Delta\varepsilon$ are kept constant, with values corresponding to the average values for long periods of time. Under these conditions, equation (20) can be integrated, and the following solution is obtained:

$$\delta_L = \frac{\delta_I + Ax}{1 + Bx} + \left(\delta_0 - \frac{\delta_L + Ax}{1 + Bx}\right) exp[-(1 + Bx)t/\tau] \tag{21}$$

According to this equation, the isotopic composition of the lake water tends to a constant value, given by the equation:

$$\delta_L = \frac{\delta_I + Ax}{1 + Bx} = \frac{\delta_I(1 - h + \Delta\varepsilon) + x(h\delta_a + \Delta\varepsilon + \varepsilon/\alpha)}{(1 - x)(1 - h + \Delta\varepsilon) + x/\varepsilon} \tag{22}$$

This equation can be used to calculate the fraction x of evaporated water, assuming all other variables can be determined or evaluated. The following expression is obtained:

$$x = \frac{\delta_L + \delta_I}{A - \delta_L B} = \frac{(\delta_L - \delta_I)(1 - h + \Delta\varepsilon)}{(\delta_L + 1)(\Delta\varepsilon + \varepsilon/\alpha) + h(\delta_a - \delta_L)} \tag{23}$$

However, the uncertainty of this determination is usually relatively large due to the difficulty of evaluating with enough precision some parameters of the previous equation, and, particularly, the isotopic deviation δ_a.

Finally, in the case of a terminal lake, in which evaporation compensates exactly for the water inflow (lake with constant volume), the fraction of evaporated water x is equal to 1, and the saturation value of the isotopic composition is given by the equation:

$$(\delta_S)_L = \alpha\delta_I(1 - h + \Delta\varepsilon) + \alpha \cdot h \cdot \delta_a + \alpha\Delta\varepsilon + \varepsilon \quad (24)$$

This is the maximum isotopic enrichment that can be reached in a lake with constant water volume. When the evaporation term is negligible for the water balance, it is possible to admit $x = 0$, and δ_L becomes equal to δ_I, as can be expected. In the case of a saturated atmosphere, h is equal to 1, $\Delta\varepsilon$ is equal to zero, according equations (11) and (12), and equation (24) is transformed into $\delta_L = \alpha\delta_a + \varepsilon$. This equation shows that the liquid and vapour are in isotopic equilibrium, as must be the case due to the lack of evaporation.

REFERENCES

Bottinga, Y. and Craig, H., 1969. Oxygen isotope fractionation between CO_2 and water and the isotopic composition of marine atmosphere. *Earth Planet. Sci. Lett.*, 5, 285–295.

Craig, H. and Gordon, L.I., 1965. Deuterium and oxygen-18 variations in the ocean and the marine atmosphere. In E. Tongiorgi (Ed.), *Proc. Conf. on Stable Isotopes in Oceanographic Studies and Paleotemperatures*. C.N.R., Laboratorio di Geologia Nucleare, Pisa, Italy, pp. 9–130.

Gonfiantini, R., 1986. Environmental isotopes in lake studies. In P. Fritz and J. Ch. Fontes (Eds.), *Handbook of Environmental Isotope Geochemistry, Vol. 2, The Terrestrial Environment, B*. Elsevier, Amsterdam, pp. 113–168.

Majoube, M., 1971. Fractionement en oxygène-18 et deuterium entre l'eau et sa vapeur. *J. Chim. Phys.*, 197, 1423–1436.

ANNEX 4

Hydraulic conductivity of the borehole casing

Most of the boreholes and wells drilled for water extraction or for hydrological studies are fitted with a casing to avoid partial collapse of the inner walls or accumulation of fine sediments inside the borehole. Only very rarely can boreholes drilled in consolidated rocks be constructed without casing. To be able to carry out tracer tests to determine the hydraulic conductivity (or permeability) of different horizons (as described in Chapter 3) and the flow tests described in Chapter 8, the casing has to be perforated with holes or slots in the stretch where such tests are to be performed. Ideally, the casing should present a much higher permeability than the rock formation in such a way that the casing does not affect groundwater flow through the borehole. This is not the case in many wells, boreholes and piezometers.

Obviously, the permeability of the casing depends on the fraction or percent f of the total surface represented by the holes or slots with respect to the outer surface of the casing. Some manufacturers supply plastic casings of different diameters with 10% perforated surface. This type of casing does not present any difficulty for the permeability or flow tests because the above-mentioned condition is fully fulfilled. However, due to the cost of such casing, most boreholes drilled for investigation purposes are equipped with casings consisting of manually perforated plastic tubes. Usually, slots of about 1 to 2 mm width and 30 to 50 mm length are made using a drill fitted with a cutting disk. The fraction of drilled surface covered by these slots is commonly much lower than in the case of the commercially available casing.

The procedure for the calculation of the casing permeability described below is interesting, particularly when manmade casings are used in rock formations with high hydraulic conductivity. The calculation procedure has been taken from a paper by Koch et al. (1967). Assuming a laminar flow, the permeability of the casing can be calculated using the following equation:

$$k = f\,\frac{\rho \cdot g}{\eta} \cdot \frac{r_H^2}{S \cdot T} \qquad (1)$$

where

k = permeability, in $\mathrm{cm \cdot s^{-1}}$,
ρ = water density, in $\mathrm{g \cdot cm^{-3}}$,

g = gravity acceleration, in $cm \cdot s^{-2}$ ($980\,cm \cdot s^{-2}$),
η = dynamic viscosity of water, in $g \cdot cm^{-1} \cdot s^{-1}$,
r_H = hydraulic radius of the slot, in cm,
S = resistance coefficient to flow, which depends on the shape of the casing cross section,
T = second resistance coefficient to flow, which depends on the shape and size of the slots.

Viscosity of water depends on temperature. For a temperature of 20°C, water viscosity is $1.002 \times 10^{-2}\,g \cdot cm^{-1} \cdot s^{-1}$, which is equivalent to 1 centipoise. At 15°C, the viscosity is $1.139 \times 10^{-2}\,g \cdot cm^{-1} \cdot s^{-1}$.

The hydraulic radius of the slot is given by double the area divided by the perimeter. For example, in the case of a rectangular slot of 2 mm width and 40 mm length, the hydraulic radius is $(2 \times 40 \times 2)/(2+2+40+40) = 1.9$ mm (0.19 cm).

The resistance coefficient S has the value 8 for casings with circular cross section.

The resistance coefficient T corrects for the existence of stagnant water surrounding the casing, located in areas not facing the slot openings. This coefficient is related to the hydraulic radius of the slot and to the dimensions of the casing according to the following equation:

$$log\left[T\left(\frac{a}{r_1+r_2}\right)^2\right] = C\,log\frac{r_H}{r_2} \qquad (2)$$

where

r_1 = inner radius of the casing,
r_2 = outer radius of the casing,
a = arbitrary reference length,
C = proportionality constant, which is determined experimentally.

Figure A4.1 shows the relationship between the first term of the previous equation and the ratio r_H/r_2 for value $a = 5$ cm, obtained experimentally by Koch et al. (1967). The diagram allows the determination of the value of T for fixed values of the other parameters of equation (2).

Assuming the values $\rho = 1\,g/cm^3$ and $\eta = 1.139 \times 10^{-2}\,g \cdot cm^{-1} \cdot s^{-1}$ (water temperature equal to 15°C), equation (1) becomes:

$$k = 8.6 \times 10^4 f \frac{r_H^2}{S \cdot T} \qquad (3)$$

Very frequently, casings with about 2 inches diameter are used, being perforated with slots of 2 mm width and 40 mm length. Let us suppose a casing of 60 mm outer diameter and 52 mm inner diameter is fitted with this type of slot. As mentioned above, the hydraulic radius is 0.19 cm. Therefore, a value of 0.063 is obtained for the

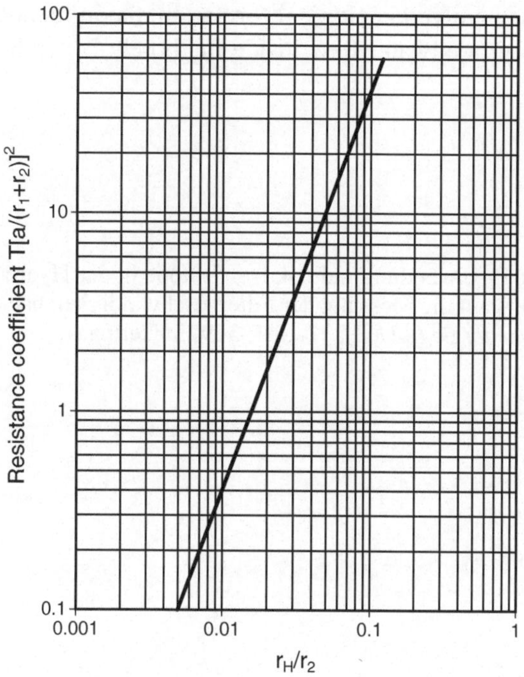

Fig. A4.1. Relationship between the parameters $T[a(r_1 + r_2)]^2$ and r_H/r_2.

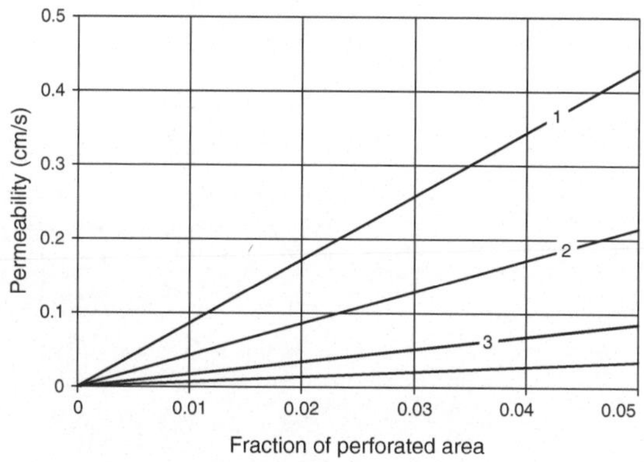

Fig. A4.2. Variation of the permeability of a casing of 60 mm outder diameter and 52 mm inner diameter, fitted with slots of 2 × 40 mm as a function of the fraction f of perforated area for different values of the parameter $r_H^2/S \cdot T$. (1): $r_H^2/ST = 10^{-5}$; (2): $r_H^2/ST = 2 \times 10^{-5}$; (3): $r_H^2/ST = 5 \times 10^{-5}$; (4): $r_H^2/ST = 10^{-4}$.

ratio r_H/r_2. Figure A4.1 provides for T a value of 125. Assuming for the coefficient S a value equal to 8, a value for the quotient $r_H^2/S \cdot T$ equal to 3.6×10^{-5} cm^2 is obtained.

On the other hand, using equation (3), the diagram of Fig. A4.2 has been elaborated, which provides the permeability of the casing as a function of the fraction f using the ratio $r_H^2/S \cdot T$ as a parameter. It is interesting to note that for values of f lower than 1%, the permeability of the casing can also be very low. Therefore, depending on the permeability of the involved geological formation, the influence of the permeability of the casing in groundwater flow inside the borehole can be considerable.

REFERENCE

Koch, A., Klotz, D. and Moser, H., 1967. Anwendung radioaktiver Isotope in der Hydrologie: VI. Der Einfluss des Filkterrohrs auf die Messung der Filtergeschwindigkeit nach der Verdünnungsverfahren. *Atomkernenergie (ATKE)*, 12, 361–369 (in German).

ANNEX 5

Estimation of the total volume of cavities in karstic aquifers from the concentration of natural tritium

Let's assume a karstic spring with an average flow-rate Q. The spring is draining groundwater stored in an aquifer system of volume V. It is assumed that any water mass entering this volume at any specific moment undergoes instantaneous mixing with the previously stored water. If this is the case, the renewal of groundwater in such volume and, therefore, the renewal of the incorporated tritium takes place according to an exponential model. If the system is in hydraulic equilibrium (same inflow and outflow), the renewal process can be illustrated by the following scheme:

$$Q;\, C_0(t) \rightarrow \boxed{\begin{array}{c} C = 0 \text{ for time } t = 0 \\ \text{Tritium concentration} = C(t) \\ \text{Water volume} = V \end{array}} \rightarrow Q;\, C(t)$$

The function $C_0(t)$ represents the input function for tritium, which corresponds to the average tritium concentration of the local precipitation or the river eventually feeding the karstic storage system (see Chapter 6 for the procedure of determination). The function $C(t)$ represents the variation of the tritium concentration in the aquifer, which is the same as the changes observed in the spring, due to the assumption of instantaneous mixing.

Under these circumstances, the tritium balance can be expressed by the following equation:

Inflow – Outflow – Radioactive decay = Change of concentration

$$Q \cdot C_0(t) - Q \cdot C(t) - \lambda \cdot V \cdot C(t) = V \cdot dC(t)/dt \qquad (1)$$

where λ is the radioactive decay constant, given by the equation:

$$\lambda = ln\, 2/t_{1/2} = 0.05668 \text{ years}^{-1}$$

with $t_{1/2}$ being the tritium half-life, equal to 12.3287 years.

This equation (1) can be rewritten in the following way:

$$V \cdot dC(t)/dt = Q \cdot C_0(t) - V(\lambda + Q/V)C(t) \tag{2}$$

The term Q/V represents the turnover time τ of groundwater stored in the volume V. The integration of equation (2) leads to the following equation:

$$C_t = \frac{1}{\tau}\int_{1953}^{t} C_0(t')exp\left[-\frac{1+\lambda\tau}{\tau}(t-t')\right]dt' \tag{3}$$

where $t$corresponds to the present year and t' to any year between 1953 and t, that is to say the period of time for which the average tritium concentrations represented by the function $C_0(t')$ have been measured or computed for local precipitation. The difference $(t - t')$ gives, therefore, the number of years since 1953.

Equation (3) can be solved in a discrete way by adding the annual tritium contributions from 1952 to the present year for a given value of the turnover time τ. The tritium concentration of the stored water and, therefore, of the spring water is calculated in this way as a function of the turnover time τ. Curves similar to the one shown in Fig. A5.1 are obtained. The almost horizontal stretch on the left side of the curve is a consequence of the almost constant tritium concentration (values corrected for radioactive decay) of the precipitation occurring after 1984, as described in Chapter 6 (Section 6.3.3). On the other hand, for concentration values between 5.5 and 10.8 T.U., two different values of the turnover time are possible. Usually, one of these two values can be neglected, taking into consideration hydrological reasons. For example, in the case of karstic aquifers, only the ascendant branch of the curve will be valid in most cases.

From the turnover time of the aquifer water corresponding to the tritium concentration measured at the spring water, the total volume of the cavities involved in

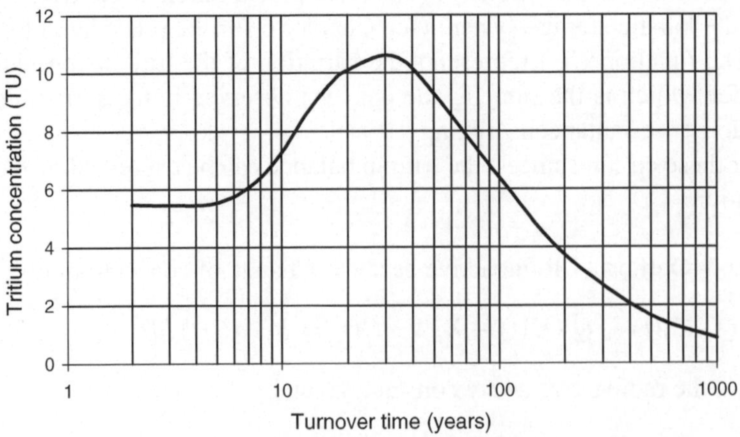

Fig. A5.1. Typical correlation between tritium concentration and turnover time obtained with the exponential or good mixing model in karstic aquifers.

the groundwater flow (cavities filled with water) can be computed by the equation $V = Q_a \cdot \tau$, where Q_a represents the average flow rate of this spring. For obvious reasons, the value obtained can be considered only a rough estimate.

Results obtained by the senior author in several karstic aquifers from different countries have proven that the exponential or good mixing model is appropriate for the explanation of the water renewal in many karstic aquifers (Plata, 1990; Plata and Braz, 1980; Plata et al., 1980, 1981, 1995).

REFERENCES

Plata, A., 1990. Parameters of carbonate rock aquifers from tracer methods. In G. Günay, A.I. Johnson and W. Back (Eds.), *Hydrogeological processes in karst terranes*. IAHS Publication No. 207, 361–380.

Plata, A. and Braz, A., 1980. Utilização de isótopos ambientais na pesquisa de recursos hídricos subterrâneos no Karst da região do Jaíba, Norte de Minas Gerais. *Rev. Bras. Geocien.*, 10, 276–291 (in Portuguese).

Plata, A., Braz, A., Gomez, F. and Barreto, L., 1980. Contribuição de diversas técnicas isotópicas no estudo do aquífero cárstico da região central da Bahia (Irecê-Iraquara). *Rev. Bras. Geocien.*, 10, 103–116 (in Portuguese).

Plata, A., Baonza, E. and Braz, A., 1981. *Estudio isotópico de un acuífero cárstico en Minas Gerais (Brasil).* Proc. Interamerican Symp. on Isotope Hydrology, Bogotá, Colombia, IAN, Bogotá, 337–357 (in Spanish).

Plata, A., Perez, E., Díaz, M.F., Luna, E. and Delgado, C., 1995. *Tiempo de renovación del agua en el acuífero cárstico de Campo de Montiel obtenido a partir del tritio de origen termonuclear.* VI Congreso de Geoquímica de España, Soria, 295–312 (in Spanish).

Author index

Subject index